Acoustic Emission - Research and Applications

Acoustic Emission - Research and Applications

Editor

Lefteris Tyler

scitus
academics

Acoustic Emission - Research and Applications
Edited by **Lefteris Tyler**

ISBN: 978-1-68117-206-4
Library of Congress Control Number: 2016934753

SCITUS Academics LLC,
www.scitusacademics.com
Box No. 4766, 616 Corporate Way,
Suite 2, Valley Cottage,
NY 10989

Printed in United States of America on Acid Free Paper

Preface

Acoustic emission is a technique used simply by fracture mechanics to evaluate structural integrity. The dynamic behavior of defects is very important because a small, increasing defect can be more important than a stable one. Acoustic emissions are the method used to control stressed defect behaviour. When an external stimulus in a structure is present, localized sources release energy in the form of stress waves that propagate to the surface and are recorded by sensors. Picometer motions can be identified and set up with the right equipment. AE sources vary from natural events such as earthquakes and rock explosions to crack initiation and growth, slip and dislocation, melting, twinning and metal transformations in phases. Cracking, breaking and debonding of matrix fibers contribute to the acoustic emissions of composites. AE has been measured and recorded amongst other materials in polymers, wood and concrete. This book shows that the AE method is constantly and intensively developed and widely used in: online monitoring of civil engineering structures, detection and location of fatigue cracks in pressure vessels and pipelines, fiber damage assessment-reinforced polymer matrix composites, welding monitoring and corrosion provision.

TABLE OF CONTENTS

CHAPTER 1 Acoustic Emission Studies on Weld Bead Defects in Nuclear Grade SS 316L Materials **1**

CHAPTER 2 Wavelet Packet Decomposition to Characterize Injection Molding Tool Damage ‡ **23**

CHAPTER 3 Identification of a Critical Time with Acoustic Emission Monitoring during Static Fatigue Tests on Ceramic Matrix Composites: Towards Lifetime Prediction† **43**

CHAPTER 4 A New Fault Location Approach for Acoustic Emission Techniques in Wind Turbines **67**

CHAPTER 5 Structural Damage Localization by Linear Technique of Acoustic Emission **89**

CHAPTER 6 The Features of Acoustic Bathymetry of Sludge in Oil Storage Tanks **101**

CHAPTER 7 Implementation of Acoustic Analogies in OpenFOAM for Computation of Sound Fields **127**

CHAPTER 8 Acoustic Emission Activity for Characterizing Fracture of Marble under Bending **153**

CHAPTER 9 Study of the Tensile Damage of High-Strength Aluminum Alloy by Acoustic Emission **171**

CHAPTER 10 Acoustic Emission in Drying Materials **189**

CHAPTER 11 Application of Acoustic Emission for Quality Evaluation of Fruits and Vegetables.......... **221**

CHAPTER 12 Otoacoustic Emissions 251

CHAPTER 13 Non-Destructive Evaluation for Corrosion Monitoring in Concrete: A Review and Capability of Acoustic Emission Technique.......... **273**

Index.......... **317**

Chapter 1

Acoustic Emission Studies on Weld Bead Defects in Nuclear Grade SS 316L Materials

Ranganayakulu S. V.[1], Shiva Raju J.[1], Kuchedludu A.[1], Ramesh Kumar B.[2]

[1]*Center for Non Destructive Evaluation, Gurunanak Institutions Technical Campus, Hyderabad, India*

[2]*Institute for Plasma Research [IPR], Bhatt, Gujarat State, India*

ABSTRACT

This paper contributes about the behaviour of Acoustic Emission (AE) signatures of implanted weld defects of SS 316L materials. Lack of penetration and lack of side fusion defects were implanted in weld bead region of the materials. Tungsten Inert Gas Welding (TIG) is adopted to weld the Stainless Steel (SS316L) nuclear grade materials. The material is fabricated with dimensions of 140 × 16 × 10 mm and AE signatures are studied under preload conditions. Mechanical Jig is fabricated to maintain constant load in concentrated weld region. When external load is applied on the weld region, the deformed specimen experiences acoustic emission signals form the weld defect region which are potential source of releasing stress energy. Liner Location Technique (LLT) is adopted for AE singal studies and the generated signal is processed by 2-channel USB—AE node and AE-WIN software. The tests are conducted on two different samples having each defect. A conventional NDT method i.e. X-ray Radiography is conducted on the samples to know the defect ranging and correlated with AE signatures. This study will be helpful to standardize the AE signals for different implanted weld defects of SS 316L materials and it is found that, the parameter "counts vs. amplitude" has given the widest distinction with respect to the type of defects.

Keywords: Acoustic Emission signatures, X-ray Radiography, Lack of side fusion, Lack of penetration, Tungsten Inert Gas welding (TIG)

1. INTRODUCTION

Acoustic Emission is a powerful tool which is useful for the study of structural health monitoring of materials. Acoustic Emission (AE) is the study and practical use of elastic waves generated by a material subjected to an external stress [1]. The phenomenon was recognized by early miners who exited a mine when the rocks or supporting timbers started groaning. Tin cry, the sound produced when a tin bar is bent, was known soon after the production of metallic tin. Allen Green and a group at Aero Jet Corporation started using AE in the testing of Saturn Rocket Propellant tanks. They used a form of triangulation based on the arrival times of the acoustic pulse at several acoustic emission sensors. According to ASTM, Acoustic Emission refers to the generation of transient elastic waves during the rapid release of energy [2] from localized sources within a material. Other sources of Acoustic Emission are melting, phase transformation, thermal stresses cool down cracking and stress build up.

The basic idea in source location is to cover a surface with a network of sensors. If one can determine the arrival times of an emission signal at several sensors, then, knowing the acoustic velocity [3] , it is possible to triangulate back to the location of the source of that emission. Most acoustic emission mechanisms involve a permanent change in the micro structure [4] of the material. Once a micro fracture occurs, it will not happen again unless there is some sort of healing mechanism. The ratio of the load value, when emission starts on subsequent loadings, to the maximum load value reached on the previous loading, known as the Felicity ratio, indicates possible damage induced by the previous loading. Many NDT tests of FRP structures apply the test load in a series of steps, returning to zero between each step. The appearance of Felicity ratios much less than 1.0 is a good indication that significant damage occurred in previous loadings.

2. EXPERIMENTAL SETUP OF MECHANICAL JIG

The samples were tested using the Linear Localization method. In order to make the defects active, an external force needs to be applied to the specimens. Samples were tested in a Mechanical Jig [Figure 2] specifically designed for this purpose. This special jig was designed for ease of loading and to keep the unwanted external noise to the minimum. This Jig has provision for the placement of the samples on an anvil at the lower side and in the centre. This anvil has two vertical supports with rounded tops on which the sample can be placed and pressed to obtain the deformations. A mandrill attached to a screw loading system can be lowered on to the anvil for pressing the samples. The samples are placed on the anvil in a flat position and the mandrill is lowered slowly by rotating the lever attached to the screw for loading the samples.

Two broad banded acoustic sensors having operating frequency range of 100 KHz - 1000 KHz, are placed on either side of the sample and Acoustic Emission data was collected while slowly loading the samples. Linear Location technique method was used while collecting the data. Suitable filters were put to avoid unwanted noise from various sources. AE signatures are recorded in terms of energy, count, and amplitude.

Two samples were tested from each of the weld category [Figure 3]. The specimens were bent to less than 45 degree angle between their two legs. Acoustic emission data was collected while slowly loading the specimens. copious emission activity was detected from both good and defective specimens. However, it was noticed that, the emission activity started much earlier in the defective specimens than from the good specimens. Also the activity from the defective samples was higher and some of the signals were noticed to be distinctly sharper. Some signals showed higher frequencies possibly from crack formation [6] during loading/deformation. It was observed that, the embedded defects showed up extra acoustic emission activity compared to specimens from good welds. The AE data was recorded for Lack of Penetration (LOP) and Lack of Side Fusion (LSF) defective samples of SS316L. NDT evaluation was also carried out on these defect specimens and the results are being correlated with AE response of the specimens.

2.1. Experimental Setup of AE Hardware and Software

4-Channel Acoustic Emission test setup and AE-Win Software is procured from Physical Acoustic Corporation, USA for Acoustic Emission studies. The USB AE node is a single channel Acoustic Emission (AE) digital signal processor with full hit and time based features, including waveforms. The USB-AE node system is capable of performing any AE application [Figure 4 & Figure 5] and is an excellent field survey tool. When a structure is subjected to an external stimulus (change in pressure, load, or temperature), localized sources trigger the release of energy, in the form of stress waves, which propagate [7] to the surface and are recorded by sensors. With the right equipment and setup, motions on the order of picometers (10^{-12} m) can be identified. The plastic component of the strain often takes longer. Some deformation [8] is immediate but some of it is delayed [Figure 6]. Counts vs. time are plotted [9] under pre load conditions.

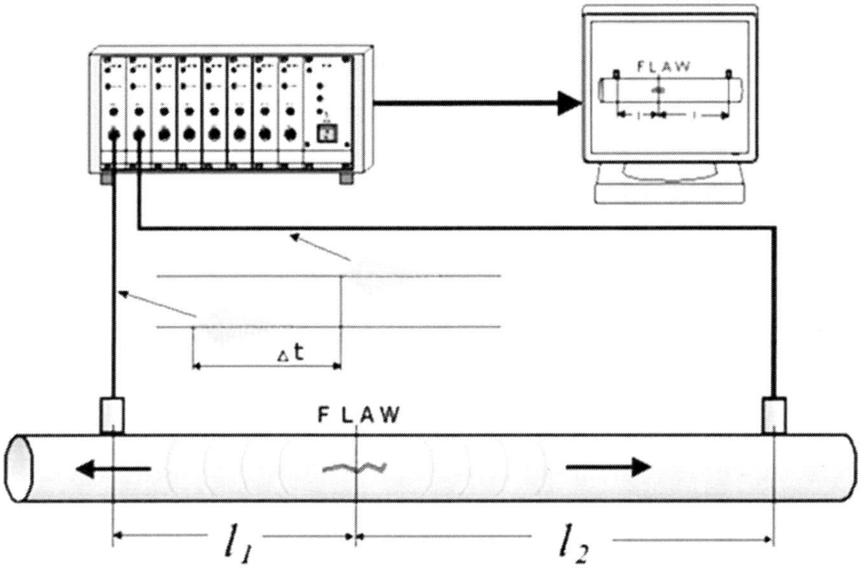

Figure 1. Set up for linera localization of AE from a flaw.

Figure 2. Sensors attached with sample in Mechanical Jig.

Figure 3. Bended sample under applied constant load.

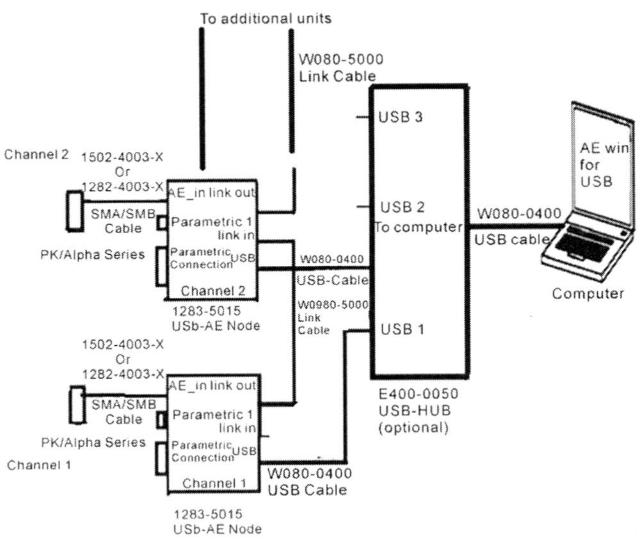

Figure 4. Typical Daisy-Chain configuration of USB-AE node and it accessories.

2.2. Linear Location Technique

One of the commonly used computed-source location techniques is the Linear Location technique [Figure 1] and it is adopted in these studies. When the source is located at the midpoint of the weld region of the sample, the time of arrival difference [10] for the wave at the two sensors is zero. If the source is closer to one of the sensors, a difference in arrival times is measured. To calculate the distance of the source location from the midpoint, the arrival time is multiplied by the wave velocity. Whether the location lies to the right or left of the midpoint is determined by which sensor [11] first records the hit. This is a linear relationship and applies to any event sources between the sensors.

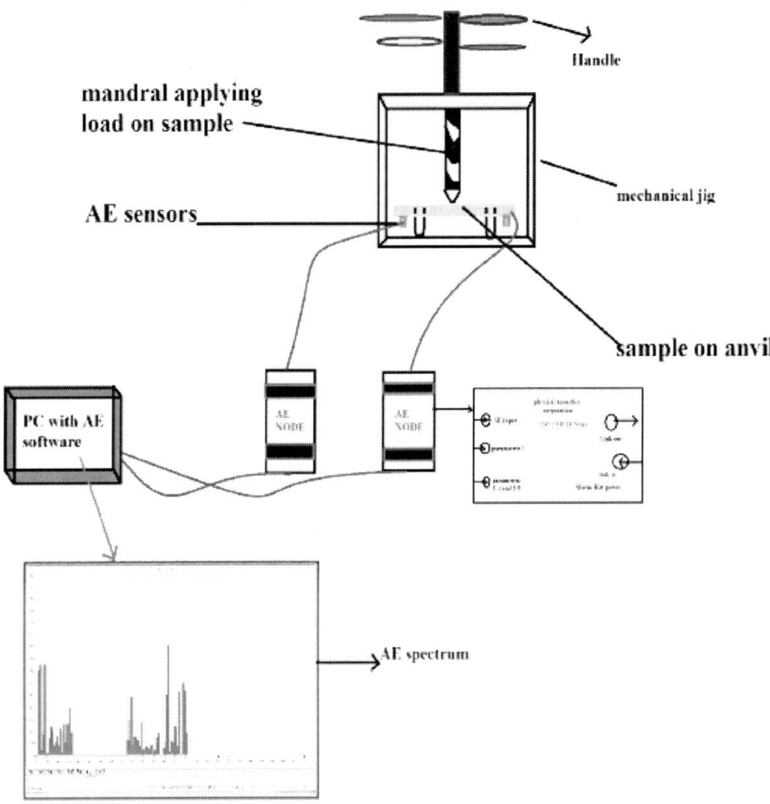

Figure 5. Schematic diagram of Acoustic Emission testing.

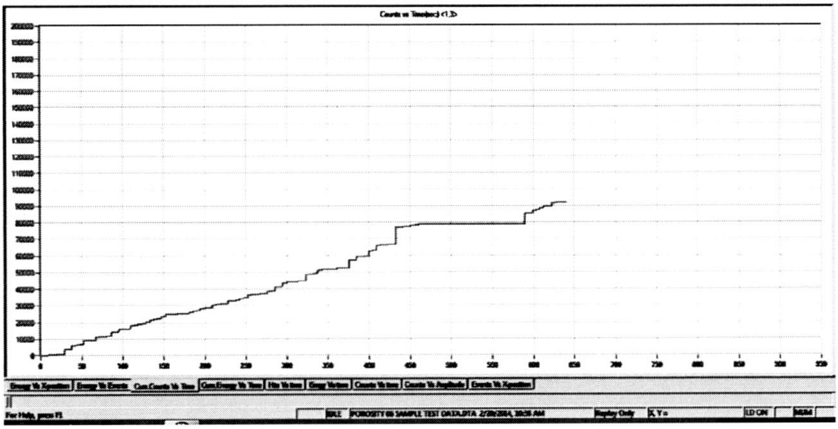

Figure 6. Acoustic Emission signals during load indicates instability.

3. SAMPLE PREPARATION DETAILS

SS316L material is adopted for Acoustic Emission testing for experimentation because it uses in construction of nuclear reactors [12] which are having high strength. Two different types of welded specimens are taken into consideration for vast studies of signals in surrounding areas of welded region. In order to monitor the weld defects, it is necessary to create such defects in experimental welds. For this, a number of weld plates were selected with suitable dimensions and Tungsten Inert Gas (TIG) as well as manual metal arc welding was carried out as per the need.

Following are the details of the parameter adopted for welding-

Plate material: SS316 L

Plate thickness: 10 mm

Plate length: 300 mm

Plate width (before welding): 70 mm*

Welding method used: TIG (Welding-current 110 to 130 Amperes)

Some plates are made with good welds using proper welding methods and matching electrodes and other plates were made by deliberately introducing Lack of Penetration (LOP) and Lack of Side Fusion defects (LSF) during their welding. The plates after welding [Figure 7] had 140 mm length, 16 mm width and 10mm thickness. The different types of defected weld samples used for the studies are 1) Lack of Penetration (LOP) and 2) Lack of Side Fusion (LSF).

4. EXPERIMENTAL PROCEDURE OF RADIOGRAPHY

4.1. Radiography Testing

Radiography method is adopted as a NDT tool for identification of defects location in welded specimens. The procedure is discussed below. It involves exposing a test object to penetrating radiation so that the radiation passes through the object being inspected and a recording medium placed against the opposite side of that, object [13] [14] .

Figure 7. SS316L samples containing (a) Lack of penetration; (b) lack of side fusion for AE study under constant load.

4.2. Radiographic Source

The Radiography testing techniques shall demonstrate that, the required radiography sensitivity has been obtained. Iridium -190 is used as source in this test carried out because the recommended minimum thickness for which Radio-active isotopes when steel is 0.75 m. The minimum recommended thickness limitation may be reduced when the radiography techniques are used to demonstrate that the required radiographic testing sensitivity has been obtained.

4.3. Radiographic Film

Commercially available industrial radiography film (Kodak type AA 400, FUJI 100 or AGFA D7) is used [15] in accordance with SE 1815 (ASTM) standard test method. D4 film is used for SS316L sample for inspection of defects. Radiography test is conducted to correlate itself with acoustic emission test results. Two samples consist of lack of penetration [Figure 8(a)] and lack of side fusion [Figure 8(b)] defects separately [16] taken into test cycle. Gamma ray is used as source. Wire type and Hole type Penetrameter is used [17] to check the test accuracy. The defects in the films are observed easily and AE test is conducted for further analysis.

4.4. Experimental Procedure of Acoustic Emission Testing

This experiment was carried out with cyclic loading using fixed load. It was found that, the primary emission from active crack growth occurred only at the peak load levels. In fact, the emission that occurred at the peak load levels, secondary emission and noise that occurred at lower load levels were gated out. At first when the crack was still small, not every cycle produced emission. But later as the crack approached the critical length for unstable propagation, every cycle produced emission. This fits well with the behavior of statically loaded specimens, that insignificant flaws tend to show the Kaiser Effect while structurally significant flaws tend to show the Felicity Effect. Damage assessment has been feasible because AE activities are function of parameters such as stress level in the crack Zone. AE activity can be directly related to fracture mechanics parameters which can be further related to crack growth rate and fatigue failure. Locating the source of significant acoustic emissions is often the main goal of an inspection. Although the magnitude of the damage may be unknown after AE analysis, follow up testing at source locations can provide these answers. Continuous emission [18] during deformation indicates that, weld region get fractured and open to the surface. Multi channels [19] are used to test large structures and to evaluate the emissions from the location and measurement of emission rates. An analysis can be done either in real time [20] (if the prevention of structural failure is the test objective) or in post-test analysis.

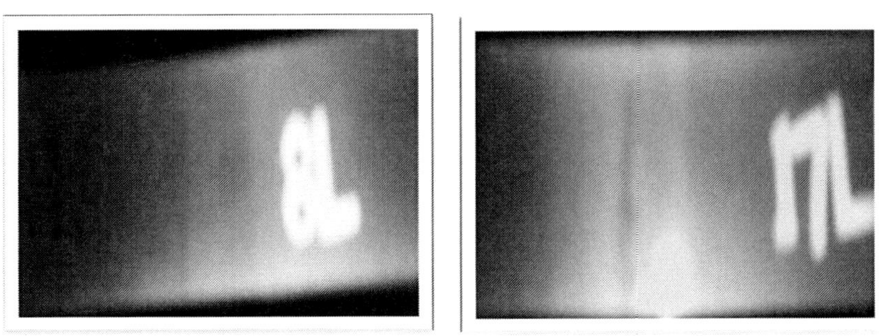

(a) (b)

Figure 8. Radiographic film shows (a) Lack of penetration and (b) Lack of side fusion weld defects.

5. RESULTS AND DISCUSSIONS

Testing of two samples of Lack of Penetration and Lack of Side Fusion defects are taken for Acoustic Emission Studies. Linear Localization technique was used for obtaining the data. While the samples are being loaded, the events were recorded with respective to the position of transducers placed on either end of the samples. Initially, we have attempted to get clear acoustic emission signals by getting higher signal to noise ratios with good results and then shifted to the monitoring of different parameters. The first data monitored was the "Events vs. the X- position". Apart from this, the data was monitored in different configurations. A graph of the hits, events, counts or absolute energy vs. the load or time is essential. The graph of the cumulative hits, etc. over the graph of the rate of hits, etc. is preferred. Because, it is visually easier to differentiate a cumulative graph than to integrate a rate graph. The classic parameter vs. time graph has been used almost from the start of AE testing. A located events count is sensitive to emission [21] from actual flaws but may miss regions where only one or two sensors are activated. Counts or absolute energy per event will give the most information about the state of flaw growth but can become confusing when there are multiple flaws. Located events count is sensitive to emission from actual flaws [22] but may miss regions where only one or two sensors are activated. Changes in the slope of the cumulative graph [23] indicate changes in the acoustic emission rate. A steady rate in the absence of external noise indicates controlled flaw growth, while an increasing rate indicates uncontrolled flaw growth. When the slope of the curve appears to increase toward an exponential, or shows a distinct knee, structural failure is usually approaching.

Following are the details of the various modes in which data was obtained, compiled and analyzed.

6. OBSERVATIONS MADE FROM THE DATA ANALYSIS

6.1. Energy vs. X-Position

It is observed that, all the samples except one recorded a good number of events coming from the centre position [Figure 9] of the sample. This must be an aberration of the experiment rather than a characteristic feature of the defect. Events from lack of penetration are high but the events formed from lack of side [Figure 10] fusion having higher counts.

6.2. Cumulative Counts vs. Time

Lack of penetration and lack of side fusion defects showed very high rate of count cumulation. The lack of penetration defect showed medium rate of count cumulation [Figure 11]. It was found that lack of penetration and lack of side fusion emitted [Figure 12] plastic deformation under the load.

6.3. Cumulative Energy vs. Time

It was found that, more variation in cumulative energy with respect to time for Lack of Penetration [Figure 13] and Lack of Side Fusion defects. Lack of penetration showed continuous cumulative emission. In the case of lack of side fusion defect, the time for deformation is very low [Figure 14]. It indicates that, high energy is emitted due to the formation of new surface.

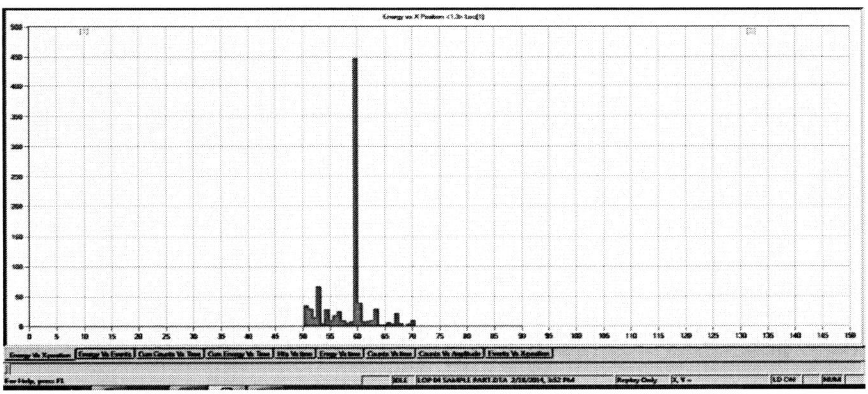

Figure 9. Energy vs. X-position for Lack of penetration defect of SS316L.

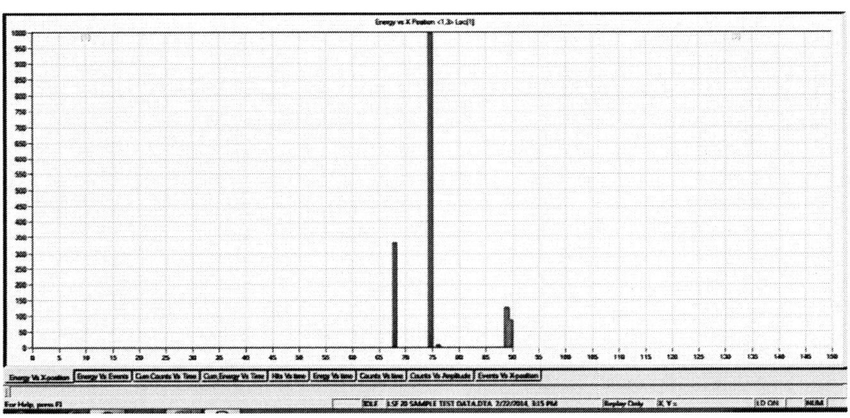

6.4. Hits vs. Time

Lack of penetration recorded the highest hits than lack of side fusion defect. After observing the complete deformation, a high emission [Figure 15] is seen in lack of penetration. Owing to lack of side fusion the highest hits is emitted [Figure 16] during deformation, when new surface is formed.

6.5. Energy vs. Time

It was found that, the defects like lack of penetration emitted higher energy [Figure 17] than lack of side fusion. This is due to incomplete penetration or welding. The crack initiates and grows in very short time compared to side fusion [Figure 18] so it produces highest energy in that order with respect to time.

6.6. Counts vs. Time

It was found that, samples having lack of side fusion showed highest number of counts with respect to time. The lack of penetration defect [Figure 19] showed more counts than lack of side fusion defect [Figure 20]. This is due to deformation of old and formation of new surface in short time.

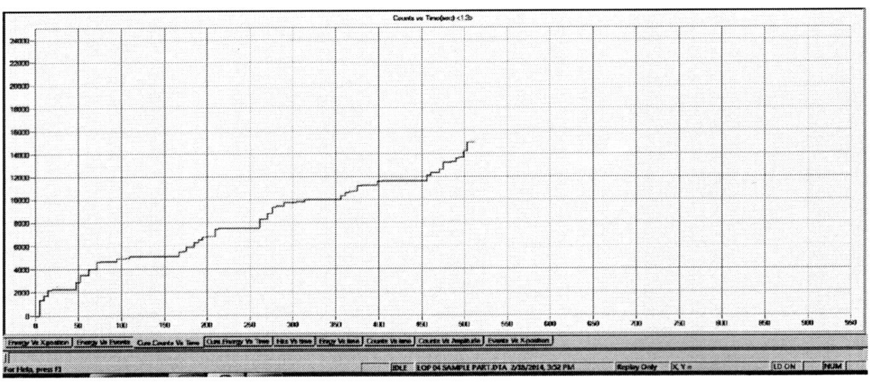

Figure 11. Cumulative Counts vs. Time for Lack of penetration defect of SS316L.

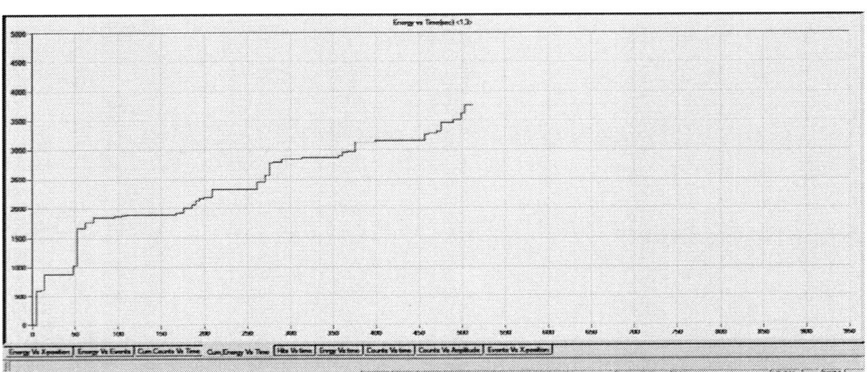

Figure 13. Cumulative Energy vs. Time for Lack of penetration defect of SS316L.

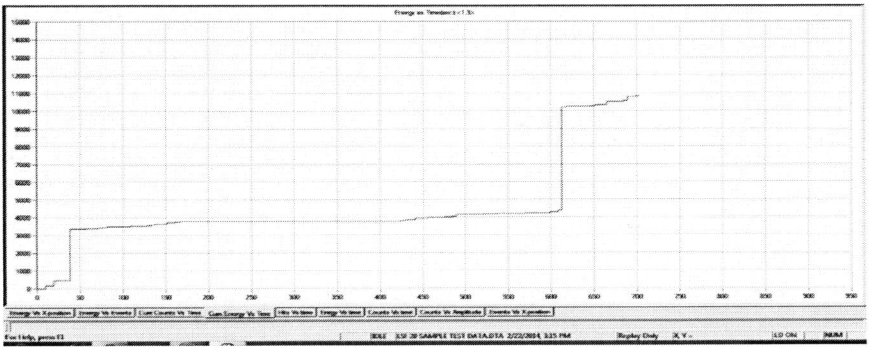

Figure 14. Cumulative Energy vs. Time for Lack of side fusion defect of SS316L.

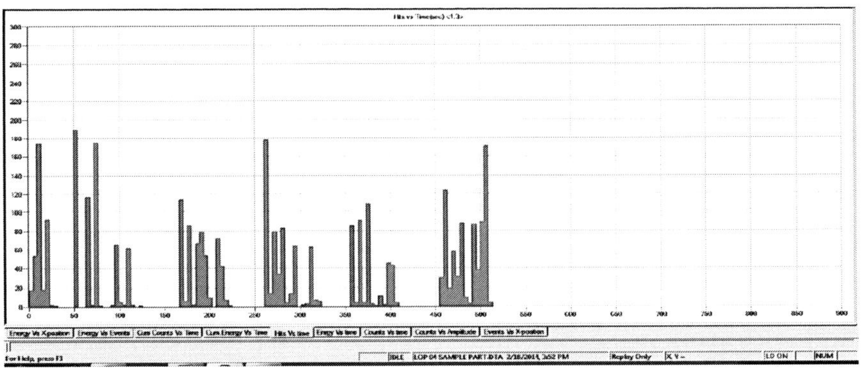

Figure 15. Hits vs. Time for Lack of penetration defect of SS316L.

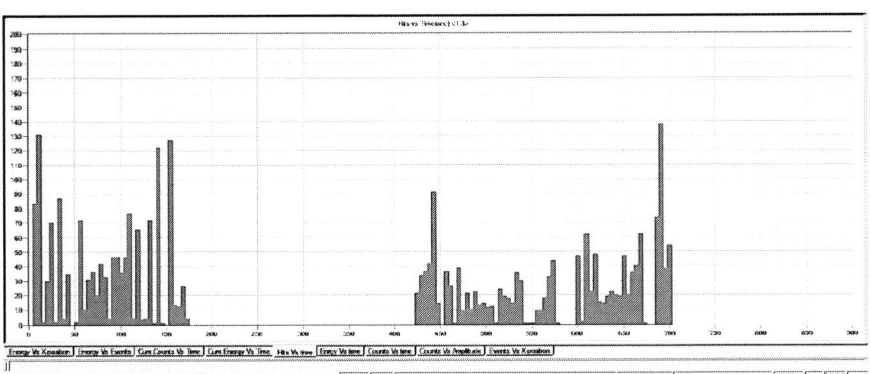

Figure 16. Hits vs. Time for Lack of side fusion defect of SS316L.

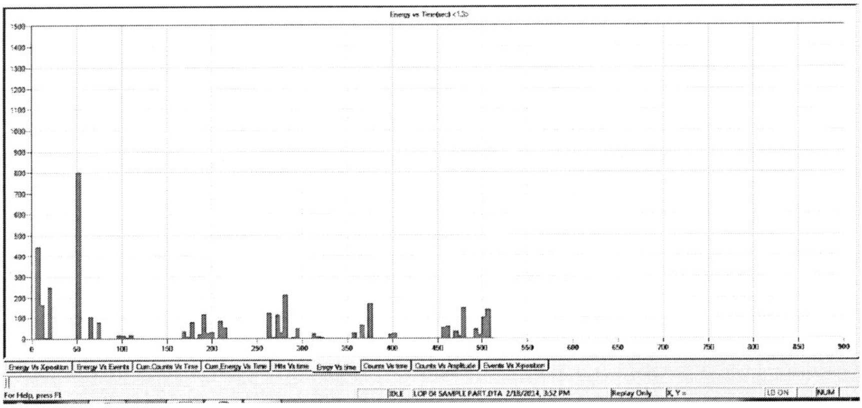

Figure 17. Energy vs. Time for Lack of penetration defect of SS316L.

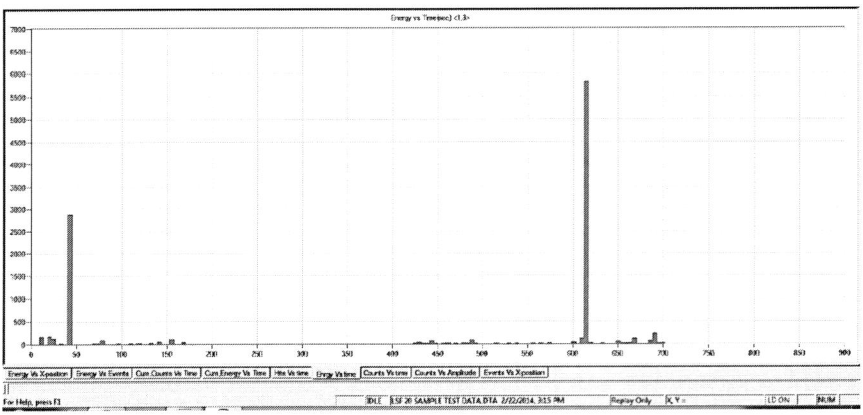

Figure 18. Energy vs. Time for Lack of side fusion defect of SS316L.

6.7. Counts vs. Amplitude

This particular parameter showed very distinctive differences for different defects. This particular parameter is dependable for segregation and characterization of different defects [Figure 21] in conjunction with other parameters. Defects like lack of penetration and lack of side fusion [Figure 22] recorded the highest count amplitude.

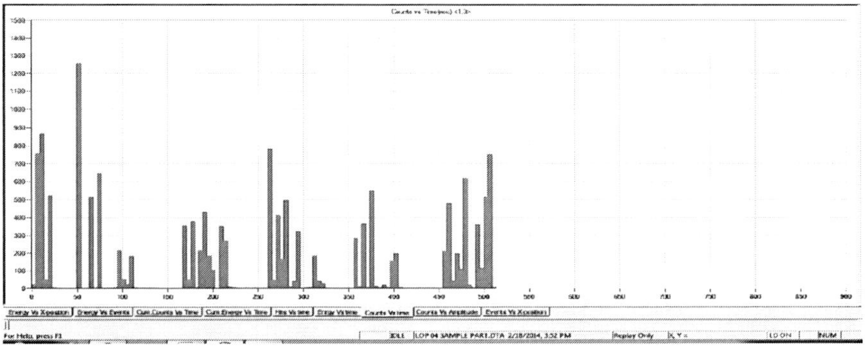

Figure 19. Counts vs. Time for Lack of penetration defect of SS316L.

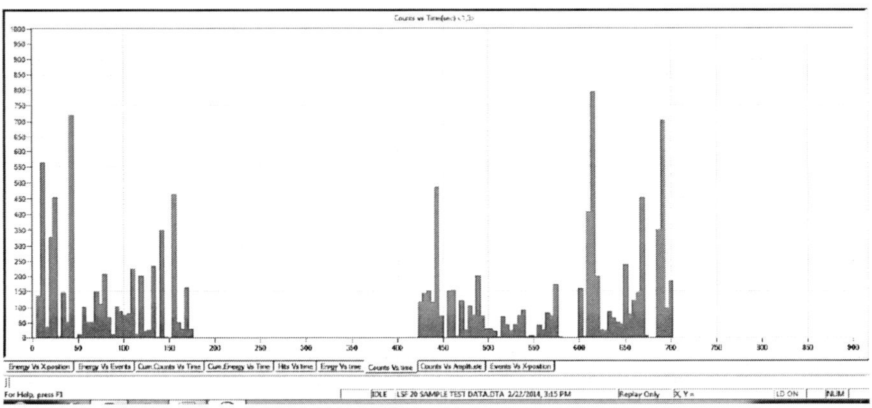

Figure 20. Counts vs. Time for Lack of side fusion defect of SS316L.

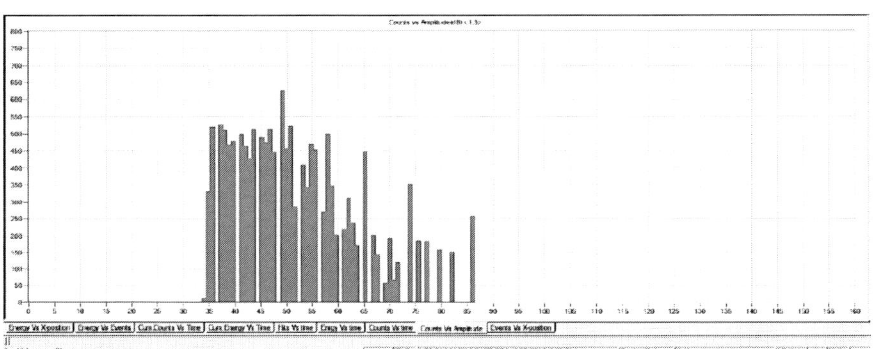

Figure 21. Counts vs. Amplitude for Lack of penetration defect of SS316L.

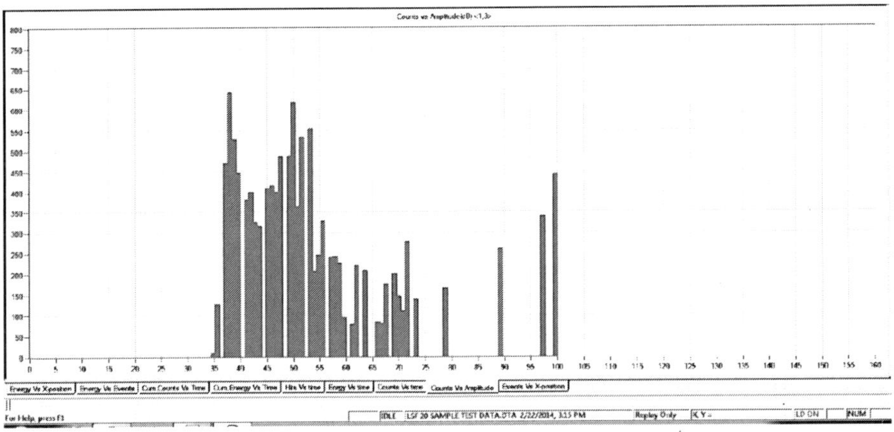

Figure 22. Counts vs. Amplitude for Lack of side fusion defect of SS316L.

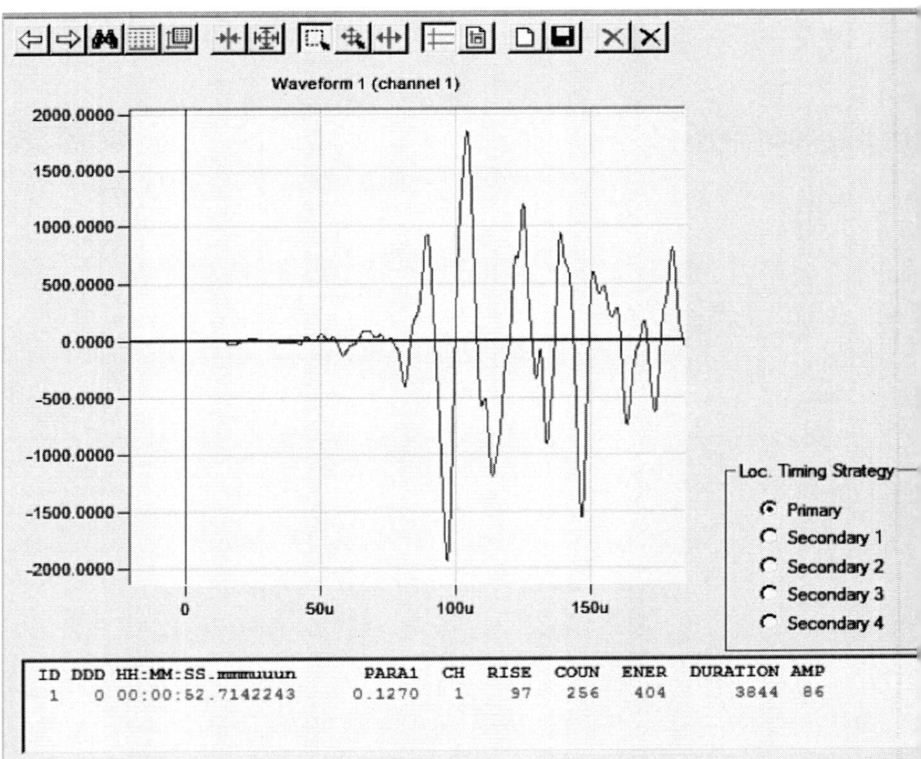

Figure 23. Amplitude vs. Time for Lack of penetration defect of SS316L.

6.8. Waveforms vs. Amplitude

Wave forms were taken for a few events having high energy from two tested samples [Figure 23]. These wave forms are plotted amplitude vs. time. It is noticed that, there are definite characteristic features in the wave forms from different defects. The defect is having higher counts showed higher energy [Figure 24] like lack of side fusion.

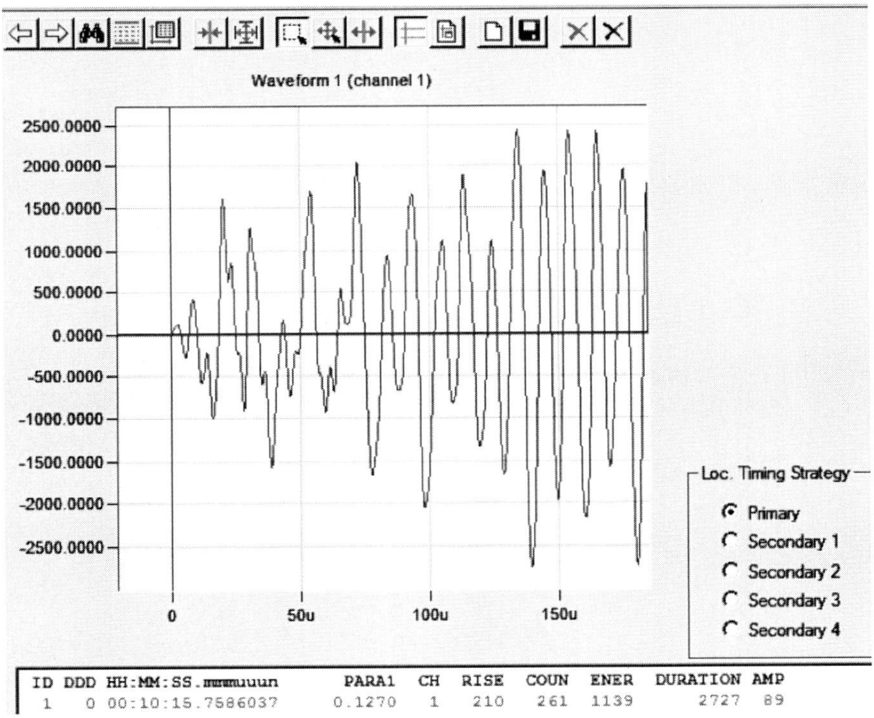

Figure 24. Amplitude vs. Time for Lack of side fusion defect of SS316L.

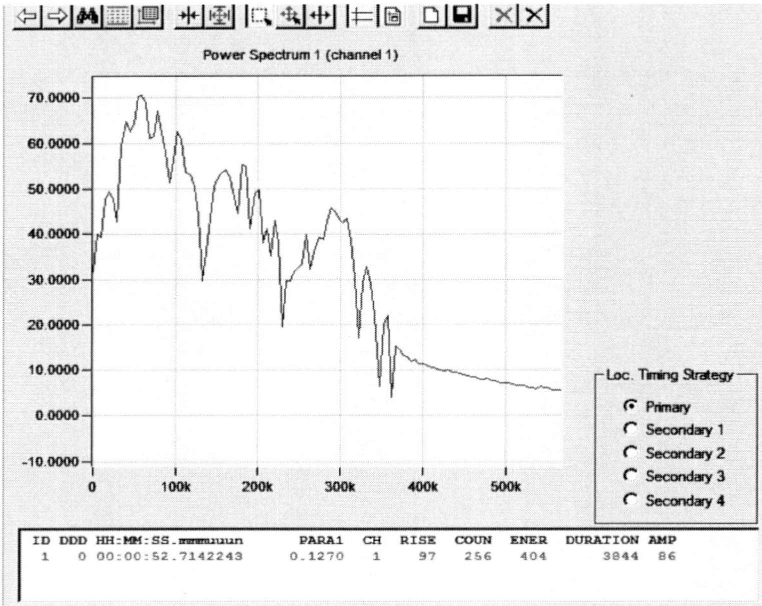

Figure 25. Frequency Spectrum for Lack of penetration defect of SS316L.

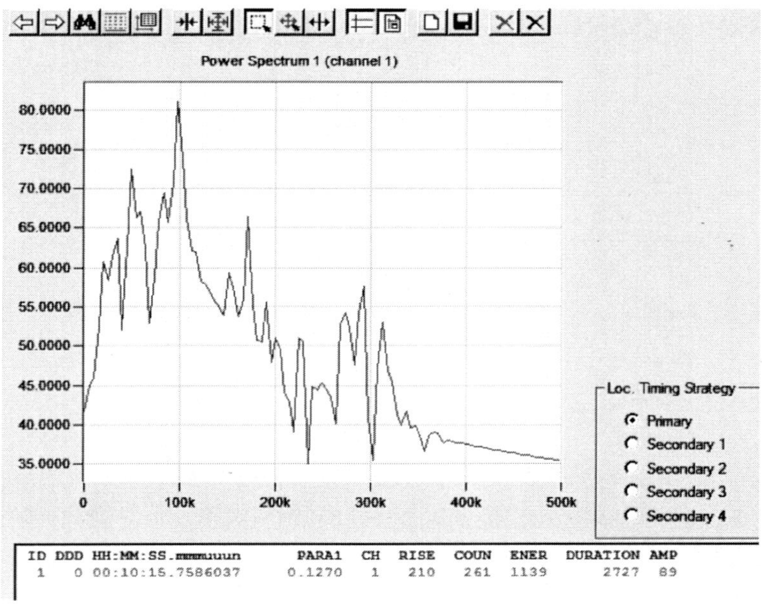

Figure 26. Frequency Spectrum for Lack of side fusion defect of SS316L

6.9. Frequencies vs. Amplitude

In the amplitude vs. frequency graphs, LOP defect has higher frequency [Figure 25] with higher amplitude content than LSF [Figure 26].

7. CONCLUSION

Data analysis was carried out for various parameters obtained from testing of two defects of samples. Comparison of various AE parameters showed that, there is a definite pattern with respect to the type of weld defects. It is observed that, the parameter "counts vs. amplitude" has given the widest distinction with respect to the type of defects. The defects like, lack of penetration have highest expression when compared to lack of side fusion defect in any of the AE parameters monitored. This lack of penetration (LOP) showed the highest counts vs. amplitude and also the highest amplitude with respect to frequency of the events. Energy raised during deformation was observed high with lack of side fusion defect.

ACKNOWLEDGEMENTS

Authors are grateful to the Board of Research in Fusion Science & Technology of Institute for Plasma Research academic research project programme for grant in aid (Project no: NFP-MATF12-01) to pursue the project.

Gratefully acknowledgements due to "Dr. H. S. Saini, Managing Director and Sardar G. S. Kohli, Vice Chairman, Guru Nanak Institutions, Ibrahimpatnam for cooperation and encouragement.

REFERENCES

1. Acoustic Emission Testing. Nondestructive Testing Handbook. 3rd edition, Vol. 6. American Society for Non-des- tructive Testing, 2005.
2. McLaskey, G.C. and Glaser, S.D. (2012) Acoustic Emission Sensor Calibration for Absolute Source Measurements. Journal of Nondestructive Evaluation, 31, 157-168.http://dx.doi.org/10.1007/s10921-012-0131-2
3. Raj, B. and Jayakumar, T. (1990) Acoustic Emission during Tensile Deformation and Fracture of Austenitic Alloys. Acoustic Emission Current Practices and Future Directions-ASTM STP 1077, 218-241.
4. Camerini, C.S., Rebello, J.M.A. and Soares, S.D. (1992) Relationship between Acoustic Emission and CTOD Testing for a Structural Steel. NDT & E International, 25, 127-133.
5. Dia, S., Monnier, T., Godin, N. and Zhang F. (2012) Primary Calibration of Acoustic Emission Sensors by the Method of Reciprocity, Theoretical and Experimental Considerations. 30th European Conference on Acoustic

Emission Testing & 7th International Conference on Acoustic Emission University of Granada, 12-15 September 2012.

6. Brunner, A.J., Tannert, T. and Vallée, T. (2010) Waveform Analysis of Acoustic Emission Monitoring of Tensile Tests on Welded Wood-Joints. Journal of Acoustic EMISSION, 28, 59-67.
7. Keprt, J. and Benes, P. (2007) Determination of Uncertainty in Calibration of Acoustic Emission Sensors. 4th International Conference on NDT, Chania, 11-14 October 2007.
8. Armentrut D.L. and Carpenter S.H. (1997) An Investigation of Luders Band Deformation and the Associated Acoustic Emission Al-4.5% Mg Alloys. Journal of Acoustic Emission, 15, 43-51,
9. Carpenter, S.H., Ono, K. and Armentrout, D. (2006) Acoustic Emission of Sensitized 304 Stainless Steel with Simultaneous Hydrogen Charging. Journal of Acoustic Emission, 24, 119-126.
10. Vahaviolos, S.J. (1998) ASTM STP 1353 Acoustic Emission Standards and Technology Update. ASTM, Florida, 76, 77.
11. Keeprt, J. and Benes, P. (2008) A Comparison of AE Sensor Calibration Methods. Journal of Acoustic Emission, 26, 60-70.
12. Romrell, D.M. (1973) Acoustic Emission Weld Monitoring of Nuclear Components. Journal of Welding Research Supplement, 81-92.
13. ASTM E94-04 (2010) Standard Guide for Radiographic Examination.
14. ASTM E390-11, Standard Reference Radiographs for Steel Fusion Welds.
15. ASTM E1815-08 (2013) Standard Test Method for Classification of Film Systems for Industrial Radiography.
16. ASTM E1032-12, Standard Test Method for Radiographic Examination of Weldments.
17. ASTM E592-99 (2009) e11, Standard Guide to Obtainable ASTM Equivalent Penetrameter Sensitivity for Radiography of Steel Plates 1/4 to 2 in. (6 to 51 mm) Thick with X Rays and 1 to 6 in. (25 to 152 mm) Thick with Cobalt-60.
18. Raasekhar, P., Bhat, M.R., Murthy, C.R.L., Phanira, C., Jayakumar, T. and Baldev Raj (2006) Discrimination of Nuclear grade Steel Samples Subjected to Different Heat Treatment Procedures Based on Acoustic Emission (AE) Data Profiling a Crack Initiation. Proceedings of National Seminar on Non-Destructive Evaluation, Hyderabad, 7-9 December 2006.
19. Aljets, D., Chong, A., Wilcox, S. and Holford, K. (2010) Acoustic Emission Source Location in Plate-Like Structures Using a Closely Arranged Triangular Sensor Array. Journal of Acoustic Emission, 28, 85-98.

20. George Varkey Strategic Electronics Group (2003) On-Line Processing of Acoustic Emission Signals. Awareness Seminar on Digital Signal Processing (ASDSP-2003), Paper 7, 1-12.
21. Manthei, G. (2010) Characterization of Acoustic Emission Sensors. Proceedings of European Conference on Acoustic Emission Testing, Venna, 8-10 September 2010.
22. Ennaceur, C., Lakshimi, A., Hervve, C., Ediouni, M. and Cherfaui, M. (2004) Acoustic Emission Technique and Potential Difference Method for Detecting the Different Stages of Crack Propagation in Carbon and Stainless Steels. Journal of Acoustic Emission, 22, 71-76.
23. Mukhopadhyay, C.K., Ray, K.K., Jayakumar, T. and Raj, B. (2000) Acoustic Emission Stress Intensity Factor Relations for Tensile Deformation of Notched Specimens of AISI Type 304 Stainless Steel. Materials Science and Engineering: A, 293, 137-145.

Chapter 2

Wavelet Packet Decomposition to Characterize Injection Molding Tool Damage ‡

Tomaž Kek [1,†], Dragan Kusić [2,†] and Janez Grum [1,*,†]

[1] *Faculty of Mechanical Engineering, University of Ljubljana, Aškerčeva 6, Ljubljana 1000, Slovenia*

[2] *TECOS Slovenian Tool and Die Development Centre, Kidričeva 25, Celje 3000, Slovenia*

ABSTRACT

This paper presents measurements of acoustic emission (AE) signals during the injection molding of polypropylene with new and damaged mold. The damaged injection mold has cracks induced by laser surface heat treatment. Standard test specimens were injection molded, commonly used for examining the shrinkage behavior of various thermoplastic materials. The measured AE burst signals during injection molding cycle are presented. For injection molding tool integrity prediction, different AE burst signals' descriptors are defined. To lower computational complexity and increase performance, the feature selection method was implemented to define a feature subset in an appropriate multidimensional space to characterize the integrity of the injection molding tool and the injection molding process steps. The feature subset was used for neural network pattern recognition of AE signals during the full time of the injection molding cycle. The results confirm that acoustic emission measurement during injection molding of polymer materials is a promising technique for characterizing the integrity of molds with respect to damage, even with resonant sensors.

Keywords: acoustic emission; injection molding; cracks; feature vector; pattern recognition

1. INTRODUCTION

Injection molding is still regarded as one of the most important and very popular manufacturing process because of its simple operation steps. A typical production cycle begins with the mold closing, followed by the injection of melt into the mold cavity. Once the cavity is filled, holding (packing) pressure is applied to compensate for the material shrinkage. In the next step, the screw turns, feeding the material for the next shot to the screw tip. This causes the screw to retract as the next cycle is almost ready. Once the molded part is sufficiently cooled, the mold opens and the part is finally ejected.

Chen and Turng [1] defined three categories of variables that characterize the hierarchy and dependencies of an injection-molding control system. The first-level variables are defined as machine variables and include barrel temperatures in several zones, pressure, sequence and motion, *etc.* The second-level variables are process-dependent variables that include melt temperature in the nozzle, runner and mold cavity, melt pressure in the nozzle and cavity, melt-front advancement, maximum shear stress, and rate of heat dissipation and cooling. The third level incorporates quality measures such as part weight and thickness, shrinkage and warpage, *etc.* Current research mostly deals with the control tasks at the first two levels. For quality control purposes, this is often achieved indirectly through online dynamic process control or offline statistical process control (SPC). The majority of commercially available in-mold sensors for monitoring second level variables are pressure and temperature sensors.

Zhang *et al.* [2] applied a mechanical wireless data transmission technique using ultrasonic waves as the information carrier for online injection mold cavity pressure measurements. Ultrasonic transmitters with specific frequency characteristics were designed, modeled, simulated, and prototyped for pressure data retrieval from an enclosed machine environment. Lijuan Zhao *et al.* [3] developed an ultrasonic monitoring technique using a high-temperature ultrasonic transducer for real-time diagnostics in polymer processing and tracking of morphologic changes in injection molding processing.

The location and propagation of a crack in the metal can be detected using the acoustic emission technique, as already reported by many researchers. The major AE signal sources detected by NDT are crack growth and plastic deformation of the steel [4]. Acoustic emission (AE) as transient elastic waves has drawn a great deal of attention because of its applicability to online surveillance and capability to acquire data with high sensitivity [5,6,7]. Cao *et al.* [8] investigated acoustic emission (AE) signals during four-point bending fatigue crack propagation in the base metal and weld of −16 Mn steel. In the first stage of fatigue damage process micro-crack initiation was a dominant AE source. During stable crack extension (intermediate stage), stacking and slipping of dislocations ahead of the crack tip are sources of AE signals. During unstable crack extension (final stage) shearing of ligaments and connectivity between dimples are AE sources. Input parameters of their neural network model are based on burst energy, peak amplitude, duration, and counts. Mukhopadhyay, *et al.* [9] investigated AE signals during fracture toughness tests of steel CT specimens. The AE generated during fracture toughness tests has been attributed to the plastic zone formation at the crack tip and initiation and/or extension of crack. They measured burst peak amplitude and *RA* value and used the differentiation of the AE burst signals' RMS voltage, count, and energy with respect to time by taking the average of the slopes of adjacent data points to evaluate fracture toughness. Drummond *et al.* [10] investigated enhancement of proof and fatigue testing procedures for wire ropes by incorporating AE signals. They suggested that to differentiate between signals from wire breaks and other sources, the most effective AE signal discriminators are peak amplitude and (cumulative) burst energy. Also, the rope's condition can be ascertained without the necessity of continuous monitoring and the AE emission is indicative of the level of damage. Ki-Bok Kim *et al.*[11] analyzed fatigue crack growth in standard CT specimens with a hydraulic loading machine. The objective of their research was the development of a neural network-based model for the prediction of stress intensity factor as a function of five basic AE parameters (ring down count, rise time, energy, event duration, peak amplitude). Their research proved that AE energy increases gradually as the number of fatigue cycles increases. Change in the AE energy with the fatigue cycle number could be one of the effective parameters to estimate the activity of the crack propagation and the stress intensity factor.

The AE method provides a huge potential for extensive *in situ* failure analysis of materials, since it provides information on sub-macroscopic failure phenomena as well as on the overall damage accumulation [12]. Generated AE signal contains useful information on the damage mechanism. One of the main issues of AE is to discriminate the different damage mechanisms from the detected AE signals. Each signal can be associated to a pattern composed of multiple relevant descriptors [13,14,15,16]. Then the patterns can be divided into clusters representative of damage mechanisms according to their similarity by the use of multivariable data analyses based on pattern recognition algorithms [17].

Piezoelectric sensors can measure ultrasound with a sensitivity of about 1 V/nm as displacement sensors and a few V/(mm/s) as velocity sensors. The dynamic amplitude range of piezoelectric sensors is around 120 dB. Another useful property when measuring ultrasound is that they measure only dynamic events and so automatically compensate for low-frequency motion of measuring objects that are caused by environmental vibrations. This intrinsic low-frequency cutoff is a consequence of the leakage of the accumulated charge. This acts as a high-pass filter that determines the low-frequency cutoff (compensational bandwidth) through a time constant given by the capacitance and resistance of the device [18]. Additionally, piezoelectric sensors are also insensitive to electromagnetic fields and radiation, enabling measurements under harsh conditions.

Correlation of basic AE burst signal parameters (peak amplitude, energy, *RMS* value, *RA* value, burst count, *etc.*) and crack propagation in steel are well described in the literature. The detection of damage in molds during the injection molding process is a topic that has not gained sufficient coverage in the literature, although injection molding is one of the most widely used production process. In this paper we present the possibility of damage detection and process steps based on time domain, frequency domain variables, and wavelet packet transform descriptors of AE bursts.

The first steps towards characterizing injection molding tool damage and process phases are the determination and calculation of AE descriptors for acquired AE bursts (Section 3.1). These descriptors are stacked to form a feature (measurement) vector *Z*. The next step is unsupervised fuzzy C means (FCM) clustering to separate the bursts characteristic for damage (cluster 5) of injection molding tool from the AE bursts characteristic for injection of the melt and holding stage of the injection molding process (Section 3.2). AE bursts of cluster 5 can be acquired only during injection molding with a damaged tool. Next we introduced a feature selection method to select the 10-dimensional feature subset from the 73-dimensional feature vector to improve suitability for classification (Section 3.3). Next, the feature subset was used for neural network pattern recognition of AE signals during the full time of the injection molding cycle (Section 3.4). At the end of the paper we present a classification of all input vectors and an evaluation of the proposed method to characterize injection molding tool damage and process phases.

2. EXPERIMENTAL SECTION

Acoustic emission signals were captured during the production cycle of standard test specimens, designed for shrinkage evaluation of various plastic materials. We used an injection molding tool (TECOS, Celje, Slovenia) with one cavity for production of standard D2 ISO test specimens with square dimensions of 60 × 60 × 2 mm^3, which can be used for a variety of tests according to ISO 294-3. The injection molding tool is made of OCR12VN steel. A polypropylene material isofil H40 C2 F NAT manufactured by Sirmax (Cittadella, Italy), mainly used in the automotive industry, was employed. A Vallen-Systeme GmbH acoustic emission measurement system AMSY-5 (Vallen-Systeme GmbH, Icking, Germany) was used to capture and analyze the AE signals at 5 MHz sampling frequency. Two piezoelectric AE sensors VS150-M with a measuring range between 100 and 450 kHz and resonance at 150 kHz were mounted on both sides of the tool via waveguides, as shown in Figure 1. These sensors are widely used in AE testing and monitoring and are low cost. The sensor also covers frequencies characteristic for AE signals generated during plastic deformation and fatigue [19]. We used an 18-kHz high pass filter (Vallen-Systeme GmbH, Icking, Germany). AE waveguides protect the sensors against the high surface temperature of injection molds during the production cycle, possible fumes exhaled out of the mold, and damage in case operator intervention is needed to help remove the specimen. The length of the waveguide is determined based on experience to protect sensors against the aforementioned hazards. PZT sensors were connected to the AMSY-5 measurement system via two AEP4 preamplifiers (Vallen-Systeme GmbH, Icking, Germany) with a fixed gain of 40 dB. The amplitude threshold was set at 40 dB. Experiments were conducted on a hydraulic injection molding machine KM 80 CX-SP 380 (KraussMaffei Group, Munich, Germany), for a broad spectrum of injection molding process parameters.

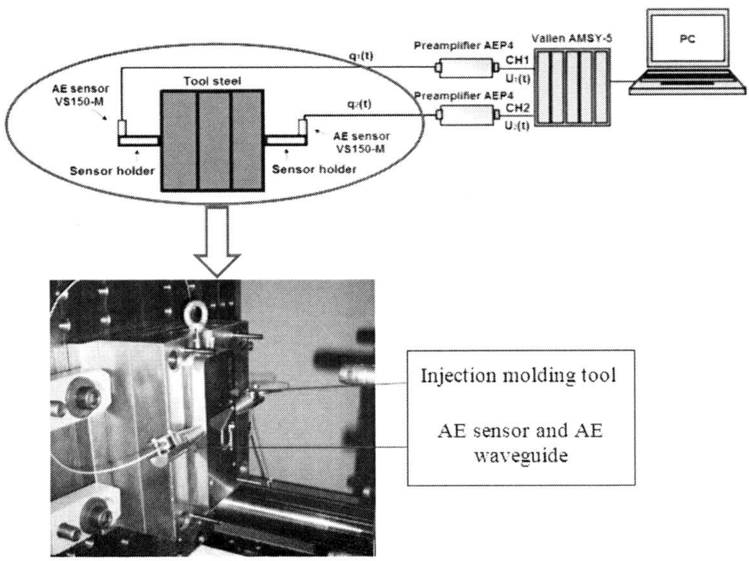

Figure 1. Experimental setup for AE monitoring during the injection molding process.

3. RESULTS AND DISCUSSION

AE monitoring is a very promising method for the detection of injection mold defects. Namely, the rapid increase of mold pressure during the filling and holding stages of the injection molding process can trigger defects (acoustic emission sources) in the tool steel. Acoustic emission testing can detect dynamic processes associated with the degradation of structural integrity.

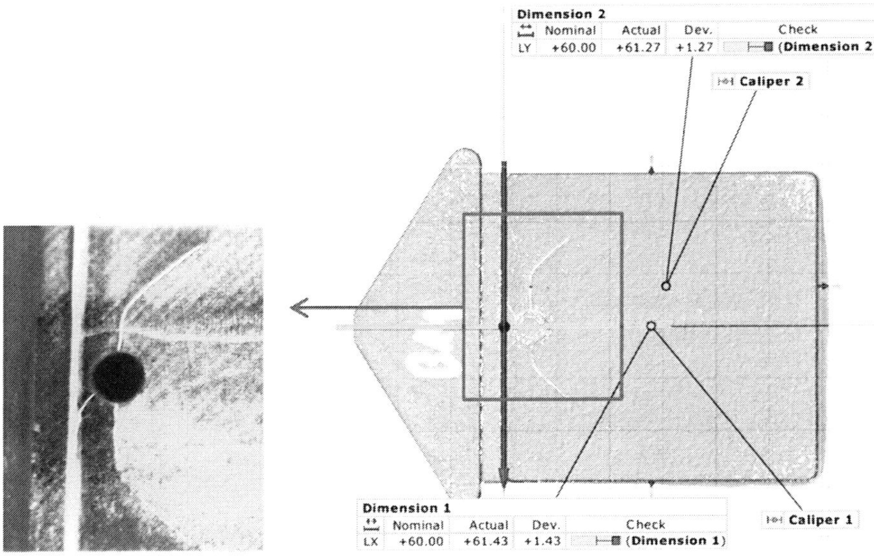

Figure 2. Cracks in the injection molding tool and accompanying PP specimen.

Experiments were conducted using a new injection mold tool and a damaged tool, respectively. Surface cracks were generated by successive local laser surface hardening of the tool steel. Both tools were tested using magnetic particle testing with fluorescent magnetic suspension. The cracks detected on the damaged tool are presented in Figure 2. After 24 h, the standard injection-molded polypropylene test specimens were scanned with an optical ATOS II SO 3D-digitizer. The ATOS II SO 3D digitizer is based on the triangulation principle whereby a sensor unit first projects different fringe patterns onto the scanned object's surface, which is then recorded by two cameras. Each measurement generates up to 4 million data points. In order to digitize a complete object, several individual measurements from different angles are required. The cracks on the damaged tool are also reflected on the surface of injection molded specimens. Shape deviations (Figure 2) in two dimensions from the ideal shape are stated in mm.

Figure 3 shows peak amplitudes and energy of AE bursts during the manufacturing of test specimens produced in the new and damaged mold, respectively. Absence of damage in the new tool offers us the opportunity of classifying measured AE bursts as process-orientated signals. On the other hand, the presence of a damaged area in the injection molding tool causes a considerable increase in the number of AE bursts detected during the filling and holding phases, above all in an amplitude range between 40 and 60 dB. Consequently, we can ascribe the additional AE bursts measured during the filling and holding phase to active AE sources from injection mold defects. A rapid increase in mold pressure during the filling phase could stimulate plastic zone formation, stacking, and slipping of dislocations ahead of the crack tip and crack extension [8,9] in the tool steel. Plastic deformations in the vicinity of the crack in the steel, mostly related to sliding dislocations, and crack growth cause distinctive AE bursts [20]. We connected AE signals with propagation of damage in the insert but this incorporates, based on the research of other authors, several physical phenomena that we described in the article as possible sources. The injection molding cycle ends after 16 s when the part is ejected out of the mold.

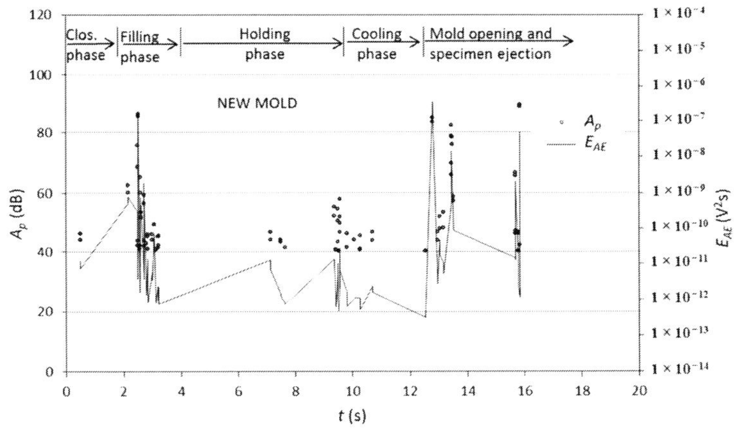

(a)

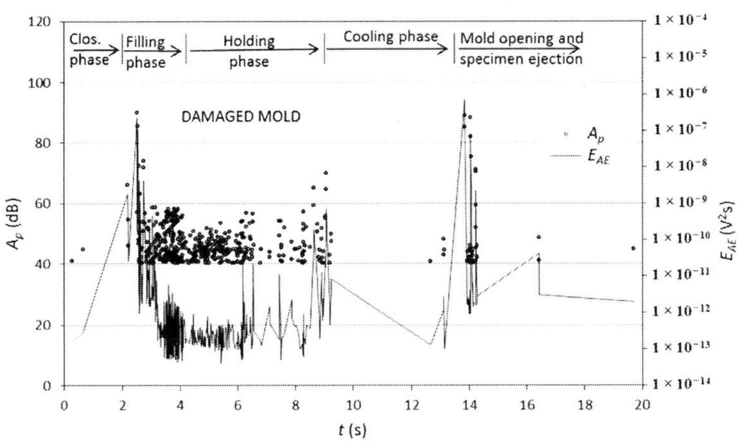

(b)

Figure 3. AE bursts' amplitude and energy during injection molding with (**a**) a new mold and a (**b**) damaged mold.

3.1. AE Burst Descriptors

Measured AE bursts are used for derivation of AE descriptors. These descriptors are stacked to form a feature (measurement) vector $Z = (z_1, z_2, ..., z_{73})$, representing acquired AE bursts during different phases of injection molding process. The union of 704 feature vectors, in this paper, composes feature space. AE burst is extracted from acquired AE waveform based on defined rearm time; in our research this was 0.46 ms. An AE burst is over when the rearm time elapses due to the absence of threshold crossings.

Calculated time domain descriptors based on acquired AE bursts are: peak amplitude value A_p, duration d, rise time Rt, and RA value. RA value is the ratio of rise time and amplitude of AE burst [15]. We calculated frequency domain descriptors. The P_L parameter characterizes the lower part of the power spectrum of each AE burst (a = 90 kHz, b = 190 kHz). Parameter P_H characterizes the higher part of the power spectrum (a = 250 kHz, b = 350 kHz), and parameter P_S covers all power spectrum frequencies (a = 50 kHz, b = 550 kHz). The parameters P_L, P_H, and P_S carry information about the energies at various frequencies in a sensor's frequency spectrum:

$$g_i = \frac{\overline{y y i}}{m} \tag{1}$$

$$m = \max\left(\mathbf{Y} \cdot \overline{\mathbf{Y}}\right) \tag{2}$$

$$P_L = \sum\nolimits_{a=90}^{b=190} g_i \tag{3}$$

$\mathbf{Y} = (y_1, y_2, ..., y_n)$ represents the vector of the discrete Fourier transform using FFT.

We introduced the kurtosis and skewness parameters describing the amplitude distribution of recorded AE burst. Kurtosis is a measure of the "peakedness" of the probability distribution for a real-valued random variable. In our case we calculated the kurtosis of measured AE burst amplitude values. Kurtosis was used in our research to characterize infrequent extreme deviations of amplitude values in AE burst as:

$$K = \frac{E(a_i - \overline{a})^4}{s^4} \tag{4}$$

where s is a sample standard deviation, a_i is an amplitude value.

Skewness S is defined as a third central moment of amplitude values in AE burst about its mean value. Skewness is in our research was used as a measure of asymmetry about the mean amplitude value of AE burst and is defined as:

$$S = \frac{E(a_i - \overline{a})^3}{s^3} \tag{5}$$

Wavelet Packet Analysis Descriptors

Wavelet analysis overcomes the pitfalls of Fourier methods and is suitable for processing AE signals during injection molding. The wavelet packet analysis is a subtler multiresolution analysis, which decomposes signal into multi-levels in the whole frequency band [21]. The detail of the signal is decomposed as well, so the frequency resolution is improved [22]. The wavelet packets include information on signals in different time windows at different resolutions. Each packet corresponds to some frequency band. Some packets contain important information while others are relatively unimportant.

Wavelet packet transform of energy limited signal $X(t)$ can be computed using the algorithm below:

$$P_0^1(t) = X(t) \tag{6}$$

$$P_j^{2i-1}(t) = HP_{j-1}^i(t) \tag{7}$$

$$P_j^{2i}(t) = GP_{j-1}^i(t) \tag{8}$$

where Pij(t) is the ith packet on j^{th} resolution, with time parameter $t = 1,2,\ldots,2^{J-j}$, $i = 1,2,\ldots,2^j$, $j = 1,2,\ldots,J$, $J = \log_2 N$. Operators H and G are convolution sum defined as:

$$H\{\cdot\} = \sum_k h(k-2t) \tag{9}$$

$$G\{\cdot\} = \sum_k g(k-2t) \tag{10}$$

where $h(t)$ and $g(t)$ are a pair of quadrature mirror filters and a time parameter t is taken as a series of integers k ($t \rightarrow k = 1, 2, \ldots$).

The energy of each wavelet packet is defined as:

$$E_j^i = \sum_{t=1}^{2^{J-j}} \left(P_j^i(t)\right)^2 \tag{11}$$

The energy of wavelet packets on selected resolution *j* is processed by normalization, and normalization value of each packet is

$$R_j^i = E_j^i / \sum_{i=1}^{2^j} E_j^i \tag{12}$$

Based on described AE signal descriptors, we have set the feature vector as $Z = (P_H, P_L/P_H, P_L/P_S, K, S, A_p, Rt, RA, d,$ $Rt, RA, d, R_5^1, R_5^2, \ldots, R_5^{32}, E_5^1, E_5^2, \ldots, E_5^{32})$.

3.2. Unsupervised Fuzzy C Means Clustering

We used unsupervised fuzzy C means clustering (FCM) for classification of feature vectors Z_j (j = 1, n) into D = 2 clusters. FCM clustering was found to be an effective algorithm for separating AE patterns composed of multiple features extracted from the AE waveforms [17,23]. Cluster 2 represents AE bursts during injection of the melt, cluster 3 represents process-orientated AE bursts during the holding stage, and cluster 5 bursts indicate damage during the injection and holding stages of the injection molding process. We used AE bursts measured during the injection molding process with a damaged tool, since we can also acquire AE signals representing cluster 5. We used fuzzy partitioning to connect feature vector Z_j (j = 1, n) to cluster ω_i (i = 1, D) with different membership grades $u_i(Z_j)$ between 0 and 1.

With FCM we find cluster centers $V = [K_1/K_2]$ that minimize the function J:

$$J(U,V) = \sum_{j=1}^{n} \sum_{i=1}^{D} \left[u_i(Z_j)\right]^f d_e^2(Z_j, K_i) \tag{13}$$

U is the membership matrix with D lines and n columns with elements $u_i(Z_j)$ representing the membership value of the j^{th}feature vector Z_j to the cluster ω_i with a condition:

$$\sum_{1=1}^{D} u_i(Z_j) = 1 \tag{14}$$

f is a scalar representing the fuzzy degree. The square of Euclidian distance d_e between the feature vector and the cluster is:

$$d_e^2(Z_j,K_i) = (Z_j\text{-}K_i)'(Z_j\text{-}K_i) \tag{15}$$

The algorithm of FCM has the following steps [17]:

1. the membership matrix U is randomly initialized with values between 0 and 1 that represent the membership values$u_i(Z_j)$ of the n feature vectors Z_j (j = 1, n) to each cluster ω_i (i = 1, D).

2. The new cluster centers K_i are calculated according to:

$$K_i - \frac{\sum_{j=1}^{n}\left[u_i\left(x_j\right)\right]^f x_j}{\sum_{j=1}^{n}\left[u_i\left(x_j\right)\right]^f} \tag{16}$$

3. The membership matrix U is updated such that:

$$u_i\left(Z_j\right) - \left[\sum_{k=1}^{D}\left(\frac{d\left(Z_j, K_i\right)}{d\left(Z_j, K_k\right)}\right)^{2/(f-1)}\right]^{-1} \tag{17}$$

Steps 2 and 3 are iterated until the improvement over the previous iteration is below a threshold θ where β are the iteration steps:

$$\| U^{\beta} - U^{\beta-1} \| < \theta,\ 0 < \theta < 1 \tag{18}$$

Figure 4 shows examples of signals designated as process-orientated AE burst (cluster 3) and AE burst indicating damage (cluster 5). Figure 5 shows wavelet packet coefficients associated with all 32 nodes on 5th resolution of the wavelet packet tree for signals represented on Figure 4. With a red color, coefficients associated with nodes 8, 16, and 24 on 5th resolution are emphasized.

Figure 6 show feature vectors representing process-orientated AE bursts (cluster 3) and feature vectors indicating damage (cluster 5) during the holding stage of the injection molding process. Axes of 3D plot are normalization values of wavelet packets energy at 5th resolution. A black circle indicates the position of the feature vector of the AE burst from Figure 4a and a black square the position of the feature vector of the AE burst from Figure 4b. Cluster 3 feature vectors are concentrated around the origin of the coordinate system while Cluster 5 feature vectors are spread away from the origin.

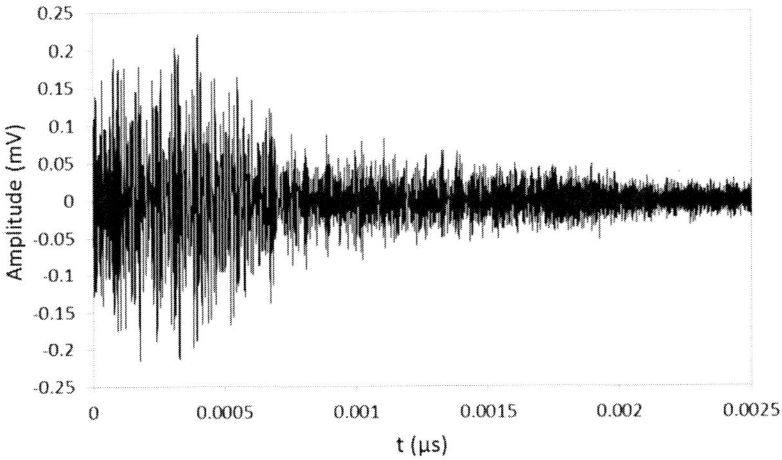

(a)

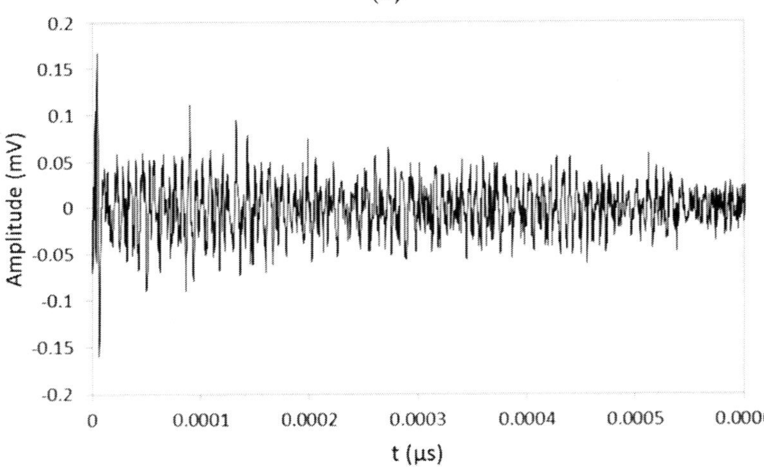

(b)

Figure 4. Signals designated as process-orientated AE burst (**a**) and AE burst indicating damage (**b**).

(a)

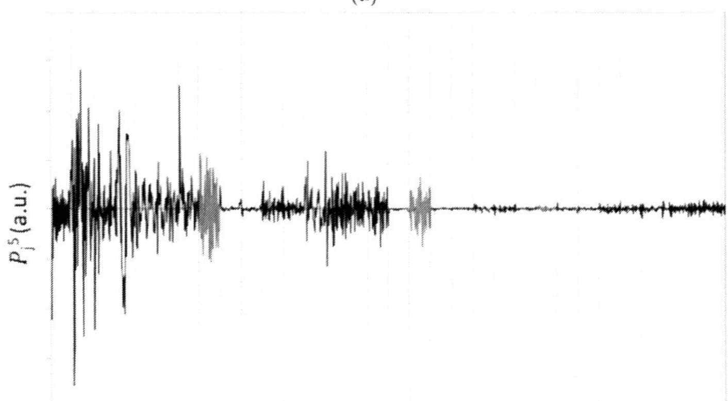

(b)

Figure 5. Wavelet packet coefficients associated with all the nodes on 5th resolution of the wavelet packet tree for above process-orientated AE burst (**a**) and AE burst indicating damage (**b**).

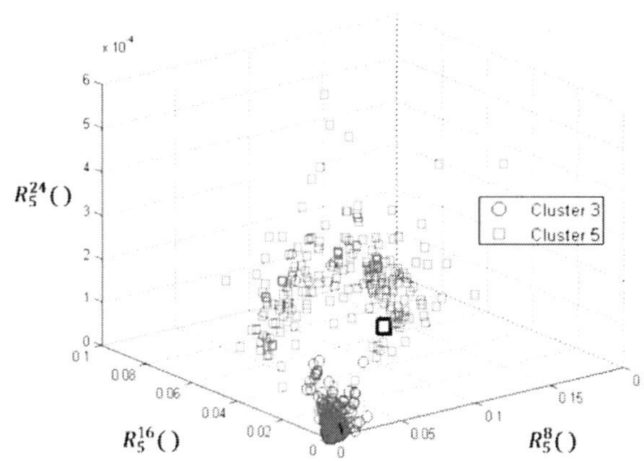

Figure 6. Visualization of feature vectors designated as cluster 3 and 5.

3.3. SFS Feature Selection

The dimensions of the feature vector cannot be arbitrary large. One reason is the computational complexity, and the second reason is that the dimension ultimately causes a decrease of performance [14,24]. Our goal is to reduce the dimensions of the data by finding a small set of important features that can give good classification performance. With feature selection we find a subset $F_i(W) = \{z_w | w=1, ..., W\}$ that outperforms all other subsets with dimension W as:

$$J(F_i(W)) \geq J(F_j(W)) \; for \;\; all \;\; j \in \{1, \ldots, q(W)\} \tag{19}$$

where $q(W)$ is the number of evaluations of performance measure $J(F_j(W))$.

We used sequential forward selection (SFS). The method starts at the bottom (the root of the tree) with an empty set and proceeds its way to the top without backtracking. At each level of the tree, SFS adds one feature to the current feature set by selecting from the remaining available measurements the element that yields a maximal increase of the performance. A structure of feature selection is shown in Figure 7, where z_n is the designation of a feature.

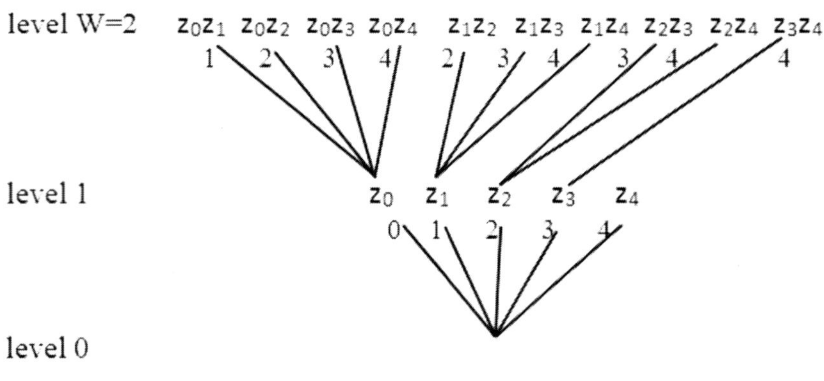

Figure 7. A bottom-up tree structure for feature selection [24].

We used a k-Nearest Neighbor (k-NN) classification algorithm as a criterion. With k-NN the input consists of the k closest training examples in the feature space. In our research k = 3. The output of the algorithm is a class membership. An object is classified by a majority vote of its neighbors, with the object being assigned to the class most common among its k nearest neighbors.

The major focus of our research was detection of damage in an injection molding tool during the injection molding process. This was the reason for using SFS with k-NN on feature vectors of AE burst signals acquired during the holding phase of the injection molding cycle on intact and damaged molds. We selected feature subset Z_s with a size of 10. Feature subset was used for neural network pattern recognition of AE signals during the full time of the injection molding cycle. The selected features based on SFS with *k*-NN are: Z_s = $(R_5^8, R_5^{16}, R_5^{24}, R_5^{21}, R_5^{18}, R_5^{20}, d, R_5^6, R_5^{32}, A_p)$.

3.4. Classification of All Input Vectors Based on AE Signals

For the purposes of presenting all feature vectors *Z*, we introduced principal component analysis (PCA) as an unsupervised feature reduction method. PCA offers us transformation of high-dimensional feature vector *Z* to a much lower dimensional vector *i.e., P*. PCA is mathematically defined [25] as an orthogonal linear transformation that transforms the data into a new coordinate system such that the greatest variance by some projection of the data comes to lie on the first coordinate (called the first principal component), the second greatest variance on the second coordinate, and so on. Figure 8shows the distribution of feature vectors *Z* in a 3D coordinate system of the first three principal components. Feature vectors*Z* are divided into five classes; Class 1—the closing of the mold, Class 2—the injection of the melt with maximum pressure, Class 3—holding (packing) stage, Class 4—opening of the tool, Class 5—AE signals characterizing damage of the tool.

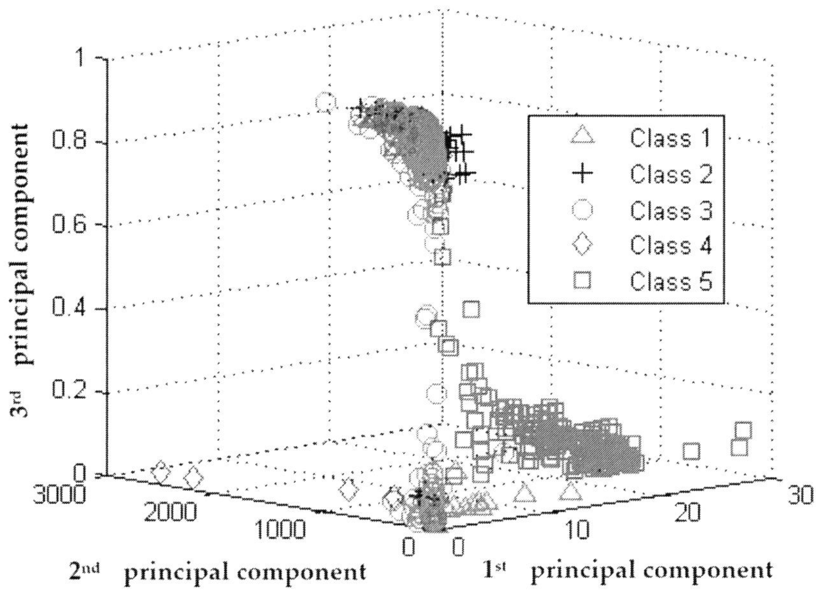

Figure 8. PCA visualization of all feature vectors *Z* during injection molding cycles.

Based on the measured AE bursts, we set 10-dimensional feature subset vector Z_S with instances that are real-valued descriptors (variables). Vector Z_S offers us defining points in appropriate multidimensional space to characterize the process. We used a classification to identify the categories or sub-populations based on a training dataset, containing instances whose category membership is known. The training data consist of a set of 704 input vectors *Zs* that have been designated based on the time moment of their acquisition during the injection molding cycle and unsupervised FCM for vectors of the damaged tool. For neural network pattern recognition we used a feed-forward network with the F Tan-Sigmoid transfer function in both the hidden and the output layer. We used 10 neurons in the hidden layer and five output neurons, because there are five target categories associated with each input vector. The network is trained with scaled conjugate gradient back-propagation.

Table 1. Classification of all input vectors in all confusion matrix.

All Confusion Matrix							
Output Class	1	8 1.1%	0	1 0.1%	0	3 0.4%	33.3%
	2	0	27 3.8%	0	4 0.6%	0	12.9%
	3	32 4.5%	0	378 53.7%	8 1.1%	4 0.6%	10.4%
	4	0	3 0.4%	1 0.1%	26 3.7%	0	13.3%
	5	2 0.3%	0	1 0.1%	0	206 29.3%	1.4%
	E	81.0%	10.0%	0.8%	31.6%	3.3%	8.4%
	o.c./t.c.	1	2	3	4	5	E
				Target Class			

E—Means error.

For training of our network, data were randomly distributed into three subsets. A training subset (70% of the data) is used for computing the gradient and updating the network weights *W* and biases *B*. In the validation subset (15% of the data) the error is monitored during the training process. The network weights and biases are saved at the minimum of the validation subset error. The testing subset (15% of the data) has no effect on training but provides an independent measure of network performance during and after training. The results of classification of all input data are shown in Table 1. The table has five output classes (o.c.) and five target classes (t.c.) with a number of vectors classified into appropriate classes.

The presented data show that we have on average 91.6% correctly classified subset feature vectors Z_S. The lowest error E = 0.8% is for Class 3 vectors, representing process-orientated AE signals during the holding (packing) stage. Detection of damage (Class 5) in the injection molding tool is 96.7% correct. Misclassified vectors representing damage are almost evenly distributed between Class 1 and 3. The highest error, 81%, is for detection of tool closing, *i.e.*, Class 1 vectors Z_S. Also, the class 4 vectors representing opening of the tool show relatively high error $E = 31.6\%$.

4. CONCLUSIONS

In this work we presented an AE monitoring system with a resonant sensor that is able to detect the damaged area in an injection molding tool. We made a close comparison of AE signals captured from a new mold and a damaged mold under the same injection molding process conditions.

The most useful information about the injection molding tool integrity based on AE bursts is gained during the filling and holding stage of the injection molding process, when the mold pressure is high enough.

We introduced time domain, frequency domain, and wavelet packet transform descriptors to classify feature vectors of AE bursts during the injection molding cycle. A 10-dimensional subset feature vector with real-valued explanatory variables is proposed for classification of AE burst signals.

Based on SFS, the feature subset consists mainly of normalization values of the energy of wavelet packets, which turned out to be the most informative for prediction of damaged areas in the injection molding tool.

Neural network pattern recognition results show very good prediction probability of tool damage and very good prediction of process-orientated signals during the filling and holding phases of the injection molding cycle.

ACKNOWLEDGMENTS

The research work was partly financed by the European Union, European Social Fund, and by the Slovenian Ministry for Higher Education, Science and Technology.

AUTHOR CONTRIBUTIONS

All the authors conceived and designed the experiments; Tomaž Kek performed the experiments, analyzed the data and wrote the paper; Dragan Kusić performed the experiments and contributed materials and injection molding machine; Janez Grum led the research work and reviewed the work.

REFERENCES

1. Chen, Z.; Turng, L.S. A Review of current developments in process and quality control for injection molding. *Adv. Polym. Technol.* **2005**, *24*, 165–182.
2. Zhang, L.; Theurer, C.B.; Gao, R.X.; Kazmer, D.O. Design of ultrasonic transmitters with defined frequency characteristics for wireless pressure

sensing in injection molding. *IEEE Trans. Ultrason. Ferroelectr. Freq. Control* **2005**,*52*, 1360–1371.

3. Lijuan, Z.; Yu, L.; Chen, P.; Cheng-Kui, J.; Kuo-Ding, W. Real-time diagnosing polymer processing in injection molding using ultrasound. *J. Appl. Polym. Sci.* **2012**, *126*, 2059–2066.
4. Pollock, A.A. Fundamentals of Acoustic Emission Testing. In *Nondestructive Testing Handbook—Acoustic Emission Testing*, 3rd ed.; Moore, P.O., Miller, R.K., Hill, E.V.K., Eds.; American Society for Nondestructive Testing: Columbus, OH, USA, 2009; pp. 31–108.
5. Davoodi, S.; Mostafapour, A. Modeling acoustic emission signals caused by leakage in pressurized gas pipe. *J. Nondestruct. Eval.* **2013**, *32*, 67–80.
6. Harri, K.; Guillaume, P.; Vanlanduit, S. On-line damage detection on a wing panel using transmission of multisine ultrasonic waves. *NDT E Int.* **2008**, *41*, 312–317.
7. Mazal, P.; Pazdera, L.; Kolar, L. Basic acoustic emission signal treatment in the area of mechanical cyclic loading. *Int. J. Microstruct. Mater. Prop.* **2006**, *1*, 341–352.
8. Cao, J.; Luo, H.; Han, Z. Acoustic emission source mechanism analysis and crack length prediction during fatigue crack propagation in 16Mn steel and welds. *Procedia Eng.* **2012**, *27*, 1524–1537.
9. Mukhopadhyay, C.K.; Sasikala, G.; Jayakumar, T.; Raj, B. Acoustic emission during fracture toughness tests of SA333 Gr.6 steel. *Eng. Fract. Mech.* **2012**, *96*, 294–306.
10. Drummond, G.; Watson, J.F.; Acarnley, P.P. Acoustic emission from wire ropes during proof load and fatigue testing.*NDT E Int.* **2007**, *40*, 94–101.
11. Kim, K.-B.; Yoon, D.-J.; Jeong, J.-C.; Lee, S.-S. Determining the stress intensity factor of a material with an artificial neural network from acoustic emission measurements. *NDT E Int.* **2004**, *37*, 423–429.
12. Baensch, F. Damage Evolution in Wood and Layered Wood Composites Monitored *in Situ* by Acoustic Emission, Digital Image Correlation and Synchrotron Based Tomographic Microscopy. PhD Thesis, ETH Zürich, Zürich, Switzerland, 17 December 2014.
13. Zaki, A.; Chai, H.K.; Aggelis, D.G.; Alver, N. Non-Destructive evaluation for corrosion monitoring in concrete: A review and capability of acoustic emission technique. *Sensors* **2015**, *15*, 19069–19101.
14. Momon, S.; Godin, N.; Reynaud, P.; Mili, M.R.; Fantozzi, G. Unsupervised and supervised classification of AE data collected during fatigue test on CMC at high temperature. *Compos. Part A* **2012**, *43*, 254–260.

15. Aggelis, D.G. Classification of cracking mode in concrete by acoustic emission parameters. *Mech. Res. Commun.* **2011**,*38*, 153–157.
16. Svečko, R.; Kusić, D.; Kek, T.; Sarjaš, A.; Hančič, A.; Grum, J. Acoustic emission detection of macro-cracks on engraving tool steel inserts during the injection molding cycle using PZT sensors. *Sensors* **2013**, *13*, 6365–6379.
17. Marec, A.; Thomas, J.H.; El Guerjouma, R. Damage characterization of polymer-based composite materials: Multivariable analysis and wavelet transform for clustering acoustic emission data. *Mech. Syst. Signal Process.* **2008**,*22*, 1441–1464.
18. Požar, T.; Možina, J. Detection of Subnanometer Ultrasonic Displacements. In *Fundamentals of Picoscience*; Sattler, K.D., Ed.; CRC Press: Boca Raton, FL, USA, 2014; pp. 553–578.
19. Kumar, J.; Punnose, S.; Mukhopadhyay, C.K.; Jayakumar, T.; Kumar, V. Acoustic emission during tensile deformation of smooth and notched specimens of near alpha titanium alloy. *Res. Nondestruct. Evaluation* **2012**, *23*, 17–31.
20. Hase, A.; Mishina, M.; Wada, M. Correlation between features of acoustic emission signals and mechanical wear mechanisms. *Wear* **2012**, *292–293*, 144–150.
21. Kunpeng, Z.; San, W.Y.; Soon, H.G. Wavelet analysis of sensor signals for tool condition monitoring: A review and some new results. *Int. J. Mach. Tools Manuf.* **2009**, *49*, 537–553.
22. Wang, X.H.; Zhu, C.M.; Mao, H.L.; Huang, Z.F. Wavelet packet analysis for the propagation of acoustic emission signals across turbine runners. *NDT E Int.* **2009**, *42*, 42–46.
23. Oskouei, A.R.; Heidary, H.; Ahmadi, M.; Farajpur, M. Unsupervised acoustic emission data clustering for the analysis of damage mechanisms in glass/polyester composites. *Mater. Des.* **2012**, *37*, 416–422.
24. Van der Heijden, F.; Duin, R.P.W.; de Ridder, D.; Tax, D.M.J. *Classification, Parameter Estimation and State Estimation*; John Wiley & Sons Ltd: West Sussex, UK, 2004.
25. Jolliffe, I.T. *Principal Component Analysis*, 2nd ed.; Springer: New York, NY, USA, 2002; pp. 10–28.

Chapter 3

Identification of a Critical Time with Acoustic Emission Monitoring during Static Fatigue Tests on Ceramic Matrix Composites: Towards Lifetime Prediction†

Nathalie Godin *, Pascal Reynaud, Mohamed R'Mili and Gilbert Fantozzi

INSA de Lyon, MATEIS (UMR CNRS 5510), 7 avenue Jean Capelle, 69621 Villeurbanne Cedex, France

ABSTRACT

Non-oxide fiber-reinforced ceramic-matrix composites are promising candidates for some aeronautic applications that require good thermomechanical behavior over long periods of time. This study focuses on the behavior of a SiC_f/[Si-B-C] composite with a self-healing matrix at intermediate temperature under air. Static fatigue experiments were performed below 600 °C and a lifetime diagram is presented. Damage is monitored both by strain measurement and acoustic emission during the static fatigue experiments. Two methods of real-time analysis of associated energy release have been developed. They allow for the identification of a characteristic time that was found to be close to 55% of the measured rupture time. This critical time reflects a critical local energy release assessed by the applicability of the Benioff law. This critical aspect is linked to a damage phase where slow crack growth in fibers is prevailing leading to ultimate fracture of the composite.

Keywords: ceramic matrix composite; static fatigue; lifetime assessment estimation; mechanical behavior; acoustic emission; damage identification

1. INTRODUCTION

Ceramic Matrix Composites (CMCs), and more particularly SiC/SiC composites are very attractive candidates for many high-temperature structural applications, because of their excellent creep resistance, high-temperature strength and light weight. Damage tolerance is achieved through the use of low shear strength fiber coatings that deflect cracks along the interfaces [1,2,3,4,5,6,7]. The multi-layered matrix was introduced in the new generations of SiC_f/[Si-B-C] composites in order to improve the lifetime under medium and high temperatures thanks to the formation of sealant glasses [8,9].

Future engine applications in civil aircrafts are foreseen for such composites. These applications require that the material can resist both severe conditions (temperatures up to 1200 °C under air) and very long static loadings at intermediate temperatures (400–700 °C during tens of thousands of hours). Various authors have studied the mechanical behavior of SiC_f/SiC composites and the damage mechanisms occurring at high temperatures [1,2,3,4,5,6,7]. Now more information is needed at intermediate temperatures: the oxidation kinetics of the different constituents are complex, and the effect of matrix sealing on the lifetime of the composite has to be examined. Expected lifetimes in use conditions are several thousands of hours, which can hardly be reached by laboratory tests for practical reasons. Therefore, a real-time prediction of the remaining lifetime during tests is necessary. It requires the monitoring of damage evolution for which AE measurement is a suitable technique. In fact, the AE technique consists in the recording and analysis of elastic waves created during material damage. It provides real-time data on initiation and evolution of damage in terms of location and mode.

Common analysis of AE is based on diagrams of cumulative hits or counts or histograms of amplitudes. Nevertheless, in the case of composite material, this approach is inadequate in order to identify the acoustic emission signature of several damage mechanisms. It was improved by grouping signals of similar shapes into clusters using classifier parameters [10]. Many works [11,12,13,14,15,16] have shown that AE techniques and multivariable classification techniques are the basis of pattern recognition tools. Kostopoulos [12], Godin [13,14], Moevus [15,16] have identified different classes of AE signals which were attributed to damage modes in oxide/oxide, glass/polyester and SiCf/[Si-B-C] composites. Moevus [16] analyzed the AE data using an unsupervised multi-variable clustering technique. This analysis showed that several types of signals may be distinguished, and that the obtained clusters were consistent with the expected damage modes in 3D SiC/SiC composites. In this way, a careful analysis of acoustic emission signals can lead to the discrimination of the different damage mechanisms occurring in a composite material. It is a possible solution for identification of damage during service with a view to component lifetime control.

A second analysis is based on a global AE analysis, on the investigation of liberated energy, with a view to identify a critical point. Many researchers investigated the elastic energy release during the failure process of materials [17,18,19,20,21,22,23,24]. They all observed that the energy release accelerated following a power law. These models are representative of an avalanche behavior very similar to that observed in seismicity.

This study focuses on static fatigue experiments realized at intermediate temperature under air on a SiC_f/[Si-B-C] composite with a self-healing matrix. A lifetime diagram is presented, and the evolution of several damage indicators is discussed. The objective of this approach is to propose a method based on acoustic energy in order to evaluate the remaining lifetime during long-term mechanical tests. This approach is based on the determination of energy released and identification of a critical point, in energy release during mechanical test. Therefore, two criteria have been defined which allow predicting the end of a static fatigue test knowing the AE activity emitted during the first half of the test. Therefore, beyond this characteristic point the criticality can be described by a power-law in order to evaluate time to failure. Moreover, a supervised classification method was used to differentiate the signals generated during fatigue tests performed on composites at intermediate temperature and build a specific library [16]. This library was used to identify the damage modes generated during fatigue tests performed at various temperatures, in order to establish a link between this critical time and the damage mechanisms.

2. EXPERIMENTAL PROCEDURE

2.1. Material

The material, manufactured by SAFRAN-HERAKLES Group (Bordeaux, France), had a multilayered [Si-B-C] reinforced matrix obtained by chemical vapor infiltration on SiC fibers and a PyC interface layer (Figure 1). This 2.5D woven composite is composed of Hi-Nicalon fibers, a pyro-carbon interphase layer and a self-healing matrix. This matrix has been processed by means of several chemical vapor infiltrations with various compositions, based on the ternary Si-B-C system. The external surface is protected by a seal-coat. The composite contains a volumic fraction of fibers equal to 35 vol.% and 20 vol.% of porosity. Dog-bone shaped specimens were used with the following dimensions: thickness 4 mm, width 16 mm, working length 40 mm with a constant cross section of 64 mm^2.

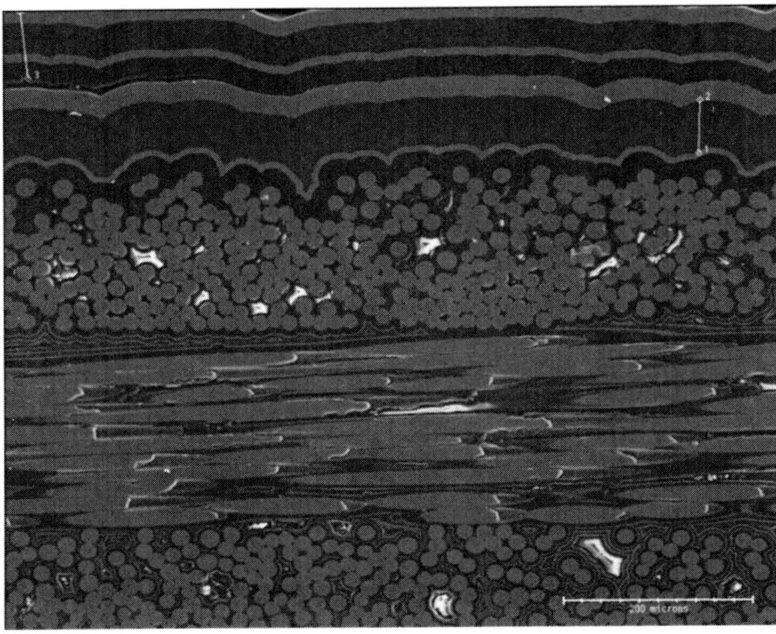

(a)

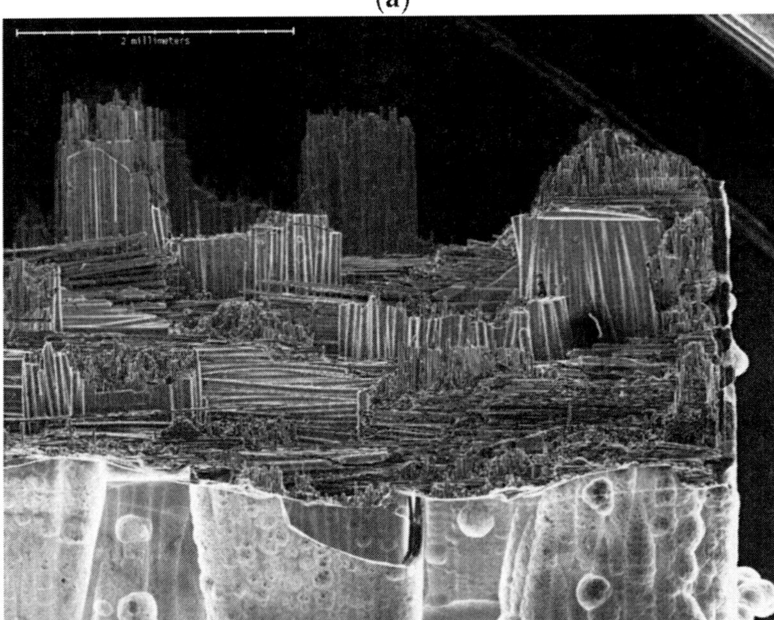

(b)

Figure 1. (**a**) SEM micrograph of the cross-section of SiC_f/[Si-B-C] composites; (**b**) Typical fracture surface of SiC_f/[Si-B-C] composite.

2.2. Static Fatigue Tests

Static fatigue tests were performed under uniaxial tensile loading with one direction of fibers parallel to the loading axis. Tests were carried out at 450 °C, 500 °C and 560 °C under air, on a pneumatic testing machine. Before loading, specimens were heated up to the testing temperature at a rate of 50 °C/min and held during 30 min to get a uniform temperature distribution in the gauge section. Specimens were loaded at a constant rate of 600 N/min up to a constant stress chosen in the range of 44% to 95% of the rupture stress. Specimen elongation was measured using a high temperature extensometer.

Seldom unload-reload cycles were applied every twelve hours in order to measure the secant elastic modulus changes during static fatigue and to evaluate the damage parameter D (Figure 2). E_0 is the elastic modulus in the undamaged state.

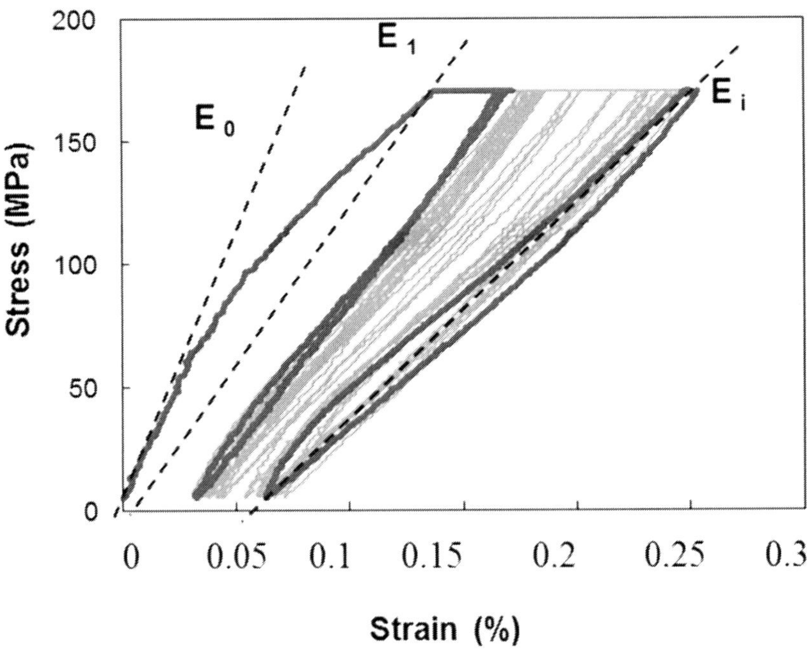

Figure 2. Schematic hysteresis loop and measure of the hysteresis loops modulus, E_0 is the initial Young's modulus, E_i is the secant elastic modulus for the unloading-reloading cycle *i*.

2.3. Acoustic Emission Monitoring

To detect acoustic emission, two MICRO-80 sensors (MISTRAS Group, Sucy en Brie, France) were put on the specimen inside the grips with vacuum grease as a coupling agent. Acquisition parameters were set as follows: preamplification 40 dB, threshold 32 dB, peak definition time 50 μs, hit definition time 100 μs, hit lockout time 1000 μs. AE signal descriptors such as amplitude, energy, duration, counts, average frequency, rise time and location as well as time, load and strain were recorded in real time by a MISTRAS 2001 data acquisition system.

For AE sources localization, AE wave velocity Ce_0 was determined before tests using a pencil lead break procedure on as received composites. This velocity was measured equal to 10,000 m/s. Since the elastic modulus decreases as damage occurs in the material, it is important to take into account the evolution of Ce during the mechanical test in order to better evaluate the location of the AE sources. As proposed by Morsher [25], the elastic secant modulus during unloading $E_i(\varepsilon)$ was measured during a cycled tensile test, where hysteresis loops were obtained at different strains. The velocity $Ce(\varepsilon)$ was then determined by using the Equation (1):

$$\frac{Ce(\varepsilon)}{Ce_0} = \sqrt{\frac{E_i(\varepsilon)}{E_0}} \tag{1}$$

where Ce_0 and E_0 are respectively the velocity and the elastic modulus in the undamaged state and $C_e(\varepsilon)$, $E_i(\varepsilon)$ respectively the velocity and the elastic modulus under a maximum strain ε. At the end of the tensile test, the velocity on SiC/SiC composite was found to be equal to 6480 m/s, instead of 10,000 m/s on the undamaged state. This decrease of wave velocity is thus not negligible. The location of sources has been calculated using the difference in times of arrival on each sensor. Only the signals coming from the working length of the specimens are then analyzed.

3. ACOUSTIC EMISSION ANALYSIS

3.1. Definition of the Acoustic Energy

Energy of an AE signal includes the energy released by the source at crack initiation and is affected by various parameters: wave propagation distance, damage attenuation, sensor/material surface coupling and sensor frequency response. Wave theory states that the energy of an acoustic wave decreases exponentially with the increase of propagation distance. Therefore, the following equation was proposed to describe the energy of recorded AE signals (for instance, at sensor 1) received from the source *n* [26,27]:

$$E_1(n) = E_s(n) \cdot A_1 \cdot e^{-B(L+x(n))} \tag{2}$$

$E_s(n)$ is the energy released at source n in the form of elastic waves. Due to differences in coupling between sensors and material surface or in sensors frequency responses, for a source at equal distance, the sensors may record significantly different amounts of energy. Thus, A_1 is the proportion of source energy that is recorded by sensor 1. It is a constant characteristic of sensor. $L + x(n)$ is the propagation distance from source *n* to the sensor 1 (2L is the distance between two sensors). The attenuation coefficient *B* is linked to the propagation medium, which may change with damage evolution. Similarly, AE signal energy received at sensor 2 is expressed as:

$$E_2(n) = E_s(n) \cdot A_2 \cdot e^{-B(L-x(n))} \tag{3}$$

The source energy is then defined as the square root of the product of the amounts of energy received at both sensors for each source:

$$E(n) = \sqrt{E_1(n) \cdot E_2(n)} \tag{4}$$

3.2. Attenuation Coefficient B

In order to evaluate energy attenuation [28], the ratio of AE signal energies recorded at both sensors is calculated for each source *n*. For an easier identification of attenuation coefficient *B*, *x(n)* is defined as the natural logarithm of this ratio. From Equations (2) and (3), it comes:

$$X(n) = \log\frac{E_1(n)}{E_2(n)} = \log\frac{A_1}{A_2} - 2 \cdot B \cdot x(n) \tag{5}$$

The second term corresponds to the effect of propagation, related to source location *x(n)* and to attenuation coefficient B. Both parameters can be estimated from *x(n)* for various sources (1,2,..., n,...) since *x(n)* is a linear function of *x*. The coefficient of attenuation B can be determined from the slope of the linear fit.

Because attenuation varies with damage, parameters estimation is done at successive time intervals. For a given time interval, values of *x(n)* are analyzed in many space intervals in order to consider uncertainties in localization of AE sources. Estimation of attenuation constants was performed as follows. Attenuation is evaluated for each time interval with the median values of *x(n)* in all space interval (width: 10 mm, overlapping: 5 mm). Each time interval is defined by 2000 AE successive sources. Each median value of *x(n)* corresponds to a few hundred AE sources located in the same space-time interval. The linear approximation is repeated in consecutive time intervals. At every increment of 500 sources, a new time interval is considered. The overlapping is set in order to accurately monitor evolution of both attenuation parameters.

3.3. Coefficient of Emission R_{AE}

The coefficient of emission RAE is defined [29], as the increment of energy ΔE recorded during a time increment Δt, divided by the total energy emitted during the initial loading of the sample:

$$R_{AE}(t) = \frac{1}{\mathrm{E_{loading}}} \frac{\Delta \mathrm{E}}{\Delta \mathrm{t}} \tag{6}$$

where $E_{loading}$ is the cumulative AE energy for all the signals recorded during the initial loading up to the nominal load of the test, ΔE is the cumulative AE energy for all signals recorded during the interval $[t; t + \Delta t]$.

3.4. Power Law

Benioff law [30], suggested initially for precursory phenomena of earthquakes, has been applied originally on composites [29]. Based on Equation (7), the increase of AE during fatigue is analyzed.

$$\Omega(t) = \sum_{i=1}^{N(t)} \sqrt{E_i} = \Omega_R + B\,(t_R - t)^{1-\gamma} \tag{7}$$

where E_i is the energy of the *i*th AE signal detected and *N(t)* is the number of AE signals recorded and located along the gauge length until time *t*.

ΩR is the value of $\Omega(t) = \sum_{i=1}^{N(t)} \sqrt{E_i}$ when $t = t_R$, t_R is the failure time. $B = -\frac{\phi}{1-\gamma}$ is negative, 1−γ is an exponent and ϕ is a constant.

The optimum circle method (OCM) [31], also developed in seismology, is used here to assess the relevance of the Benioff law. In addition, it determines which AE sources should be considered in order to achieve the best approximation. The OCM method was used to evaluate the time t_{start} when the AE energy is well simulated by the Benioff law. Two approximations are carried out on the energy release resulting from each time interval $[t_{start}; t_r]$: a power-law approximation using the Benioff law and a linear approximation ($\Omega(t) = \alpha t + \beta$)

used as reference. The *c*-value is defined by the ratio of the root mean square error of the approximation by the Benioff law over that of the linear fit. When the *c*-value is lower than 1, there is a positive contribution of the Benioff law since the approximation error is lower than that of a linear fit. It is a relative validation of the relevance of the approximation by the Benioff law. Therefore, to ensure quality of the approximation, only *c*-values lower than 0.5 are considered to be relevant.

3.5. Identification of Damage Mechanisms with Supervised Clustering

In such CMCs, matrix cracking occurs in several modes of cracking [16]. First, cracks initiate in the external seal-coat and in macropores inside the composite, then propagate through the inter-yarn matrix. Afterwards cracks propagate inside the transverse yarns through fiber/matrix interfaces. Multiple matrix cracking finally occurs inside the axial yarns. These cracks are deviated by the fiber/matrix interphase layer, yielding to fiber debonding and overloading. Some fiber breaks occur under high stresses. They rapidly lead to unstable fracture of entire yarns and of the composite. Hence, the sources of AE are the various matrix cracks, fiber/matrix interfacial debonding, individual fiber fractures and yarn ruptures.

The supervised classification technique requires a data base of signals that have been labelled: the training set. This training set was created by merging data collected during tensile test [16]. As described in the previous paper [16], the analysis of AE signals, observation of microstructures and analysis of the mechanical behavior of the composite led to the identification of 4 types of AE signals and to the following labelling of classes:

Class A: Signals of cluster A are assigned to two damage mechanisms which are chronologically separated: seal coat cracking and tow breaks.

Class B: cluster B contains also signals from two damage mechanisms which are chronologically well separated: longitudinal matrix cracking and individual fiber breaks in the fracture zone just before failure.

Class C: Cluster C contains signals with relatively short duration, short rise time and low amplitude when compared to the others: transversal matrix cracking.

Class D: this cluster is the last one to be activated and it becomes more active as strain increases and the D-type signals have a longer rise time when compared to other signals: sliding at fiber/matrix interfaces, fiber/matrix debonding.

In order to establish the training set of labelled signals for the supervised analysis, the same amount of signals (500 signals) in each class (A, B, C and D) was used. This training set included all the damage modes that may operate in this composite.

The supervised method consists in comparing each detected signal to those of the library and to assign it to the class of the K nearest neighbors [32] in the library. First, it is necessary to determine the number K of nearest neighbors required to obtain the best classification. For this purpose, a self-validating procedure was used which consists in comparing each signal of the training set to all the others. The classification error rate was estimated for all the values of K using the "leave-one-out" method [33]. The optimal value of K corresponded to the lowest value of error, and in the present study $K = 11$, and error = 2.5%. Then, for each AE signal of a given test, the Euclidean distance to all signals of the training set was calculated and the signal was labelled according to the class the most frequent class among its 11 nearest neighbors. The supervised clustering procedure was carried out on the different fatigue static tests.

4. RESULTS AND DISCUSSION

4.1. Mechanical Analysis

The lifetime diagram, obtained at several temperatures and under various stresses, of the composite for static fatigue at intermediate temperature is plotted in Figure 3a. All the points are aligned in the logarithmic representation and follow the power-type law:

$$t \cdot \sigma^{n} = A \tag{8}$$

where *t* is the lifetime, σ is the applied stress, *A* and *n* are constants depending on the material, the temperature and the environment. The stress exponent *n* is estimated at 3.2 ± 0.3. A comparison between our results on composite and on the fibers in literature [34] under static fatigue at 500 °C (Figure 3b) shows that at intermediate and high stresses the lifetime of the composite is longer than the lifetime of the fibers. This is characterized by two different values of *n*, *n* is equal to 3.2 for the composite and to 8.45 for the fibers [34].

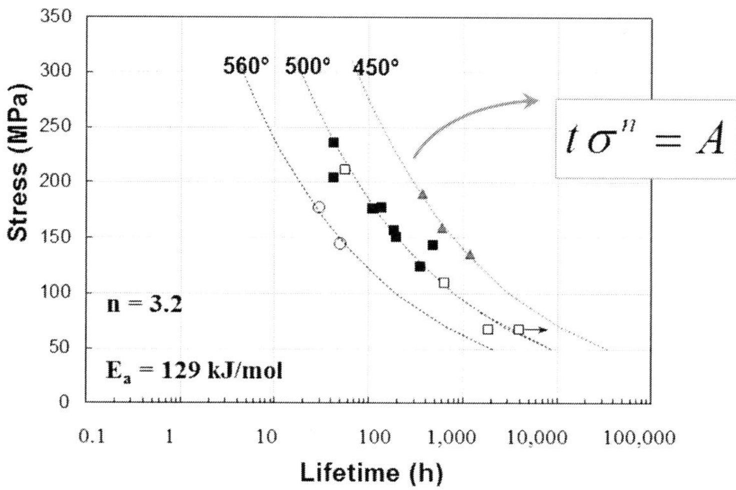

(a)

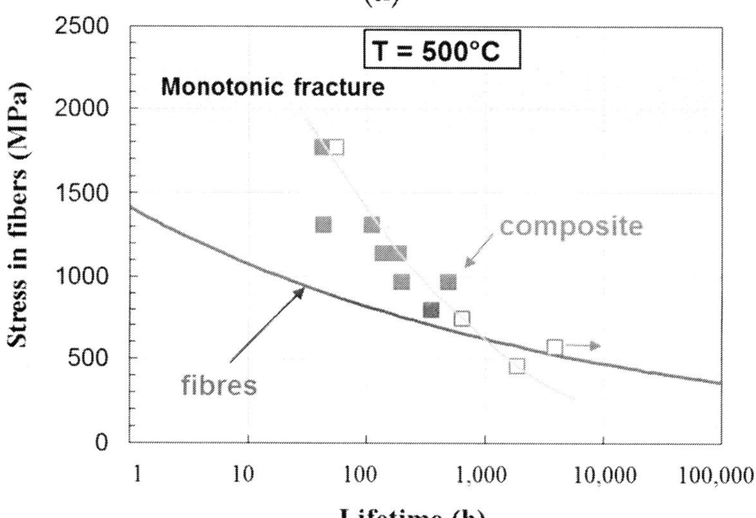

(b)

Figure 3. (**a**) Lifetimes obtained during static fatigue at 450 °C, 500 °C and 560 °C under air on SiC_f/[Si-B-C] composites for several applied stresses; (**b**) Lifetimes on the fibers bundles [34] and on SiC_f/[Si-B-C] composites in static fatigue at 500 °C under air.

For this range of applied stress, slow crack growth of the fibers needs the diffusion of oxygen but is controlled by the chemical reaction at the crack tip. As shown by Gauthier and Lamon [34], the activation energy 129 kJ/mol corresponds to the reactivity of carbon (present in the fibers) to oxygen. In the composite the diffusion of oxygen toward the fiber surfaces is slackened by the oxygen consumption due to reaction with self-healing matrices and with interphases. Under lower stresses, the lifetimes of the composite and of the fiber bundles are similar. For this low level of stresses reaction kinetics with fiber surfaces are slower than diffusion kinetics of oxygen into the matrix crack. Hence the subcritical crack growth kinetics in dry bundles and in the fibers of the composites are similar.

Stress as a function of strain is shown in Figure 4. The grey curve represents the monotonic tensile curve whereas the other curves correspond to the static fatigue curves for various ratios of stress and with some loading and unloading sequences. One may notice that strain increases significantly during the test, and that hysteresis secant loops evolve as well: the mean elastic modulus decreases whereas the loops width increases. The increase in the loops area is linked to a change in the interfacial shear stress, suggesting that some debonding occur during static fatigue. The evolution of the Young's modulus was evaluated during the tests by measuring the hysteresis loops modulus E_i. Then the damage parameter $D = 1 - E_i/E_0$ was calculated, E_0 being the initial Young's modulus of the composite. The evolution of D is plotted *versus* time inFigure 5, for several σ/σ_R ratios. The biggest increase of D occurs during the loading step and the first 24 h of static fatigue. Then D is observed to rise monotonically up to the failure of the specimen. The general trend is that the highest the σ/σ_Rratio, the highest the D parameter for similar durations of test. The final D values obtained at failure are not exactly the same: they vary in the range 0.6–0.8, and it seems that this value decreases when the σ/σ_R ratio decreases. However it seems difficult to define a failure criterion based on the damage parameter.

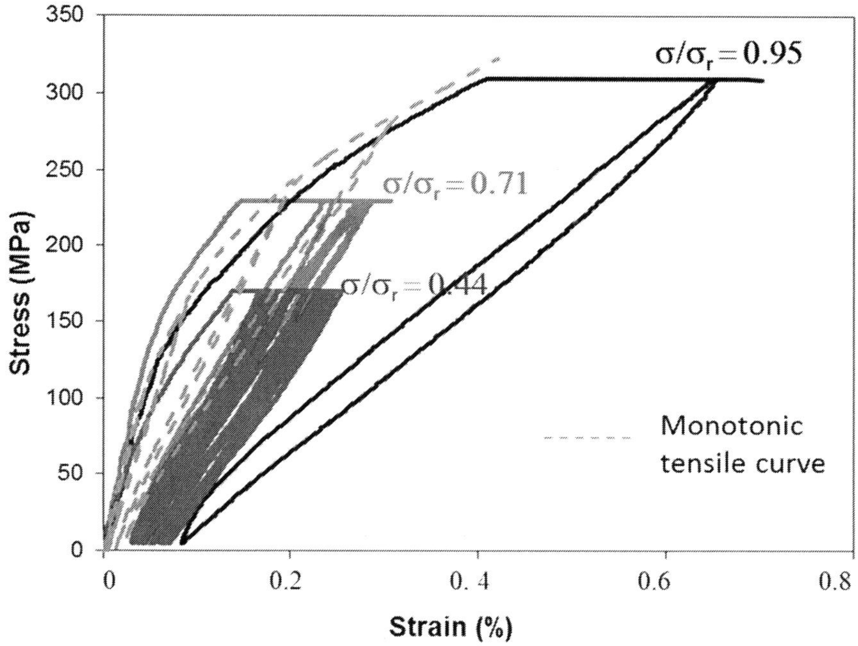

Figure 4. Stress *versus* strain during static fatigue experiments for different σ/σ_R ratios: 0.44, 0.71 and 0.95 at 500 °C on SiC_f/[Si-B-C] composites. The grey curve represents the monotonic tensile curve.

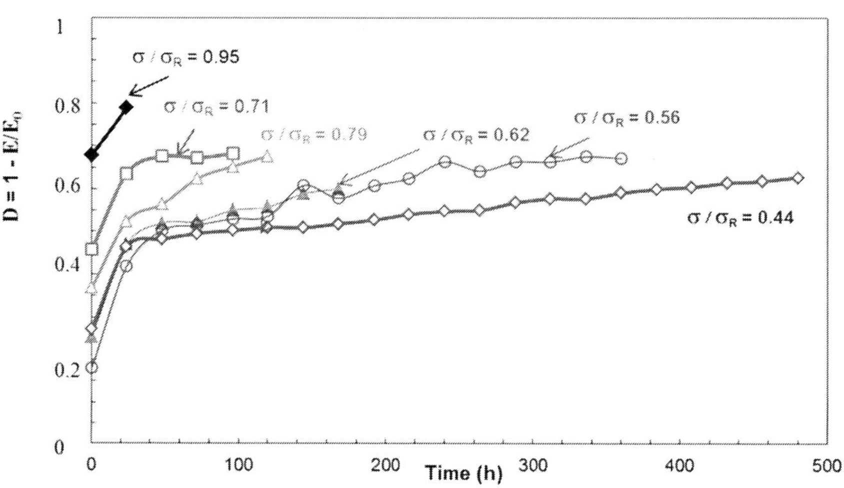

Figure 5. Evolution of the damage parameter *D* during static fatigue experiments for various σ/σ_R ratios 0.44, 0.56, 0.62, 0.71, 0.79 and 0.95 at 500 °C on SiC_f/[Si-B-C] composites.

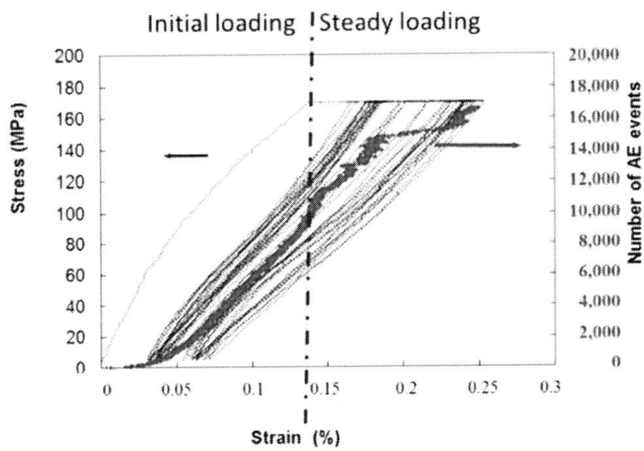

(a)

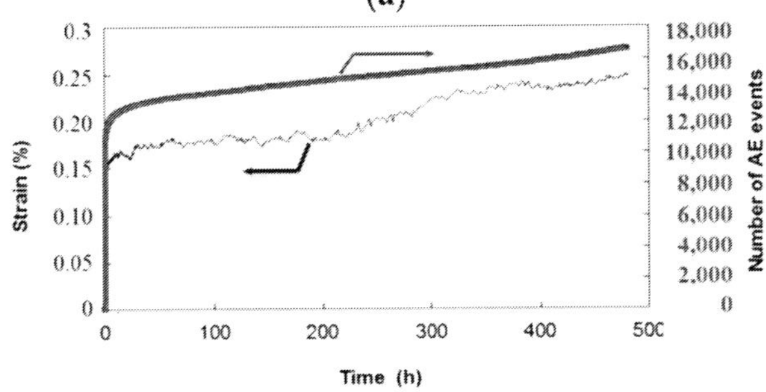

(b)

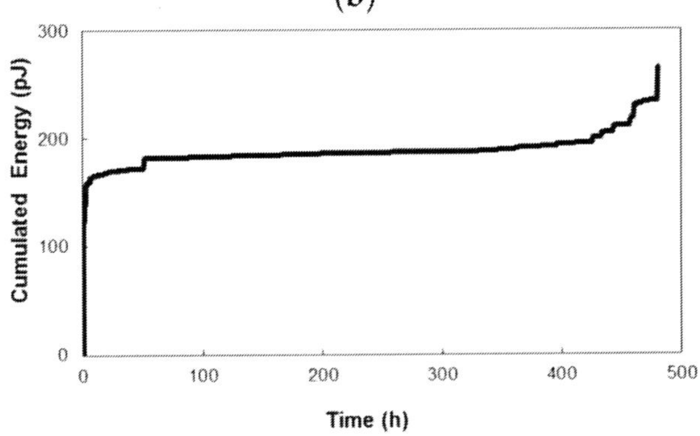

(c)

Figure 6. (**a**) Stress-strain curve and cumulated number of acoustic emission (AE) events recorded during initial loading up to the nominal load, and during the static loading at a constant load ($\sigma/\sigma_R = 0.44$ and $T = 500$ °C); (**b**) Strain and cumulated number of AE events *versus* time; (**c**) Cumulated acoustic energy *versus* time.

The evolution of strain and acoustic emission events is shown in Figure 6 for a specimen loaded at $\sigma/\sigma_R = 0.44$. Under the steady loading the strain rate is important at the beginning then decreases before reaching a constant value. At this time, strain rises monotonically up to the final failure for the majority of the specimens (Figure 6b). The AE activity recorded during the constant load hold is also plotted in Figure 6b. The acoustic emission activity evolves with the same trend. Figure 6c shows the evolution of the cumulated acoustic energy. The main activity occurred at the beginning of the test and then reached a plateau. The renewal of AE activity (Figure 6c) just before failure seems to be an interesting way to anticipate the fracture but was not always noticeable on the AE energy *vs.* time plot.

The main activity occurs at the beginning of the static fatigue. Then the AE rate decreases without stopping. During the initial loading up to the nominal load, the number of events increases with the applied stress, which is consistent with the increasing number of cracks in the matrix (the main source of acoustic emission) as the applied stress increases. Then, during the static loading, the number of events before failure is lower under high stress than under low stress (Figure 7): damage needs to progress more under low stress to cause the final failure of the composite, than under high stress where the matrix already contains a lot of cracks.

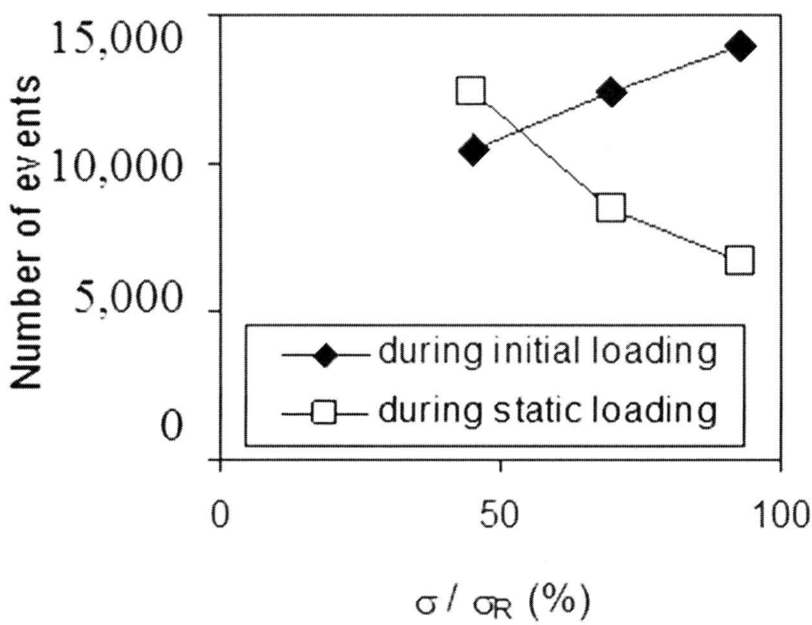

Figure 7. Number of AE events recorded during initial loading up to the nominal load and during the static loading at a constant load.

4.2. Identification of Critical Time

The evolution of RAE is plotted *versus* time in Figure 8 for three experiments. This coefficient is observed to decrease first, then to reach a minimum value for $t = t_m$, and then to increase up to the failure of the composite at time t_R. This ultimate increase in AE activity was not always visible on the curves of AE energy *versus* time, but it was revealed by this representation. The ratio t_m/t_R appears to be quite reproducible: $t_m/t_R = 0.57 \pm 0.06$ (Figure 9). Therefore, the RAE ratio may be used as a criterion for predicting the remaining lifetime of a specimen under static loading. During static fatigue test, attenuation coefficient *B* growths significantly during the first part of tests up to a plateau value near 50% of the rupture time (Figure 10). For each test, the slight increase of *B* observed during initial loading is very low compared to the one occurring during static fatigue. Therefore, the increase of attenuation coefficient *B* may be related to matrix crack opening and to the recession of interfaces.

The detection of the plateau during the typical evolution of attenuation coefficient B may be considered as an indicator for lifetime prediction. As the minimum of coefficient R_{AE}, the value plateau of *B* indicates the beginning of the critical damage phase and provides an estimation of the remaining lifetime.

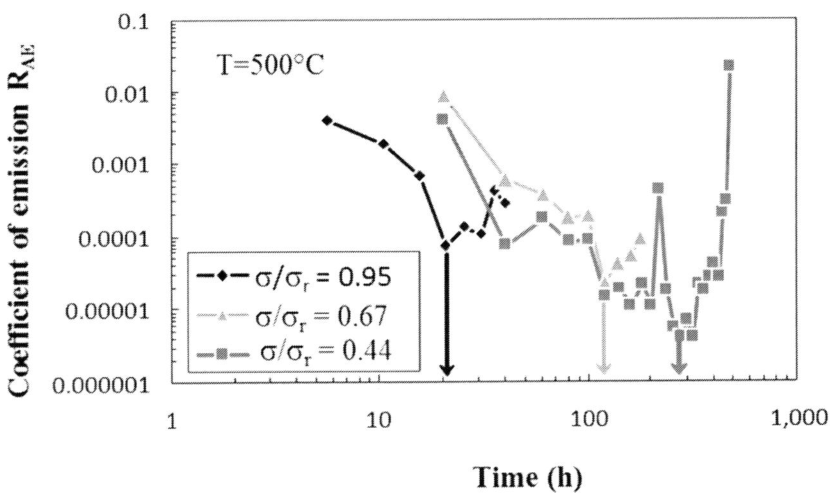

Figure 8. Evolution of the R_{AE} coefficient during the static load hold on SiC_f/[Si-B-C] composites.

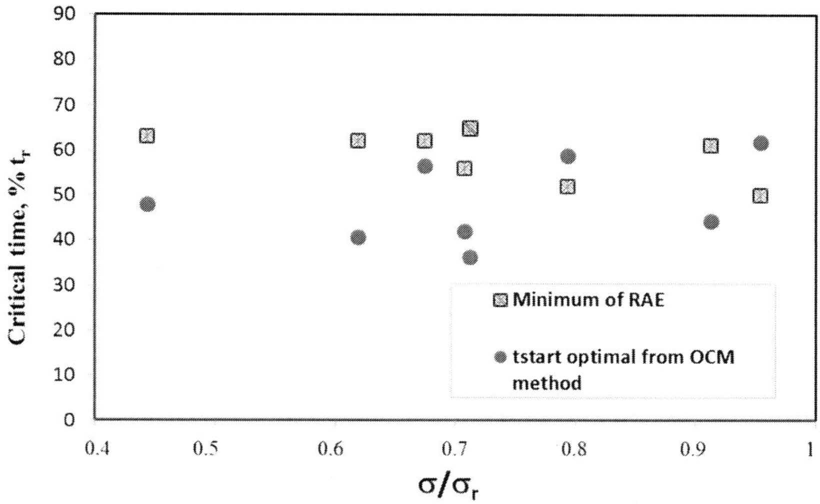

Figure 9. Critical times obtained with the minimum value of R_{AE} and the optimumt_{sart} for the Benioff law obtained with the OCM method.

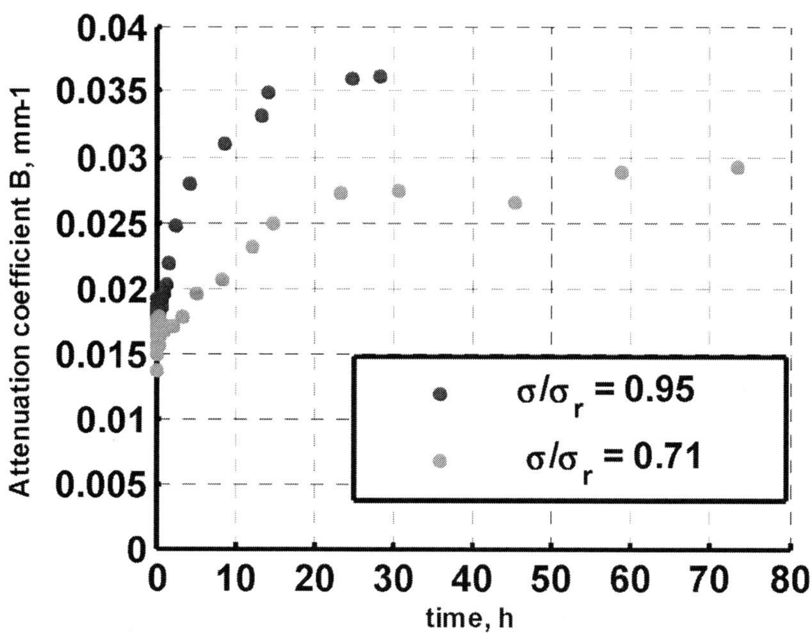

Figure 10. Evolution of the attenuation coefficient B during the static load hold (σ/σ_R = 0.95, σ/σ_R = 0.71 and T = 500 °C).

The critical damage phase prior to final rupture may be attributed to the avalanche fibers breaks, controlled by fibers oxidation and by recession of interfaces. In order to confirm this hypothesis, the coefficient R_{AE} is calculated for several damage mechanisms identified with clustering analysis of AE data. The Figure 11 show the activities of several damage mechanisms identified with the supervised analysis. During static fatigue, clusters B and D are more active. The evolution of the coefficients *B* obtained for the two classes A and B go through a minimum, contrary to those of classes C and D (Figure 12). For the B class (mainly fiber failure at the end of the tests), the minimum value of the coefficient *B* is observed around 65% of the lifetime. For the A class (mainly yarn fractures or collective fiber breaks at the end of the tests), the minimum is also observed at 65%. One may noticed that the minimum is observed only for clusters A and B corresponding to fibers breaks during the second part of the test. If the growth of the attenuation coefficient *B* is linked to matrix crack opening, coefficient B also allows considering the plateau observed on the evolution of attenuation coefficient *B* shows that matrix crack opening leads to an equilibrium state near 40%–50% of the rupture time. The significant increase of matrix crack opening pointed out before 50% of the rupture time is linked to carbon oxidation in the interphases provoking an increase in length of the debonded region of fibers in the vicinity of matrix cracks. Beyond 50% of the rupture time, the oxygen flux, determined by the degree of matrix crack opening, controls the rate of fibers break by slow crack growth. This critical time corresponds to the beginning of a second damage phase where slow crack growth in fibers is prevailing, leading to the ultimate fracture of the composite.

The minimum of coefficient RAE and the value plateau of B appear to be a promising tool to perform "short" static fatigue experiments which could be stopped once the minimum value of RAE is reached, rather than waiting the final failure of the specimen. In this way the test duration could be divided by 1.5 to 2.

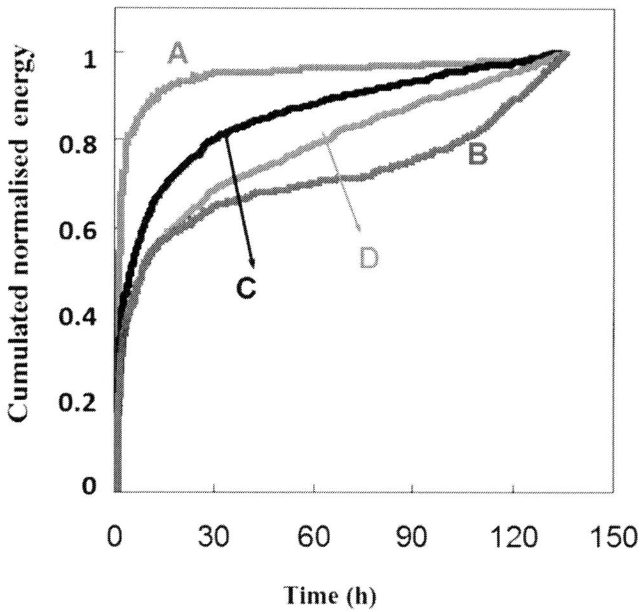

Figure 11. Activities of several clusters during static fatigue test at 500 °C and $\sigma/\sigma_R = 0.79$.

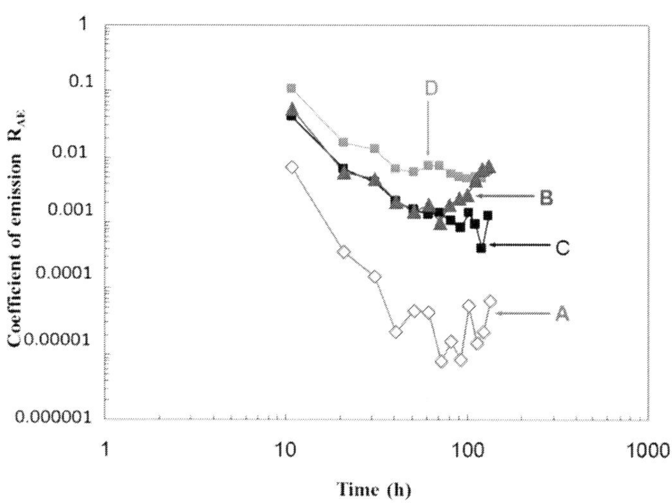

Figure 12. Evolution of the coefficient of emission for several clusters at 500 °C and $\sigma/\sigma_R = 0.79$.

4.3. Toward Lifetime Prediction

The Benioff law was applied to AE data recorded during the static fatigue tests in order to evaluate the lifetime of the studied material. This law is based on a power law description of the cumulative energy (Ω) released by the composite (Equation (2)) and characterized by 4 parameters (Ω_r, γ, ϕ and t_r).

Momon [11] studied the applicability on AE signals recorded for C/SiC and SiC/SiC composites. Data beyond the minimum of R_{AE} were in better agreement with the Benioff law than those before the minimum. This point was confirmed with the estimation of the *c*-value.

The Optimum Circle Method was applied to the fatigue tests. Optimum values (leading to the minimum *c*-value) of t_{start} are reported on the Figure 9, which means that the approximation by the Benioff law was relevant for the data collected behind this time. A value of t_{start} was taken every hour in time interval of 10% to 90% of rupture time t_r. The minimum *c*-value appears clearly around on average 40%–50% of rupture time. The energy release prior to rupture under static fatigue exhibits a critical evolution at 50% of rupture time regardless of the applied stress level. Thus, the Benioff law, initially used to study the activation of earthquakes, may also be applied to the damage of composites. However, this procedure requires preliminary tests until rupture to determine γ and ϕ for the studied material.

5. CONCLUSIONS

Static fatigue experiments were performed at intermediate temperatures under air to assess lifetimes of a SiC_f/[Si-B-C] composite. The lifetime as a function of applied stress follows a power-type law, which can be used to predict lifetimes. Additional information is obtained from strain measurement and AE monitoring during the tests. Two criterion based on the AE cumulative energy has been defined, which can be used to predict the final failure of a specimen if the AE activity during the first half of the test is known. This could be a way to shorten such static fatigue experiments, and to divide their duration by 1.5. The coefficient R_{AE} and the attenuation coefficient *B* confirm the existence of two distinct phases during damage of CMCs in static fatigue at intermediate temperatures. The first phase being mainly attributed to interfacial changes and the second one to the predominance of subcritical crack growth in fibers. Beyond this characteristic point, energy release may be modelled with the Benioff law in order to extrapolate AE activity and hence evaluate time to failure. The same analysis is in progress for the behavior during cyclic fatigue tests. Future works will focus on the use of the Benioff law as a predictive model.

ACKNOWLEDGMENT

The authors gratefully acknowledge Snecma Propulsion Solide, CNRS and DGA for supporting this work in the frame of the research program: "Modélisation, extrapolation, validation de la durée de vie des composites à matrice céramique".

AUTHOR CONTRIBUTIONS

All the authors have contributed to perform mechanical tests, to analyze data and to write this paper.

REFERENCES

1. Fantozzi, G.; Reynaud, P.; Rouby, D. Thermomechanical behaviour of long fibres ceramic-ceramic composites. *Silic. Ind.* **2002**, *66*, 109–119.
2. Reynaud, P.; Rouby, D.; Fantozzi, G. Effects of temperature and of oxidation on the interfacial shear stress between fibres and matrix in ceramic-matrix composites. *Acta Mater.* **1998**, *46*, 2461–2469.
3. Naslain, R.; Christin, F. Sic-matrix composite materials for advanced jet engines. *MRS Bull.* **2003**, *28*, 654–658.
4. Lamon, J. Micromechanics-based approach to the mechanical behavior of brittle matrix composites. *Compos. Sci. Technol.* **2002**, *58*, 2259–2272.
5. Ohnabe, H.; Masaki, S.; Onozuka, M.; Miyahara, K.; Sasa, T. Potential applications of ceramic matrix composites to aero-engine components. *Compos. A* **1999**, *30*, 489–496.
6. Lamon, J. CVI SiC/SiC composites. In *Handbook of Ceramics Composites*; Bansal, N.P., Ed.; Springer: Berlin, Germany, 2005; pp. 55–76.
7. Fantozzi, G.; Reynaud, P. Mechanical behaviour of SiC fiber reinforced ceramic matrix composites. In *Comprehensive Hard Materials*; Sarin, V.K., Ed.; Elsevier: Philadelphia, PA, USA, 2014; pp. 345–366.
8. Quemard, L.; Rebillat, F.; Guette, A.; Tawil, H.; Louchet-Pouillerie, C. Self-healing mechanisms of a SiC fiber reinforced multi-layered ceramic matrix composite in high pressure steam environments. *J. Eur. Ceram. Soc.* **2007**, *27*, 2085–2094.
9. Forio, P.; Lamon, J. Fatigue behavior at high temperatures in air of a 2D SiC/SiBC composite with a self-sealing multilayered matrix. In *Ceram Transactions*; Bansal, N.P., Singh, J.P., Lin, H.T., Eds.; American Ceramic Society: Westerville, OH, USA, 2001; Volume 128, pp. 127–141.
10. Ono, K.; Huang, Q. Pattern Recognition Analysis of Acoustic Emission Signals. *NDT and E Int.* **1997**, *30*, 109.

11. Yan, T.; Holford, K.; Carter, D.; Brandon, J. Classification of acoustic emission signatures using a self-organization neural network. *J. Acoust. Emiss.* **1999**, *17*, 49–59.
12. Kostopoulos, V.; Loutas, T.H.; Kontsos, A.; Sotiriadis, G.; Pappas, Y.Z. On the identification of the failure mechanisms in oxide/oxide composites using acoustic emission. *NDT E Int.* **2003**, *36*, 571–580.
13. Pappas, Y.Z.; Kontsos, A.; Loutas, T.H.; Kostopoulos, V. Failure mechanisms analysis of 2D carbon/carbon using acoustic emission monitoring. *NDT E Int.* **1998**, *31*, 571–580.
14. Godin, N.; Huguet, S.; Gaertner, R. Integration of the Kohonen's self-organising map and *k*-means algorithm for the segmentation of the AE data collected during tensile tests on cross-ply composites. *NDT E Int.* **2005**, *38*, 299–309.
15. Moevus, M.; Rouby, D.; Godin, N.; R'Mili, M.; Reynaud, P.; Fantozzi, G.; Fayolle, G. Analyse of damage mechanisms and associated acoustic emission in two SiC/[Si-B-C] composites exhibiting different tensile curves. Part I: Damage patterns and acoustic emission activity. *Compos. Sci. Technol.* **2008**, *68*, 1250–1257.
16. Moevus, M.; Godin, N.; Rouby, D.; R'Mili, M.; Reynaud, P.; Fantozzi, G.; Farizy, G. Analyse of damage mechanisms and associated acoustic emission in two SiC/[Si-B-C] composites exhibiting different tensile curves. Part II: Unsupervised acoustic emission data clustering. *Compos. Sci. Technol.* **2008**, *68*, 1258–1265.
17. Smith, R.L.; Phoenix, S.L. Asymptotic distributions for the failure of fibrous materials under series-parallel structure and equal load-sharing. *J. Appl. Mech.* **1981**, *48*, 75–82.
18. Curtin, W.A. Theory of mechanical properties of ceramic-matrix composites. *J. Am. Ceram. Soc.* **1991**, *74*, 2837–2845.
19. Newman, W.I.; Phoenix, S.L. Time dependent fiber-bundles with local load sharing. *Phys. Rev. E Stat. Nonlin. Soft Matter Phys.* **2001**, *63*, 021507.
20. Turcotte, D.; Shcherbakov, R. Can damage mechanics explain temporal scaling laws in brittle fracture and seismicity?*Pure Appl. Geophys.* **2006**, *163*, 1031–1045.
21. Guarino, A.; Garcimartin, A.; Ciliberto, S. An experimental test of the critical behaviour of fracture precursors. *Eur. Phys. J. B* **1998**, *6*, 13–24.
22. Nechad, H.; Helmstetter, A.; El Guerjouma, R.; Sornette, D. Andrade and critical time-to-failure laws in fiber-matrix composites: Experiments and model. *J. Mech. Phys. Solids* **2005**, *53*, 1099–1027.

23. Deschanel, S.; Vanel, L.; Vigier, G.; Godin, N.; Ciliberto, S. Experimental study of crackling noise: Conditions on power law scaling correlated to fracture precursors. *J. Stat. Mech. Theory Exp.* **2009**.
24. Shaira, M.; Godin, N.; Guy, P.; Vanel, L.; Courbon, J. Evaluation of the strain-induced martensitic transformation by acoustic emission monitoring in 304L austenitic stainless steel: Identification of the AE signature of the martensitic transformation and power-law statistics. *Mater. Sci. Eng. A* **2008**, *492*, 392–399.
25. Morscher, G.N. The velocity and attenuation of acoustic emission waves in SiC/SiC composites loaded in tension.*Compos. Sci. Technol.* **2002**, *62*, 1171–1180.
26. Maillet, E.; Godin, N.; R'Mili, M.; Reynaud, P.; Lamon, J.; Fantozzi, G. Analysis of Acoustic Emission energy release during static fatigue tests at intermediate temperatures on Ceramic Matrix Composites: Towards rupture time prediction. *Compos. Sci. Technol.* **2012**, *72*, 1001–1007.
27. Maillet, E.; Godin, N.; R'Mili, M.; Reynaud, P.; Lamon, J.; Fantozzi, G. Determination of acoustic emission sources energy and application towards lifetime prediction of ceramic matrix composites. In *Mechanical Properties and Performance of Engineering Ceramics and Composites VII*; Singh, D., Salem, J., Halbig, M., Mathur, S., Eds.; Wiley: Hoboken, NJ, USA, 2012; Volume 33, pp. 15–25.
28. Maillet, E.; Godin, N.; R'Mili, M.; Reynaud, P.; Fantozzi, G.; Lamon, J. Real-time evaluation of energy attenuation: A novel approach to acoustic emission analysis for damage monitoring of ceramic matrix composites. *J. Eur. Ceram. Soc.***2014**, *34*, 1673–1679.
29. Momon, S.; Moevus, M.; Godin, N.; R'Mili, M.; Reynaud, P.; Fantozzi, G.; Fayolle, G. Acoustic emission and lifetime prediction during static fatigue tests on ceramic-matrix-composite at high temperature under air. *Compos. A Appl. Sci. Manuf.* **2010**, *41*, 913–918.
30. Bufe, C.G.; Varnes, D.G. Predictive modeling of the seismic cycle of the Greater San Francisco Bay Region. *J. Geophys. Res.* **1993**, *98*, 9871–9883.
31. Bowman, D.D.; Ouillon, G.; Sammis, C.G.; Sornette, A.; Sornette, D. An observational test of the critical earthquake concept. *J. Geophys. Res.* **1998**, *103*, 24359–24372.
32. Likas, A.; Vlassis, N.; Verbeek, J. The global *k*-means clustering algorithm. *Pattern Recognit.* **2003**, *36*, 451–461.
33. Lee, M.S.; Keerthi, S.; Jin Ong, C.; Decoste, D. An efficient method for computing leave one out error in support vector machines with gaussian kernels. *IEEE Trans. Neural Netw.* **2004**, *15*, 750–757.

34. Gauthier, W.; Lamon, J. Delayed failure of Hi-Nicalon and Hi-Nicalon S multifilament tows and single filaments at intermediate temperatures. *J. Am. Ceram. Soc.* **2009**, *92*, 702–709.

Chapter 4

A New Fault Location Approach for Acoustic Emission Techniques in Wind Turbines

Carlos Quiterio Gómez Muñoz * and Fausto Pedro García Márquez

Ingenium Research Group, Department of Business Management, University of Castilla-La Mancha, Ciudad Real 13071, Spain

ABSTRACT

The renewable energy industry is undergoing continuous improvement and development worldwide, wind energy being one of the most relevant renewable energies. This industry requires high levels of reliability, availability, maintainability and safety (RAMS) for wind turbines. The blades are critical components in wind turbines. The objective of this research work is focused on the fault detection and diagnosis (FDD) of the wind turbine blades. The FDD approach is composed of a robust condition monitoring system (CMS) and a novel signal processing method. CMS collects and analyses the data from different non-destructive tests based on acoustic emission. The acoustic emission signals are collected applying macro-fiber composite (MFC) sensors to detect and locate cracks on the surface of the blades. Three MFC sensors are set in a section of a wind turbine blade. The acoustic emission signals are generated by breaking a pencil lead in the blade surface. This method is used to simulate the acoustic emission due to a breakdown of the composite fibers. The breakdown generates a set of mechanical waves that are collected by the MFC sensors. A graphical method is employed to obtain a system of non-linear equations that will be used for locating the emission source. This work demonstrates that a fiber breakage in the wind turbine blade can be detected and located by using only three low cost sensors. It allows the detection of potential failures at an early stages, and it can also reduce corrective maintenance tasks and downtimes and increase the RAMS of the wind turbine.

Keywords: acoustic emission; wind turbine; fault detection and diagnosis; macro-fiber composite; non-destructive testing

1. INTRODUCTION

The renewable energy industry is undergoing continuous improvement to cover the current demands of electricity, wind energy being one of the most important. The new technologies, communication systems and advances in mathematical models for signal processing aid in achieving that goal [1]. The complexity of these devices causes a reduction of the reliability, availability, maintainability and safety of the system (RAMS) and increases the maintenance costs due to the occurrence of non-monitored failures [2,3,4].

Nowadays, fault detection and diagnosis (FDD) by non-destructive testing (NDT) is employed in maintenance management [5,6,7], for example in structural health monitoring (SHM) [8]. SHM enables identifying and diagnosing the fault and its location by detecting changes in the static and dynamic features of the structure [9,10]. SHM can be remotely managed, reducing the costs of manual inspections and the time between the fault occurrences, and this has been noted [11,12]. This will lead to an increase in the productivity, reducing the potential downtimes for the wind farms and increasing the RAMS of the wind turbine [13,14,15].

The purpose of this paper is to design an FDD model for the SHM of a wind turbine blade [16,17,18]. The case study proposes a novel localization method using signals from macro-fiber composite (MFC) sensors. Three MFC sensors are strategically located along a blade section to detect incipient breakages in the structure [19,20]. The case study involves some considerations, e.g., the appearance of the scattering phenomena, the orientation of the sensors when the excitation is received, *etc*. However, it will be demonstrated that the proposed method can set the location with high accuracy. The analysis identifies a single point obtained from a graphical method that is analytically set by nonlinear equations.

The accuracy of this method depends on the transducer sensitivity, the type of composite material, irregularities in the material, the environmental noise, *etc*. The localization precision of the emission source will be affected by the type of composite material, the sensitivity of the materials, environmental noise, false positives due to impacts on the piece, *etc*. Moreover, in real working conditions, considering environmental conditions, e.g., rain or hail, or impacts on the blade, it can cause false alarms.

In working conditions, it would be possible to distinguish between the frequencies associated with the vibration of the blade (low frequencies) and the frequencies associated with the acoustic emission of the fiber breakage (frequencies within the audible range and the ultrasonic range) [21]. It is possible to filter the frequencies associated with the vibration from the collected signal. The authors demonstrated this in [22].

2. EXPERIMENTS

The experiments are done in a section of the wind turbine blade. The fragment, shown in Figure 1, is made of glass fiber-reinforced polymer (GFRP), with dimensions of 100 × 79.5 cm. The section is composed of a honeycomb central layer embedded between two fiberglass layers made of polyester resin. This type of material has good structural properties, resistance to fatigue and other advantages. The attenuation of the acoustic emission in the blade is high, and it depends on the material, wave frequency and travelling distance between the failure source and the sensor location [23].

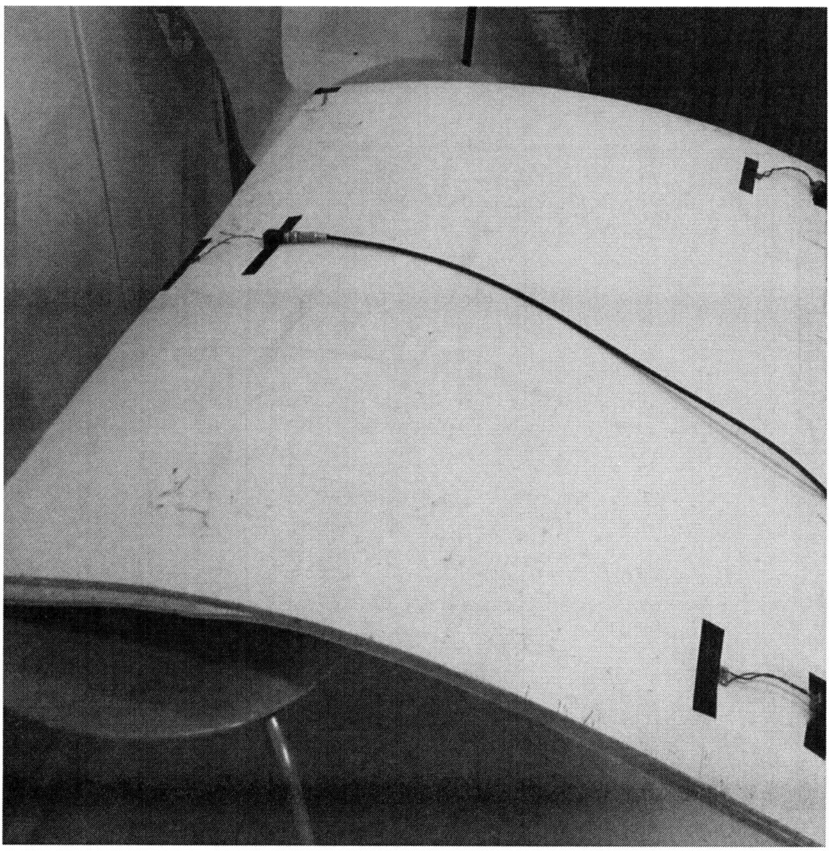

Figure 1. Wind turbine section with sensors for acoustic emission location.

The waves with the same velocity form a circular wave front when they propagate through an isotropic material. The velocity generally does not depend on the direction of propagation, but in anisotropic materials, e.g., the composite materials of the wind turbines, the velocity depends on the direction of propagation. A slowness factor could be introduced in order to consider the propagation direction, e.g., it has been observed that the configuration of layers (+45/−45) for a composite has a strong dependency of the direction of propagation. However, it has been demonstrated that the direction of propagation does not affect the velocity in the blade section studied in this paper. Therefore, the slowness factor has not been introduced in these experiments.

3. LOCATION OF THE FIBER BREAKAGE BY THE TRIANGULATION METHODOLOGY

The SHM on wind turbine blades is employed to detect the defect online and to locate it with accuracy [24]. The wind turbine blades are becoming larger and more complex, and this requires setting the exact location of a fiber breakage to reduce the maintenance cost and the productivity.

The glass fiber breakages of a wind turbine blade have been simulated in the laboratory on a real blade. A novel location method by triangulation has been developed. The aim of the paper is to locate the acoustic emission source in four different points on the blade section. The acoustic emission produced by the division of the glass fibers is simulated by breaking the tip of the lead from a mechanical pencil [25,26,27]. Three MFC transducers (A, B and C) were used to detect the acoustic emissions. The three transducers are used as sensors that collect the wave front of the mechanical wave produced by the acoustic emission. These signals received by the sensors present a low amplitude, and therefore, they need to be pre-amplified before being acquired by the oscilloscope [28].

In working conditions, there are many factors that could influence the configuration of the arrangement of the sensors on a blade, for example the length of the blades, the intensity of the acoustic emission, the accuracy of the sensors, the background noise, attenuation, *etc*. Depending on these factors, many groups of three sensors would be established, as they are required to cover the entire blade.

The propagation velocity of the acoustic emission (see Figure 2) has been experimentally calculated by breaking a pencil lead and measuring the delays in the excitement of the sensors S1 and S2 (Figure 3).

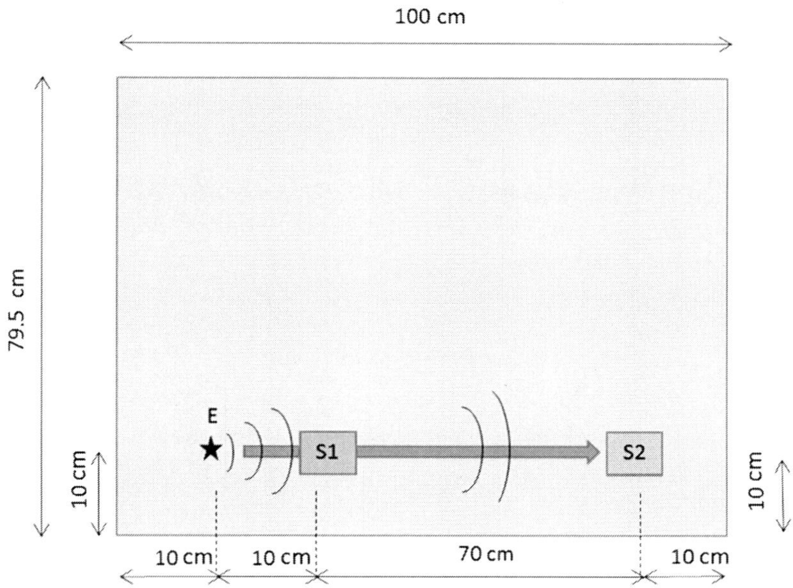

Figure 2. Measuring the experimental propagation velocity in the composite material.

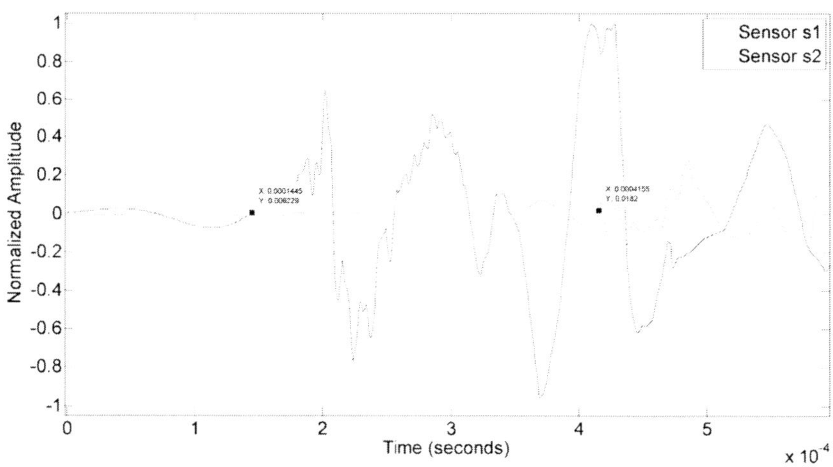

Figure 3. Peak detection of the acoustical emission collected by Sensor 1 (**blue**) and Sensor 2 (**green**) to obtain the experimental propagation velocity in the composite material.

The delay between Signals 1 and 2 is 271 μs (542 samples), and the propagation velocity for the composite material is 2583 m/s.

Four experiments have been conducted at four different locations of the acoustic emission. Twelve tests have been done applying the same force, angle of inclination and length (1 mm approximately). The main objective is to get similar signals for all of the case studies. The data are also filtered for the signal processing, where undesired frequencies are filtered [29]. The peak detection algorithm identifies the wave front of each signal. This process is complex because the waves are compounded by a large number of frequencies. Moreover, there are multiple elements in the blade that could affect the scattering of the acoustic signal, such as the edges of the geometry, the junction with the beam, adhesives, *etc*.

The signal processing consists of a pass band filter that eliminates low and high frequencies and carries out a comparison of the peaks of the wave front in the same frequency range of Signals A, B and C, generated by the above-mentioned MFC sensors, A, B and C (Figure 4).

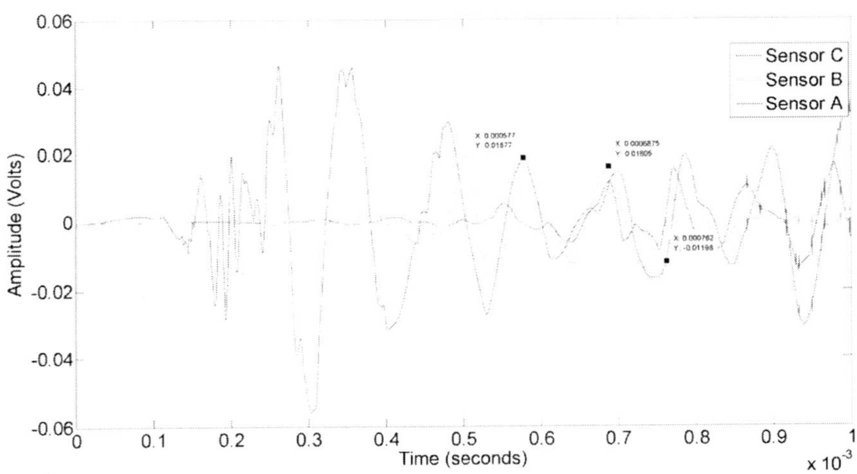

Figure 4. Pre-processing of the signal. Wave front collected by Sensors C (**blue**), B (**green**) and A (**red**).

The MFC Sensors A, B and C are placed as an equilateral triangle (see Figure 5). D is the location of the emission source by the breaking, D (D is known in this experiment).

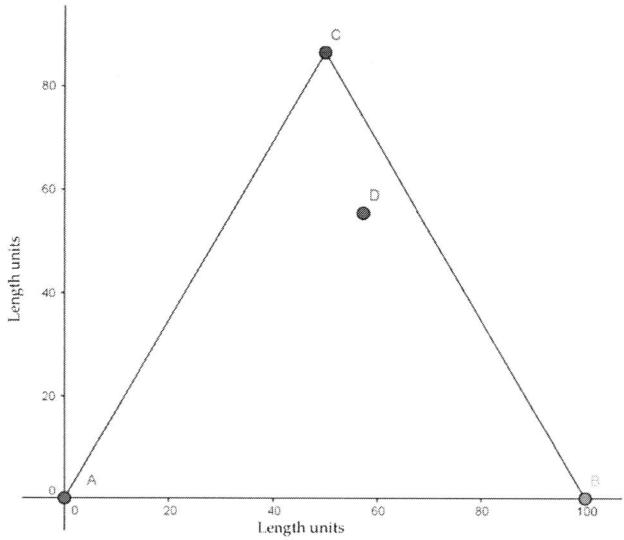

Figure 5. Location of Vertices A, B and C and the, defect D.

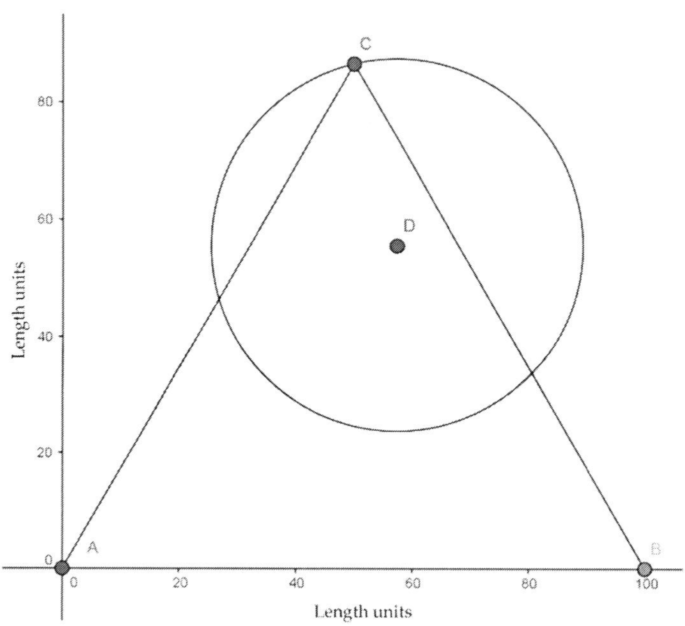

Figure 6. Wave front of the acoustic emission collected by the nearest Sensor C.

The nearest Sensor C is the first to be excited due to the wave front coming from the acoustic emission (Figure 6).

The delay between the excitation of the first Sensor C and the second closest Sensor B to the defect, D, is given by the distance from E to B (Figure 7). The delay time and the speed of the wave propagation on the blade is calculated by Equation (1):

$$D_{EB} = v \times t_{CB} \quad (1)$$

where D_{EB} is the distance between E and B, v is the propagation velocity of the wave (obtained experimentally) and t_{CB} is the time delay between the excitation of Sensors C and B.

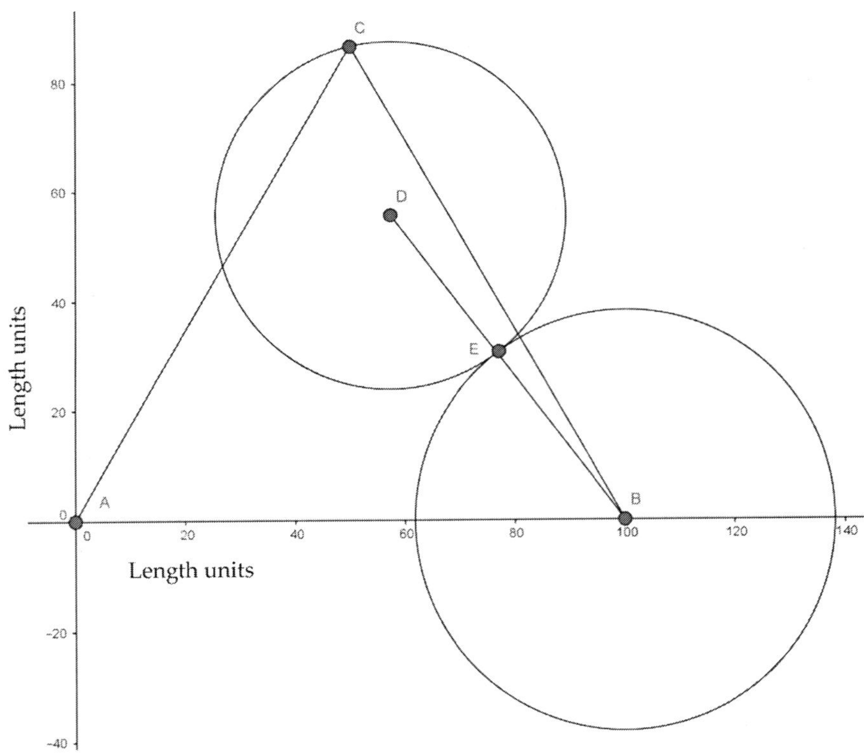

Figure 7. Location of Point E, set by the delay between the excitation time in Sensors C and B.

The delay between Sensor C and Sensor A, the farthest one from Defect D, is given by the distance from F to A D_{FA} in Equation (2).

$$D_{FA} = v \times t_{CA} \quad (2)$$

where t_{CA} is the time delay between the excitation of Sensor C and Sensor A. Figure 8 shows the scheme of the triangulation approach, the delay being represented by a circle.

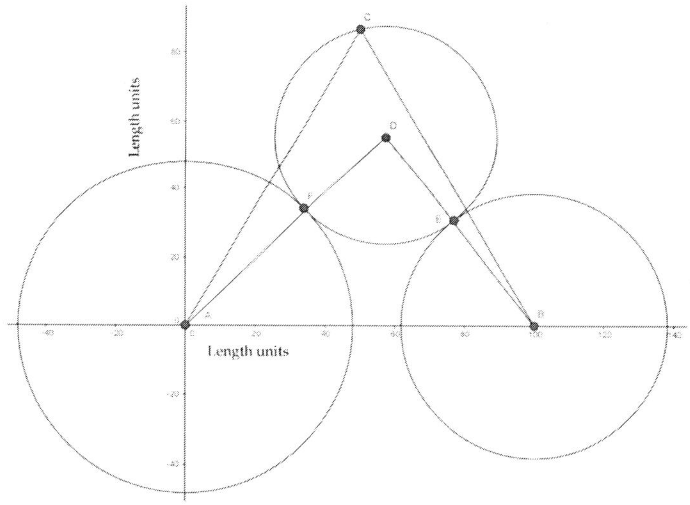

Figure 8. Scheme of the acoustic emission delays for locating the source.

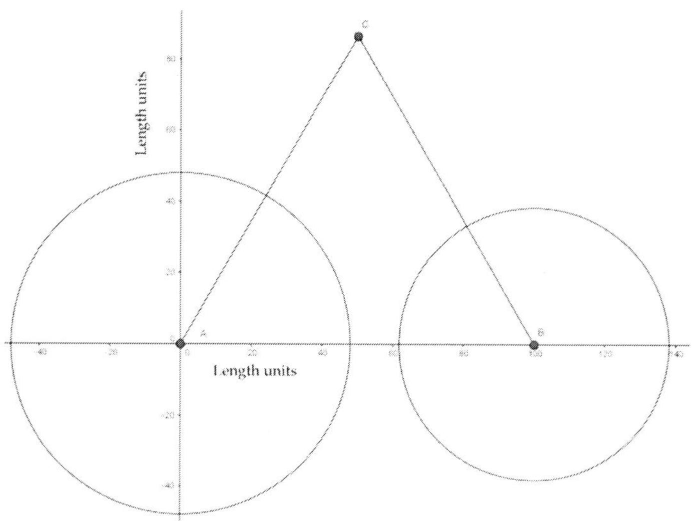

Figure 9. Initial conditions to locate the source of the acoustic emission.

In a real case study, Point D is unknown regarding the time and location, and the delays between the different sensors can be calculated. This condition is shown in Figure 9, where the circumferences represent the delays of the signal that comes to each sensor with respect to the first sensor (C).

The objective is to find the source of the acoustic emission D mentioned above. This point is the center of a circle that is tangential to two given circles and passes through Point C (see Figure 8). The solution is obtained in this paper employing a graphical method and an analytical method using a system of seven nonlinear equations.

4. TRIANGULATION EQUATIONS SYSTEM

The seven nonlinear equations to solve this problem are given by Equations (3) to (9), considering the scheme shown in Figure 4, where the MFC sensors are located at Points A, B and C, and the defect is at Point D. The coordinates and radius are:

- x_c: x-coordinate at the top of the triangle.
- y_c: y-coordinate at the top of the triangle.
- x_a: x-coordinate at the left lower corner of the triangle.
- y_a: y-coordinate at the left lower corner of the triangle.
- x_b: x-coordinate at the right lower corner of the triangle.
- y_b: y-coordinate at the right lower corner of the triangle.
- r_a: radius of the circle originated from A (delay of Sensor A).
- r_b: radius of the circle originated from B (delay of Sensor B).

The data mentioned above are known. The unknown variables are x_1, x_2, x_3, x_4, x_5, x_6 and x_7, being:

- x_1 and x_2 the coordinates of the emission Source D.
- x_3 and x_4 the coordinates of the tangency of Point F.
- x_5 and x_6 the coordinates of the tangency of Point E.
- x_7 is the radius of the circumference with the center D.

The following equations define the method analytically.

Equation (3) considers a circle with the center at D and passing through C:

$$F(1) = (x_c - x_1)^2 + (y_c - x_2)^2 - (x_7)^2 \qquad (3)$$

Equation (4) represents a circle with the center at D and passing through F:

$$F(2) = (x_3 - x_1)^2 + (x_4 - x_2)^2 - (x_7)^2 \quad (4)$$

Equation (5) sets a circle with the center at D and passing through E:

$$F(3) = (x_5 - x_1)^2 + (x_6 - x_2)^2 - (x_7)^2 \quad (5)$$

Equation (6) represents a circle with the center at A and passing through F:

$$F(4) = (x_1 - x_a)^2 + (x_4 - y_a)^2 - r_a^{\,2} \quad (6)$$

Equation (7) considers a circle with the center at B and passing through E:

$$F(5) = (x_5 - x_b)^2 + (x_6 - y_b)^2 - r_b^{\,2} \quad (7)$$

Equation (8) provides the straight line passing through Points A and F:

$$F(6) = \frac{(x_4 - y_a)}{(x_3 - x_a)} \times x_1 + (y_a - \frac{(x_4 - y_a)}{(x_3 - x_a)} \times x_a) - x_2 \quad (8)$$

Equation (9) sets the straight line passing through Points B and E:

$$F(7) = \frac{(y_b - x_6)}{(x_b - x_5)} \times x_1 + ((y_b - \frac{(y_b - x_6)}{(x_b - x_5)})) \times x_b) - x_2 \quad (9)$$

5. EXPERIMENTAL PROCEDURE AND RESULTS

The time of flight and distances are set in this section for Sensors B and A regarding C, C being the first sensor to receive the acoustic signal of the breakage. The experiments are repeated four times to take into account the deviations of the results. The algorithm gives the exact location of the defect, as well as a graphic outline, knowing the radius of the circles with centers at B and C. The dimensions of the blade section, the distribution of the sensors and the emission source (star) in the wind turbine blade are shown in Figure 10. The mathematical results obtained with the algorithm are given in Table 1,Table 2, Table 3, Table 4, Table 5, Table 6, Table 7 and Table 8.

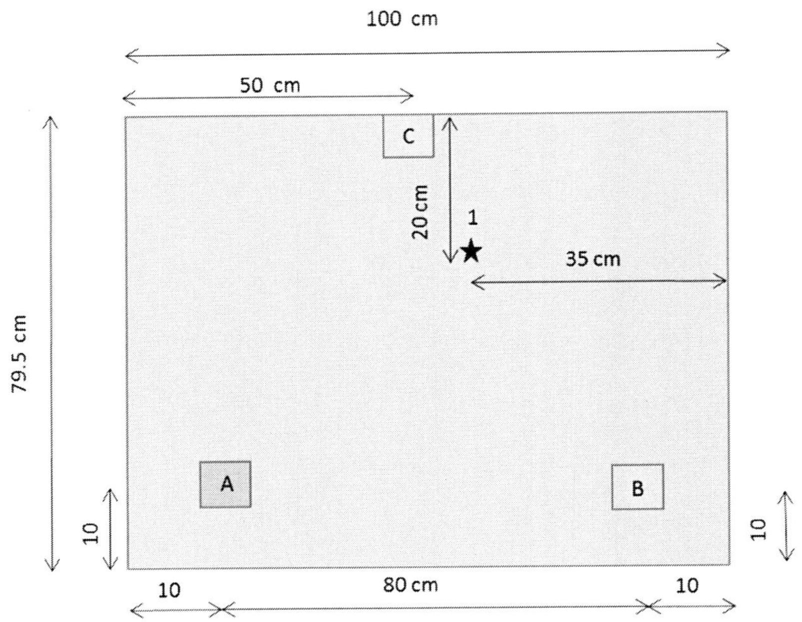

Figure 10. First experiment. Case Study 1.

Table 1. First case study: detection time; delay with C; delay; theoretical distance; experimental distance.

Sensors	Detection Time (Samples)	Delay with C (Samples)	Delay (s)	Theoretical Distance (m)	Experimental Distance (m)
C	1152	-	-	-	-
B	1381	229	1.15×10^{-4}	0.30	0.30
A	1528	376	1.88×10^{-4}	0.49	0.49

Table 2. Initial data of the first case study.

Locations	x-Coordinate (m)	y-Coordinate (m)	Radius (m)
A	0	0	0.49
B	0.8	0	0.30
C	0.4	0.69	-
1	0.55	0.495	-

Table 3. Second case study: detection time; delay with C; delay; theoretical distance; experimental distance.

Sensors	Detection Time (Samples)	Delay with C (Samples)	Delay (s)	Theoretical Distance (m)	Experimental Distance (m)
C	912	-	-	-	-
B	1063	151	7.55×10^{-5}	0.20	0.20
A	1296	384	1.92×10^{-4}	0.50	0.50

Table 4. Initial data of the second case study.

Locations	x-Coordinate (m)	y-Coordinate (m)	Radius (m)
A	0	0	0.50
B	0.8	0	0.20
C	0.4	0.69	-
2	0.65	0.495	-

Table 5. Third case study: detection time; delay with C; delay; theoretical distance; experimental distance.

Sensors	Detection Time (Samples)	Delay with C (Samples)	Delay (s)	Theoretical Distance (m)	Experimental Distance (m)
C	962	-	-	-	-
B	1298	336	1.68×10^{-4}	0.43	0.43
A	1087	125	6.25×10^{-5}	0.17	0.16

Table 6. Initial data of the third case study.

Locations	x-Coordinate (m)	y-Coordinate (m)	Radius (m)
A	0	0	0.16
B	0.8	0	0.43
C	0.4	0.69	-
3	0.2	0.445	-

Table 7. Fourth case study: detection time; delay with C; delay; theoretical distance; experimental distance.

Sensors	Detection Time (Samples)	Delay with C (Samples)	Delay (s)	Theoretical Distance (m)	Experimental Distance (m)
C	1155	-	-	-	-
B	1650	495	2.48×10^{-4}	0.64	0.64
A	1385	230	1.15×10^{-4}	0.29	0.30

Table 8. Initial data of the fourth case study.

Locations	x-Coordinate (m)	y-Coordinate (m)	Radius (m)
A	0	0	0.30
B	0.8	0	0.64
C	0.4	0.69	/
4	0.05	0.645	/

5.1. Case Study 1

The breaking of the lead is made in the following coordinates from Sensor A at Point 1 (star); see Figure 10. Sensor A is the coordinate origin.

- Coordinate x: 0.55.
- Coordinate y: 0.495.

The location of the source employing the algorithm is: Point 1: (x: 0.5533, y: 0.4920). The error in the location is: coordinate x: 3.3 mm; coordinate y: 30 mm.

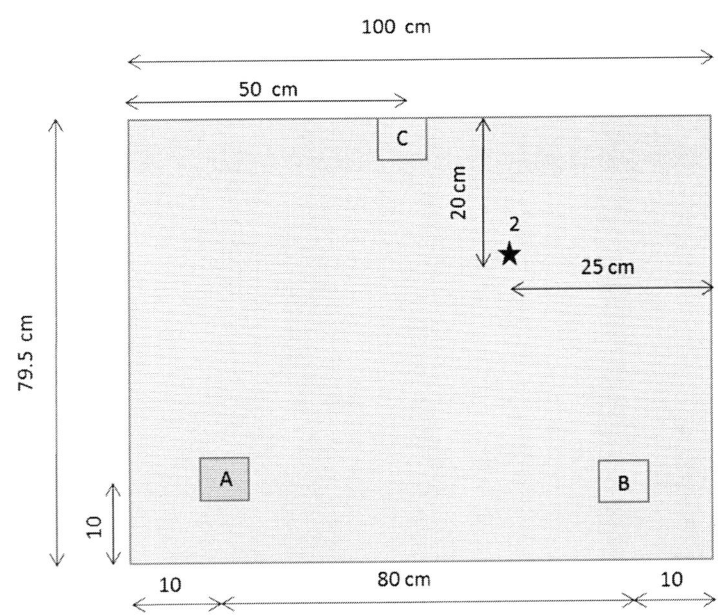

Figure 11. Scheme for Case Study 2.

5.2. Case Study 2

In this case, the emission source was generated at Point 2 (star), shown in Figure 11:

- -Coordinate x: 0.65.
- -Coordinate y: 0.495.

The location of the source employing the algorithm is: Point 2: (x: 0.6502, y: 0.4950). The errors in the location are: coordinate x: 0.2 mm; coordinate y: 0.00 mm.

5.3. Case Study 3

In this case, the emission source was generated at Point 3 (star); see Figure 12:

- -Coordinate x: 0.20.
- -Coordinate y: 0.445.

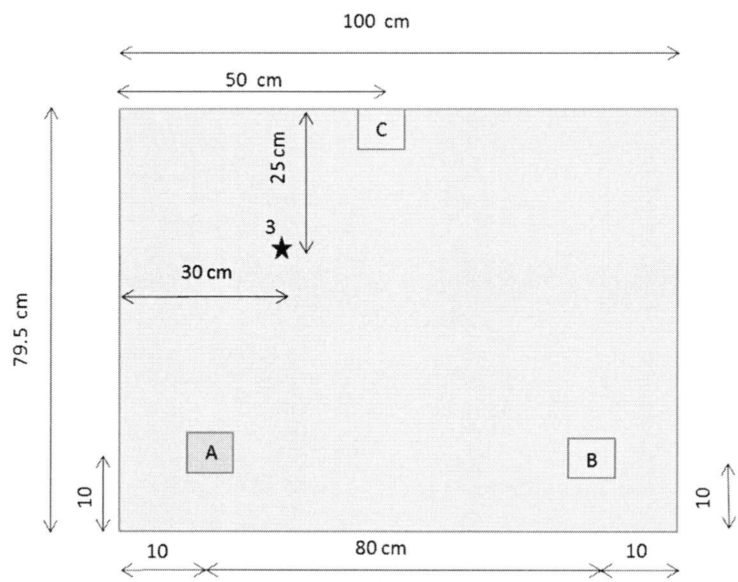

Figure 12. Scheme for Case Study 3.

The location of the source employing the algorithm is: Point 3 (x: 0.1914, y: 0.4434). The errors in the location are: coordinate x: 8.6 mm; coordinate y: 1.6 mm.

5.4. Case Study 4

In this case, the emission source was generated at Point 4 (star), and it is shown in Figure 13:

- -Coordinate x: 0.05.
- -Coordinate y: 0.645.

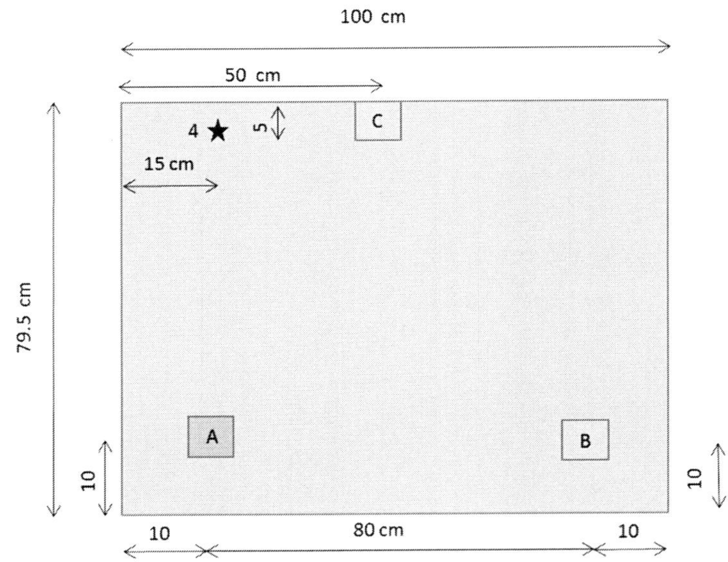

Figure 13. Scheme for Case Study 4.

The location of the source employing the algorithm is: Point 4: (x: 0.050, y: 0.6495). The errors in the location are: coordinate x: 0 mm; coordinate y: 4.5 mm.

Different waves with different speeds appear as a result of the scattering phenomena when a large number of frequencies are excited by the breakage. This makes the identification of peaks to measure the delays of the signals complicated. The orientation of the sensors, when they receive the excitation, can affect the shape of the signal collected.

It is observed that the algorithm provides correct and coherent results. It detects the location of the acoustic emission with an accuracy of two decimals (millimeters). The maximum error registered was 9 mm.

Finally, the algorithm shows the position of the acoustic emission point with the real dimensions of the blade. Figure 14shows the location of the acoustic emission for the first case study.

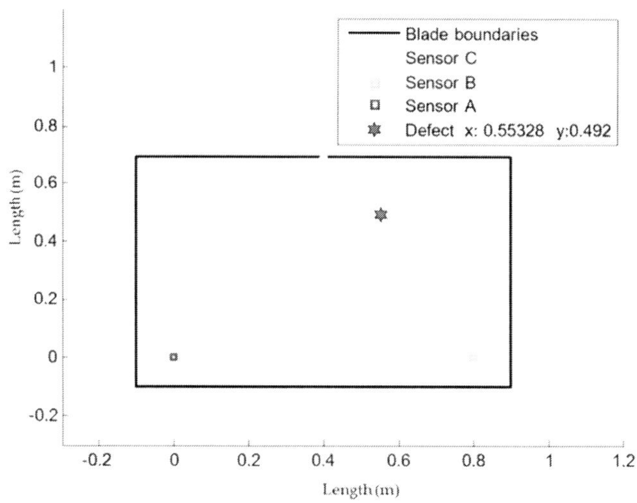

Figure 14. Scheme of the location of the acoustic emission for the first case study.

6. CONCLUSIONS

The development of a localization approach presented in this paper is set using macro-fiber composites to detect cracks in blades in an SHM system. This approach, based on NDT, automatically identifies and locates an acoustic emission source coming from a fiber's breakage in a wind turbine blade section by a novel signal processing method. It can be extrapolated to other similar structures, e.g., airplane wings.

Three sensors are strategically located in the blade. It is demonstrated that the approach is able to detect the location of the simulated defect accurately employing acoustic emissions signals. The signal processing is based on a graphical method of triangulation and seven nonlinear equations. The signals are previously filtered. Different experiments are performed to demonstrate the effectiveness of the proposed method.

The approach detects the location of the acoustic emission with high accuracy, 9 mm being the maximum error registered.

There are conditions that affect the accuracy of the emission source location, e.g., the type of composite material, the sensitivity of the transducers, environmental noise, false positives due to impacts on the piece, *etc.* The method shows the position of the acoustic emission point with the real dimensions of the blade.

ACKNOWLEDGMENTS

This project is partly funded by the EC under the FP7 framework program (Ref. 322430), OPTIMUS and the MINECO project WindSeaEnergy (Ref. DPI2012-31579), where OPTIMUS is supporting to the new condition Monitoring System, and WindSeaEnergy the novel signal processing approach presented in this research paper.

AUTHOR CONTRIBUTIONS

Carlos Quiterio Gómez Muñoz and Fausto Pedro García Marquez conceived and designed the experiments; Carlos Quiterio Gómez Muñoz and Fausto Pedro García Marquez performed the experiments; Carlos Quiterio Gómez Muñoz and Fausto Pedro García Marquez analyzed the data; Carlos Quiterio Gómez Muñoz and Fausto Pedro García Marquez contributed reagents/materials/analysis tools; Carlos Quiterio Gómez Muñoz and Fausto Pedro García Marquez wrote the paper.

REFERENCES

1. Dai, D.; He, Q. Structure damage localization with ultrasonic guided waves based on a time–frequency method. *Signal Process.* **2014**, *96*, 21–28.
2. Spinato, F.; Tavner, P.J.; van Bussel, G.; Koutoulakos, E. Reliability of wind turbine subassemblies. *IET Renew. Power Gen.* **2009**, *3*, 387–401.
3. Ajayi, O.O.; Fagbenle, R.O.; Katende, J.; Ndambuki, J.M.; Omole, D.O.; Badejo, A.A. Wind energy study and energy cost of wind electricity generation in nigeria: Past and recent results and a case study for south west nigeria. *Energies***2014**, *7*, 8508–8534.
4. Marugán, A.P.; Márquez, F.P.G. A novel approach to diagnostic and prognostic evaluations applied to railways: A real case study. *Proc. Inst. Mech. Eng. F J. Rail Rapid Transit* **2015**.
5. Chen, X.; Zhao, W.; Zhao, X.L.; Xu, J.Z. Failure test and finite element simulation of a large wind turbine composite blade under static loading. *Energies* **2014**, *7*, 2274–2297.

6. Márquez, F.G.; Roberts, C.; Tobias, A.M. Railway point mechanisms: Condition monitoring and fault detection. *Proc. Inst. Mech. Eng. F J. Rail Rapid Transit* **2010**, *224*, 35–44.
7. Pliego Marugán, A.; García Márquez, F.P.; Lorente, J. Decision making process via binary decision diagram. *Int. J. Manag. Sci. Eng. Manag.* **2015**, *10*, 3–8.
8. Marquez, F.P.G. An approach to remote condition monitoring systems management. In Proceedings of the Institution of Engineering and Technology International Conference on Railway Condition Monitoring, Birmingham, UK, 29–30 Novemmber 2006; pp. 156–160.
9. Light-Marquez, A.; Sobin, A.; Park, G.; Farinholt, K. Structural damage identification in wind turbine blades using piezoelectric active sensing. In *Structural Dynamics and Renewable Energy*; Springer: New York, NY, USA, 2011; pp. 55–65.
10. García, F.P.; Pinar, J.M.; Papaelias, M.; Ruiz de la Hermosa, R. Wind turbines maintenance management based on FTA and BDD. *Renew. Energy Power Qual. J.* **2012**. Available online: http://icrepq.com/icrepq'12/699-garcia.pdf (accessed on 7 January 2016).
11. Pedregal, D.J.; García, F.P.; Roberts, C. An algorithmic approach for maintenance management based on advanced state space systems and harmonic regressions. *Ann. Oper. Res.* **2009**, *166*, 109–124.
12. Yang, H.-H.; Huang, M.-L.; Yang, S.-W. Integrating auto-associative neural networks with hotelling T^2 control charts for wind turbine fault detection. *Energies* **2015**, *8*, 12100–12115.
13. Chen, X.; Qin, Z.W.; Zhao, X.L.; Xu, J.Z. Structural performance of a glass/polyester composite wind turbine blade with flatback and thick airfoils. In Proceedings of the American Society of Mechanical Engineers (ASME) 2014 International Mechanical Engineering Congress and Exposition, Montreal, QC, Canada, 14–20 November 2014.
14. De la Hermosa González, R.R.; Márquez, F.P.G.; Dimlaye, V. Maintenance management of wind turbines structures via mfcs and wavelet transforms. *Renew. Sustain. Energy Rev.* **2015**, *48*, 472–482.
15. Márquez, F.P.G.; Pedregal, D.J.; Roberts, C. New methods for the condition monitoring of level crossings. *Int. J. Syst. Sci.* **2015**, *46*, 878–884.
16. García, F.P.; Pedregal, D.J.; Roberts, C. Time series methods applied to failure prediction and detection. *Reliab. Eng. Syst. Saf.* **2010**, *95*, 698–703.

17. García Márquez, F.P.; García-Pardo, I.P. Principal component analysis applied to filtered signals for maintenance management. *Qual. Reliab. Eng. Int.* **2010**, *26*, 523–527.
18. Márquez, F.P.G.; Muñoz, J.M.C. A pattern recognition and data analysis method for maintenance management. *Int. J. Syst. Sci.* **2012**, *43*, 1014–1028.
19. Michaels, J.E. Detection, localization and characterization of damage in plates with an *in situ* array of spatially distributed ultrasonic sensors. *Smart Mater. Struct.* **2008**, *17*, 035035.
20. Chen, H.; Yan, Y.; Chen, W.; Jiang, J.; Yu, L.; Wu, Z. Early damage detection in composite wingbox structures using hilbert-huang transform and genetic algorithm. *Struct. Health Monit.* **2007**, *6*, 281–297.
21. Eftekharnejad, B.; Carrasco, M.; Charnley, B.; Mba, D. The application of spectral kurtosis on acoustic emission and vibrations from a defective bearing. *Mech. Syst. Signal Process.* **2011**, *25*, 266–284.
22. Gómez, C.Q.; Villegas, M.A.; García, F.P.; Pedregal, D.J. Big data and web intelligence for condition monitoring: A case study on wind turbines. In *Handbook of Research on Trends and Future Directions in Big Data and Web Intelligence*; Information Science Reference, IGI Global: Hershey, PA, USA, 2015.
23. Bohse, J. Acoustic emission characteristics of micro-failure processes in polymer blends and composites. *Compos. Sci. Technol.* **2000**, *60*, 1213–1226.
24. Márquez, F.P.G.; Pérez, J.M.P.; Marugán, A.P.; Papaelias, M. Identification of critical components of wind turbines using FTA over the time. *Renew. Energy* **2015**, *87*, 869–883.
25. Gorman, M.R. Plate wave acoustic emission. *J. Acoust. Soc. Am.* **1991**, *90*, 358–364.
26. Ruiz de la Hermosa, R.; García Márquez, F.P.; Dimlaye, V.; Ruiz-Hernández, D. Pattern recognition by wavelet transforms using macro fibre composites transducers. *Mechan. Syst. Signal Process.* **2014**, *48*, 339–350.
27. Betz, D.C.; Staszewski, W.J.; Thursby, G.; Culshaw, B. Structural damage identification using multifunctional bragg grating sensors: II. Damage detection results and analysis. *Smart Mater. Struct.* **2006**, *15*, 1313–1322.

28. Coverley, P.; Staszewski, W. Impact damage location in composite structures using optimized sensor triangulation procedure. *Smart Mater. Struct.* **2003**, *12*, 795–803.
29. Gómez, C.Q.; Ruiz de la Hermosa, R.; Trapero, J.R.; Garcia, F.P. A novel approach to fault detection and diagnosis on wind turbines. *Glob. Nest J.* **2014**, *16*, 1029–1037.

Chapter 5

Structural Damage Localization by Linear Technique of Acoustic Emission

Md. Tawhidul Islam Khan[1*], Nagafuchi Sunichi[1], Mehedi Hasan[2]

[1]Department of Advanced Technology Fusion, Saga University, Saga, Japan

[2]Department of Mechanical Engineering, CUET, Chittagong, Bangladesh

ABSTRACT

Linear source location of acoustic emission (AE) technique has been applied in the present paper for finding the source of material damage under fatigue loading. A plate type structure of ductile cast iron has been used to undergo fatigue damage in a servopulsing machine in the experiment. AE sensors were attached to the specimen for getting the time delay of AE signal propagations through the specimen. After receiving the time delay data of AE signals due to the damage initiation in the material for the provided fatigue loads, linear source location algorithm has been applied and the crack positions are identified. Before applying the technique, a series of pencil lead breaks (PLBs) tests have been conducted upon a ductile cast iron plate of same dimension for verifying the applied algorithm. According to the PLBs varification, the failure location of ductile cast iron (pearlite type) due to the fatigue loading has been characterized. In both experiments, the active ability of the proposed technique for source location of structural damage has been identified clearly and successfully.

Keywords: Structural Damage, Linear Source Location Technique, Acoustic Emission

1. INTRODUCTION

Damage localization in different sensitive equipment or structures like aircraft structure or some internal flow- paths like cooling channels of atomic reactor and so on is a great challenge in NDT (nondestructive testing) for their perfect running. Therefore, getting the perfect information of damage initiation, damage propagation as well as damage location becomes challenging topics to the present scientists and engineers. Although, several methods are also trying to apply for the mentioned problems solution, the acoustic emission (AE) technique plays an important role in finding these structural damage locations or sources as NDT technique as well. It is mentioned more critically that the source location of a damage initiation is often considered as the main goal of acoustic emission technique [1] .

Structural health monitoring (SHM) refers to the structural health assessment technique by involving some non-destructive devices consisting of sensors and relative appliances to monitor the structural condition of integrity, working ability, reliability, safety as well as the performance efficiency of the structural system together [2] . In this system, appropriate processes are involved in monitoring the structure over a period of time by extracting damage sensitive features from AE sensing systems and then identify the conditions of damage intensity, damage propagation, and damage location in an early stage as well as in matured stage [3] . There are various techniques that are involved in SHM such as, vibration monitoring technique, ultrasonic wave propagation technique, acoustic emission technique etc. Acoustic emission technique is one of these non-destructive techniques (NDT) which is widely and successfully used in monitoring structures in aerospace engineering, mechanical engineering and civil engineering fields [4] [5] .

Acoustic emission (AE) is a phenomenon of transient stress wave propagation which is caused due to the sudden release of elastic energy, resulting from mechanical deformations in materials and or due to the initiation and propagation of microcracks, dislocation movement or occurring similar irreversible changes [6] . AE sensors are used in detecting the AE acoustic waves produced by AE events at the surface of the material. The output of AE signals acquisition is featured by its representing properties [7] . Therefore, AE plays an important role in non-destructive testing.

Source location is a powerful tool in AE testing for all engineering applications. In both for laboratory as well as for industrial testing, AE source location algorithms are efficiently used [8] [9] . In laboratory based source location experiments are used to confirm the characteristics of original source and the wave propagations. By applying these information the origin of damage is localized in laboratory application along with the AE feature values.

The main objective of the present research is to clarify the source localization characteristics of acoustic emission technique (linear source location) based on the fatigue damage in ductile cast iron. As ductile cast iron contains carbon globules to increase the ductile properties of the material, the AE wave transformation through this material faces scatterings and anomalousness in propagations. Therefore, two types of experiments were conducted in characterizing the AE source location algorithm. First, the pencil lead fracture (PLF) tests are occurred, and based on that results laboratory experiments are conducted and cracking sources are localized.

Linear source location algorithm has been explained in the next section; after which, the experimental methodologies are discussed. Experimental results and discussions are explained in the following section of the methodologies. Finally, conclusions are presented before the acknowledgements and references.

2. LINEAR SOURCE LOCATION ALGORITHM

One of the important advantages of AE structural health monitoring (SHM) system is the ability to locate the damage source. Source localization is an important objective of any types of monitoring system. Therefore, the AE based SHM plays an influential role in health monitoring system as this technique can locate damage source in real time, whereas other methods such as vibration or ultrasound based monitoring techniques detect damage after it has been occurred.

There are several methods in linear source location technique. Out of them, time difference (TD) method which is widely used has applied in the present research as well. The schematics of TD algorithm is shown in Figure 1, where, AE source is mentioned by So. Similarly, s_1, s_2 and t_1, $t_2$indicate as the AE sensor 1, sensor 2 and time of arrival (TOA) to sensor 1, time of arrival to sensor 2 respectively.

Furthermore, l, l_1 and l_0 indicate the distance between two sensors (s_1, s_2), axial distance of AE source to sensor 1 and axial distance between AE source to the midpoint of two AE sensors respectively. The mathematical relations of the TD algorithm are explained as follows [9] [10] :

$$l_0 = \frac{1}{2}(t_1 - t_2) \cdot v = \frac{1}{2}\Delta t \cdot v \quad (1)$$

$$l_1 = \frac{1}{2}l - l_0 = \frac{1}{2}(l - \Delta t \cdot v) \quad (2)$$

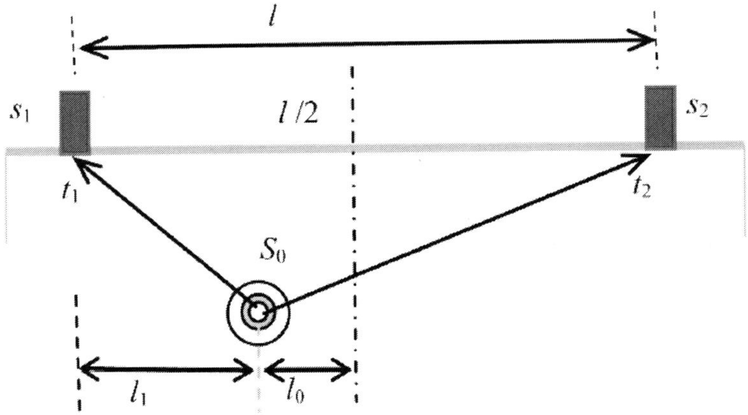

Figure 1. Modeling of linear source location technique.

where, Dt indicates the time difference between two arrival times to sensor 1 and sensor 2 and v indicates the AE wave velocity which has been measured as:

$$v = \frac{d}{t} \tag{3}$$

where, d and t indicate the AE wave travelling distance and travelling time respectively.

3. EXPERIMENTAL METHODOLOGY

To characterize the AE source localization technique in ductile cast iron, two major types of experiments have been conducted in the present research. One is the pencil lead fracture experiment, and the other one is the laboratory test under the fatigue loading. Furthermore, before entering into the source location calculations, experiments for identifying the velocity of AE waves through ductile cast iron are conducted. Details of these experiments are explained as follows.

3.1. Measurement of AE Wave Velocity

A ductile cast iron plate with same thickness of the specimen used in PLF or fatigue tests is used for AE wave velocity experiment. For the purpose of getting wave velocity inside the specimen material (described later), two AE sensors, s_1 and s_2 (detail specification is given in Table 1) are attached upon the specimen with a central distance of d. For getting AE excitation, pencil lead (uni 0.5 B Nano Dia Blended Hi-Quality pencil lead of Mitsubishi Pencil Co. LTD.)

with a length of 10 mm and 45 degrees inclination to the specimen surface is fractured. The positions of sensors and the place of PLF is shown in Figure 2. The AE wave travelling time (t) from sensor 1 to sensor 2 is calculated from the AE wave velocity experiments. The schematics of velocity measurement experiment are shown in Figure 3.

3.2. Experiments of PLF

Pencil lead fracture (PLF) tests have been conducted in a noise free environment on the same specimen as is used in fatigue tests. Therefore, the results of PLF experiments are used in calibrating the source location techniques for using in fatigue experiments as well. The details of the specimen preparation and the specimen material properties are discussed as follows.

Specimen for the experiments of PLF and laboratory fatigue tests, pearlite type ductile cast iron (PDI) block has been cast into a block and machined with appropriate size as mentioned in Figure 4. Chemical compositions of the test materials are listed in Table 2. The numerical values in this table are represented by the corresponding values in percent content of carbon (C), silicon (Si), manganese (Mn), phosphorus (P), sulphur (S) and copper (Cu) in percent mass respectively. Mechanical properties of pearlite ductile cast iron are shown in Table 3. The white cast iron is heat treated according to the method of pearlite grain determination test for steel of Japanese Industrial Standards (JISG0552) for converting the hard iron carbide structure into malleable pearlite malleable matrix. The pearlitic structure contains some combined carbons in the matrix which result higher strength and higher hardness than ferritic malleable iron matrix. The specimen is prepared by cutting a central hole and curvatures at the middle position of the specimen to control the crack growth point at the radial surface of this hole. Other endpoint holes are prepared for the attachment to the fatigue loading machines. The determination of the rate of spherical graphite is carried out in compliance with the spheroidal graphite iron castings of Japanese Industrial Standards (JISZ 5502). The average ferrite grain size (D_f), average particle size (D_m), area fraction of graphite (f_g) and other related values have been calculated accordingly as well. Then, the specimen has been mirror polished to examine the crack propagation by conducting the microstructural analysis (not mentioned in the present paper). The dimension of the prepared specimen with four AE sensor positions is shown in Figure 4. In PLF tests, lead fractures are done at both sides of the central hole accordingly as shown in this figure. As under fatigue loading, cracks growth generally start at the peripheral sides the central circular hole, in PLF test, therefore, fractures are considered to occur at the same locations. All sensors have been connected with preamplifiers, main amplifiers. Data acquisition device (digital oscilloscope) and finally, the results are saved to a PC for further analysis.

Table 1. Specification of AE sensors and preamplifiers.

Types/Parameters	Specification/Value
AE sensor	R15a (Physical acoustics)
Operating frequency	50 - 400 kHz
Resonant frequency	75 [150] kHz
Sampling frequency	1 MHz
Preamplifier	R-6-1 (40 dB)

Table 2. Chemical composition of the specimen (PDI).

	C	Si	Mn	P	S	Cu
Mass (%)	3.68	2.08	0.16	0.017	0.001	0.67

Table 3. Mechanical properties of the specimen (PDI).

Tensile strength (Mpa)	0.2% proof stress (Mpa)	Stretching (%)	Hardness (HBW)
793	483	5.9	218

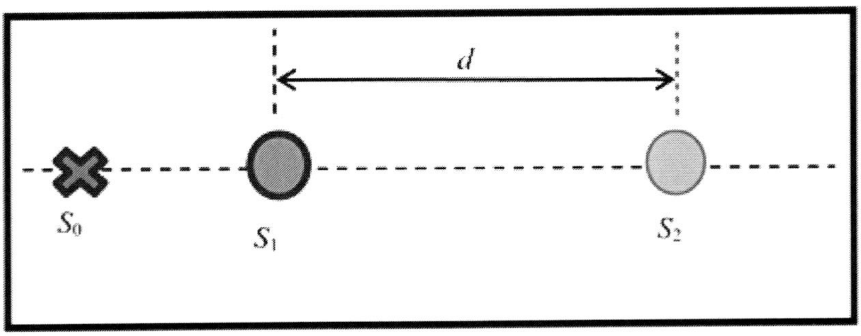

Figure 2. Schematics of sensor positions (s1, s2) and source of PLF (s0).

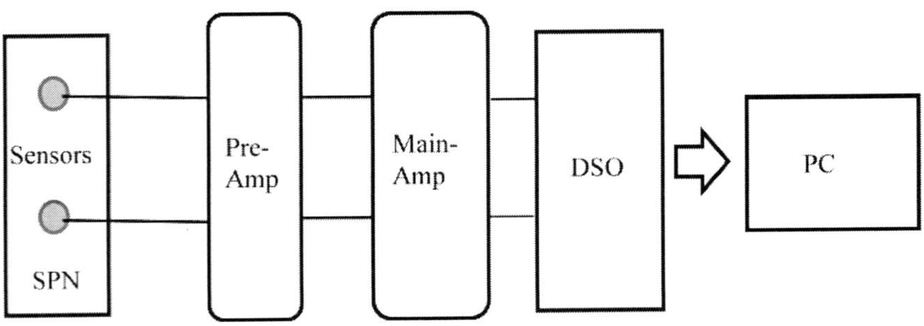

Figure 3. Schematics of experimental apparatus for the measurement of AE wave velocity.

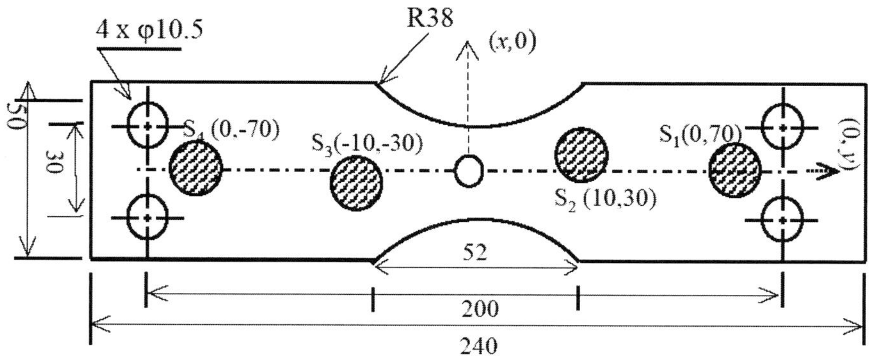

Figure 4. Dimension of specimen with 4 AE sensors (s1 - s4).

3.3. Experiments of Fatigue Test

The specimen for the fatigue test is attached to the servopulser (fatigue testing machine) for pressing required fatigue loads for attaining the adequate cracks to the specimen and continued until the crack-fracture. Four AE sensors (s_1, s_2, s_3, and s_4) are attached to the specimen (SPN). Special silica gel is used between the sensor and specimen for avoiding signal leakage as well as reducing artifacts. Each sensor is connected to a pre-amplifier (Pre.AMP) and then to the 4-channel main AE amplifier (Main.AMP). After applying adequate filtration, the AE signals are stored in a digital storage oscilloscope (DSO) from where the AE data are transferred to the personal computer for the further analysis. Total schematics of the experimental set-up are shown in Figure 5. In this figure the term servopulser is indicated with dotted line where the specimen is attached through four bolts. The controller of servopulser is attached to the system for the controlling of fatigue loads, loading cycles and the power on-off to the machine. The connections of the AE sensing systems in fatigue test are similar to the PLF tests.

In AE experiments, all sensors have been significantly coupled with the specimen for avoiding wave leakage. The preamplifier is setting to the gain of 40 dB. Main amplifier gain is setting to zero. The sampling frequency is selected as 1 MHz as explained above. For avoiding noise, the specimen in PLF tests has mounted upon the acoustic absorbing material. However, for avoiding noise in fatigue tests, special initial thresholds values (1.28 volts in trigger setting) have been applied.

4. EXPERIMENTAL RESULTS AND DISCUSSION

Structural damage localization of ductile cast iron (pearlite type) was characterized by the linear source location technique. Several experiments have been conducted to identify and validate the present algorithm.

4.1. PLF Experimental Results

According to the discussions above the PLF experimental results for linear source location have been summarized and shown in Table 4. The results that have been summarized in the table are also based on average data of ten iterations. The results show the source locations satisfactorily (exact source locations are shown in Figure 4).

4.2. Fatigue Experimental Results

Fatigue experiments have been conducted as explained before. All results corresponding to the above experiments are summarized in Table 5. Sensor settings as well as other preparations for this experiment are also based on the same fundamental preparations of PLF experiments.

AE source localization is compared in Figure 6. The comparison shows good agreements, although, some artifacts are observed in channel 2. It is considered that this difference is caused due to the receiving of higher AE energy for having the near location of the damage site in fatigue test. This artifact in channel 2 during the PLF test is relatively less compared to that of fatigue test. The reason that have been considered for getting much AE energy in sensor 2 may not consider in this case, since the intensity is found less in PLB test as well. Moreover, in PBL test the direction of pencil lead fracture may affect the result as well.

4.3. Comparisons of AE Parametric Results

Several sample AE parametric characteristics are shown in the following AE data. The amplitude characteristics of four sensors (s_1, s_2, s_3 and s_4) in PLF experiments are shown in Figure 7. It is found that the peak amplitudes of sensor 2 and sensor 3 are higher compared to that of sensors 1 and 4. It is reasoning that since the sensors 2 and 3 are placed nearer to the fracture location, it received more AE energy (near field excitation) compared to the sensors 1 and 4 (far field excitation). Same results are also observed in fatigue experiments.

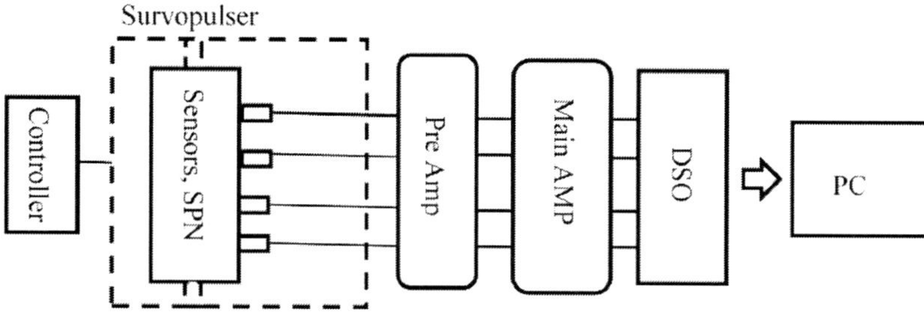

Figure 5. Schematics of experimental apparatus, in PLF tests as well as fatigue tests.

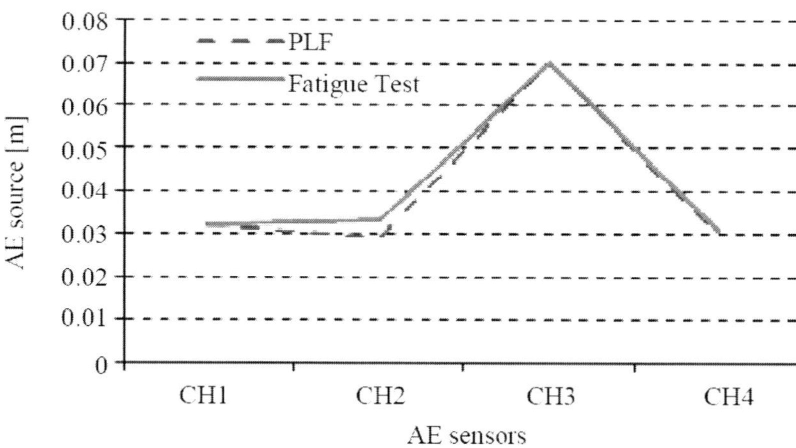

Figure 6. Comparison of AE sources in PLF and fatigue tests.

Table 4. Linear source location result in PLF technique.

Observed sources	Δt [s]	l [m]	l_1 [m]
s_2 - s_4	2.10E−05	0.10049	0.032
s_1 - s_3	2.60E−05	0.10049	0.029
s_1 - s_4	1.00E−06	0.14	0.07
s_2 - s_3	3.00E−06	0.6324	0.030

Table 5. Linear source location result in fatigue tests.

Observed sources	Δt [s]	l [m]	l_1 [m]
s_2 - s_4	2.20E−05	0.10049	0.032
s_1 - s_3	2.10E−05	0.10049	0.033
s_1 - s_4	0.00	0.14	0.07
s_2 - s_3	1.00E−06	0.6324	0.031

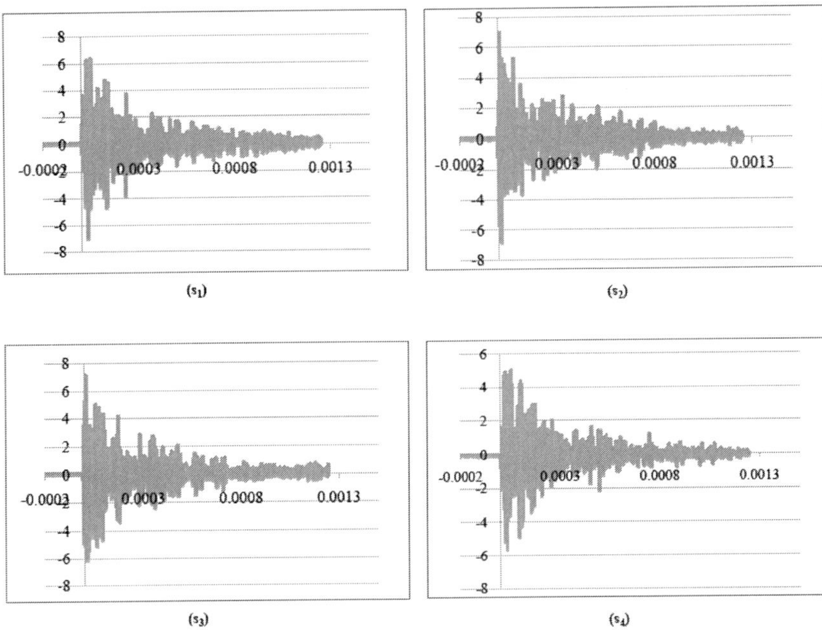

Figure 7. AE parametric results (PLF) of sensors, 1, 2, 3, 4 which show the peak amplitude characteristics in volts.

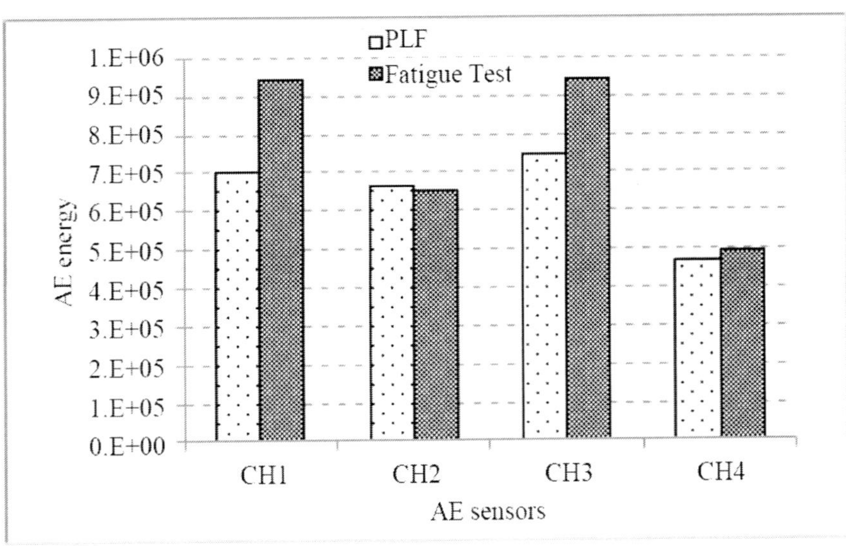

Figure 8. Comparison of AE energy in PLF and fatigue tests for four sensors.

The parametric analysis of AE energy for both PLF and fatigue tests are shown in Figure 8. It is found from this figure that AE energy received from fatigue test is higher than that of PLF test. Furthermore, the sensor located at far field (CH 4) is receiving less AE energy compared to sensors 2 or 3, and it shows the facts that the location of AE source is nearer to the location of sensors 2 or 3, and from the source location results as discussed above also satisfy these techniques validation. Moreover, it is observed that the sensor 1 at fatigue tests received excessive noise from the specimen attachments device since it is located near to the fatigue loading side to the survopulser.

4.4. Discussion

A linear source location technique has been discussed in the present paper for finding the source of fatigue damage in the material. The major advantage of the applied AE technique is its simplicity. For its application no external guided signals are required to insert in identifying internal information of the material. Moreover, as the emitted acoustic waves captured by AE sensors have been generated during the internal changes of the material damage, therefore, impedance loss in travelling is much less compared to other ultrasonic techniques. However, since the AE sensors are attached to the specimen, noise due to vibration and other dynamic effects have easily affected the results in AE technique. Therefore, similar features of data contamination with excessive noise transmissions to the sensors, particularly, to those sensors near to the specimen attachments are observed in the present experimental results (Figure 6). However, in source location of material damage, AE technique still plays important roles and the maximum contaminated noise or dynamic noise are minimized with appropriate thresholds. Thus, suitable thresholds values (maximum 2 volts) have been added in the present experiments as well to minimize the excess noise contaminations from the experimental apparatus (servopulser in fatigue loading) during fatigue experiment.

5. CONCLUSIONS

Structural damage localization technique by applying AE linear source location algorithm is successfully implemented and analyzed in the present research.

Two major experiments of PLF and fatigue test are used to discuss the present source location technique. The results of fatigue tests are successfully validated by the PLF technique.

Finally, the damage that has been occurred by fatigue loading in ductile cast iron is successfully identified by the AE source localization technique in the present research.

It is understood that the further improvements in specimen attachment technique to the fatigue loading machine is necessary for avoiding excessive noise transmission to the AE source location technique.

ACKNOWLEDGEMENTS

This investigation has been supported by a Grant-in-Aid for Scientific Research (KAKENHI 24560297) from Japanese Society for the Promotion of Science. The author is also expressed his gratitude to HINODE Ltd., Japan for supplying the test materials and other related assistance.

REFERENCES

1. Abe, S. (2006) Acoustic Emission Testing. Nondestructive Testing Series I, Japan Society for Non-Destructive Inspection, 4-22. (In Japanese)
2. Dawson, B. (1976) Vibration Condition Monitoring Techniques for Rotating Machinery. The Shock and Vibration Digest (London), 8, 3.http://dx.doi.org/10.1177/058310247600801203
3. Wilcox, P.D., Lee, C.K., Scholey, J.J., Friswell, M.I., Wisnom, M.R. and Drinkwater, B.W. (2006) Quantitative Structural Health Monitoring Using Acoustic Emission. SPIE, 6173, 1-10.
4. Meyendorf, N., Frankenstein, B. and Schubert, L. (2012) Structural Health Monitoring for Aircraft, Ground Transportation Vehicles, Wind Turbines and Pipes-Prognosis. 18th World Conference on Nondestructive Testing, Durban, 16- 20 April 2012.
5. Finlayson, R.D., Friesel, M., Carlos, Cole, P. and Lenain, J.C. (2001) Health Monitoring of Aerospace Structures with Acoustic Emission and Acousto-Ultrasonics. Insight, 43, 1-4.
6. Beghi, M.G. (2013) Modeling and Measurement for Acoustic Wave and for Acoustic Microdevices. In Tech.
7. Theobald, P. and Dar, F. (2006) AE Sensor Calibration for Out-of-Plane and In-Plane Displacement Sensitivity. Journal of Advanced Material Research, 13, 91-98.http://dx.doi.org/10.4028/www.scientific.net/AMR.13-14.91
8. Yoshioka, T. and Fujiwara, T. (1982) A New Acoustic Emission Source Locating System for the Study of Rolling Contact Fatigue. Wear, 81, 183-186.http://dx.doi.org/10.1016/0043-1648(82)90314-3
9. Baxter, M.G., Pullin, R., Holford, K.M. and Evans, S.L. (2007) Delta T Source Location for Acoustic Emission. Mechanical Systems and Signal Processing, 2007, 1512-1520.http://dx.doi.org/10.1016/j.ymssp.2006.05.003
10. Jonathan, J.S., Paul, D.W. and Michael, R.W. (2008) Two-Dimensional Source Location Techniques for Large Composite Plates. Proceedings, EWGAE 2008, 160-165.

Chapter 6

The Features of Acoustic Bathymetry of Sludge in Oil Storage Tanks

Manel Monteiro[1], Victor Svet[2], Donald Sandilands[1]

[1]*ITS Ltd., Aberdeen, UK*

[2]*Laboratory of Signal Processing and Imaging, Institute of Acoustics, Moscow, Russia*

ABSTRACT

The features of acoustic bathymetry of sludge in crude oil tanks are considered. Different parameters of crude oil and sludge which are important for the selection of acoustic parameters such as sound speed, the acoustic impedances of "oil-sludge" boundaries and sound attenuation are analyzed and compared with experimental data. The main sources of errors of sludge volume estimation are discussed.

Keywords: Crude Oil, Storage Oil Tanks, Sludge, Acoustic Bathymetry, Acoustic Profiling

1. INTRODUCTION

Large aboveground storage tanks (AST) are used as reservoirs for the storage of crude oil prior to processing. Typical construction is a steel cylinder (API 650) with a diameter up to 120 m and height up to 20 m. The tank is not pressurised and a pontoon roof with a rim seal floats on the oil surface to minimize the release of hazardous vapours to the atmosphere. When the crude oil is fully drained from the tank the pontoon roof is mechanically supported on evenly distributed leg supports. The height of the legs in the maintenance position is about 2.5 m. Crude oil is a very heterogenous raw material. Depending on the producing oilfield the composition can contain different fractions of water, gravel, sand and heavy asphaltenes, paraffin or wax hydrocarbons. Over time these mixed fractions become deposited on the bottom of the tank as a heavy viscous sludge. This sludge reduces the available tank volume, accelerates corrosion, traps water and may contain pockets of gas or other toxic impurities. On the other hand, the sludge has some value if processed, as it may contain up to 80% of usable hydrocarbons. Tank sludge cleaning is a very costly and time consuming operation, and it is therefore necessary to know not only the amount of sludge in the tank, but the spatial distribution. If in some places the local sludge height exceeds the height of the leg support then there is a risk of damage to the roof when lowered to the maintenance position. Measurement of the volume and its spatial distribution is typically performed manually using a special weighted metal tape measure. The sludge/oil interface is detected by the operator feeling the change in weight of the measurement tape as it makes contact with the higher density sludge surface. Manual measurement is time consuming, subjective and places the operator in a potentially hazardous area for long periods since H_2S gas is typically present in crude oil. Manual measurements can be made only in a few entry points so the reliability of the measured volume and character of sludge distribution over the bottom are very poor. The error of the measured volume can be up to 35% - 45%. Note that a measurement error of a few centimeters can lead to an error in sludge volume of several hundred cubic meters.

Currently the most promising method of sludge evaluation in large storage tanks is acoustic profiling. Acoustic profiling is similar to acoustic bathymetry in Oceanography but the technical implementation of suitable devices differs from marine multi-beam echo sounders due to the specific conditions of sound propagation in crude oil and the characteristics of closed tank environments that significantly influence signal processing methods. In this article we will discuss the basic physical effects associated with sound propagation in the crude oil which has a major influence on the reliability of measurements of sludge distribution and the evaluation of its volume.

2. PARAMETERS OF CRUDE OIL

The properties important for acoustics are only well understood for pure refined oils. The acoustic properties of crude cannot be tabulated because of the wide variation of crude oil produced from different oilfields. The most important parameters are sound speed velocity in the crude oil, C_{cr}, attenuation, α_{cr} and character and level of backscattering echo-signals from the boundary of the "oil-sludge". In turn, the sound speed depends on the temperature of crude oil, its density, water content and dissolved gas. Sound attenuation is determined by the absorption and additional losses due to sound scattering caused by the dissolved gas, various suspensions and water-oil emulsion layer. Another important fact is that all these parameters can significantly vary both in time and in space, i.e. their distribution can be anisotropic. This in turn causes additional refraction effects of the acoustic rays. Furthermore the oil tank is an enclosed design and its metal walls and different internal metal constructions (roof supports, heaters, mixers, and pipes) perfectly reflect the sound waves. Such reflections are a source of strong reverberation, which has a negative impact on performance.

2.1. Specific Gravity of Crude Oil (Oil Density)

In accordance with the classification of API (American Petroleum Institute) all crude oil is divided into 4 types: the "light", "medium", "heavy" and "extra-heavy" [1] . To estimate the density of the oil industry uses specific units API^0, which are associated with normal density SG = ρ_0

$$API = \frac{141.5}{\rho_0} - 131.5 \tag{1a}$$

$$SG = \rho_0 = \frac{141.5}{131.5 + API} \tag{1b}$$

Light Crude has an API gravity higher than 31.1° API, Medium Crude has an API gravity between 22.3° API and 31.1° API and Heavy Crude has an API gravity below 22.3° API. This specification is for T = 16.5°C. Figure 1 shows plots of the kinematic viscosity of oil versus temperature for different values of the API, the dotted line on these graphs shows the kinematic viscosity of crude versus density at temperature T = 20°C. Many empirical relations are offered for the calculation of oil density versus temperature. A commonly used formula is:

$$\rho_n = 3.37694 \times 10^{-4} T^2 - 7.06892 \times 10^{-1} T + 8.80964 \times 10^2 \tag{2}$$

2.2. Sound Speed

Numerous measurements of sound speed in crude oil showed that it decreases with increasing temperature. This is opposite to the temperature dependence of sound speed in water. We will not discuss in detail the physical reasons for such behavior of sound speed especially since in the literature there are many often opposing points of view. Researchers have proposed many empirical formulas of which the following is widely used, [3] -[5] :

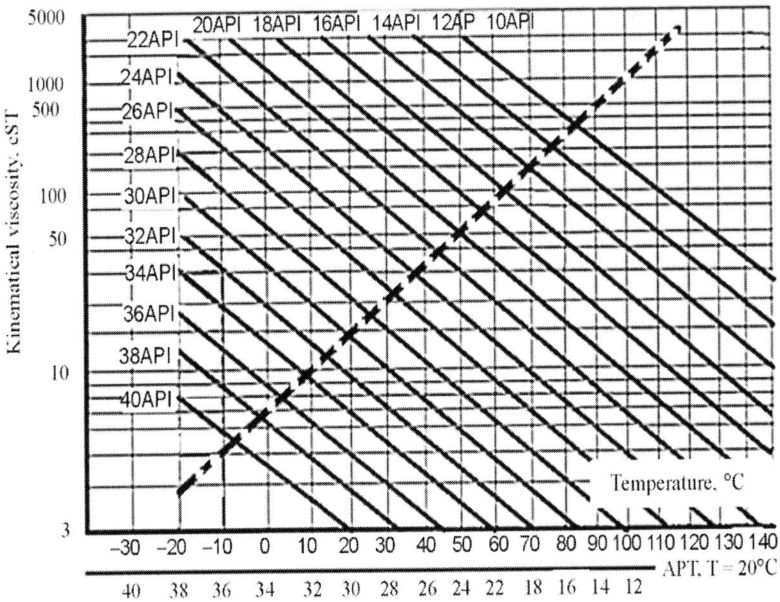

Figure 1. Oil viscosity versus gravity and temperature according ASTM Standard Viscosity Temperature Charts for Liquid Petroleum products (D34), (Courtesy of Paragon Engineering Service, Inc.).

$$C_{\text{crude oil}} = \frac{15450}{\sqrt{77.1 + \text{API}}} - 3.7T\text{þC} \tag{3}$$

The dependence of the sound velocity versus temperature for oil with different API is presented inFigure 2.

In the presence of water in of crude in the form of a water emulsion then the sound speed increases with an increasing of amount of water, Figure 3.

As follows from these graphs, the sound speed decreases almost linearly with decreasing % of water. It is interesting to compare the calculated data of sound speed with real measurements of sound speed in different types of crude. Direct measurements of sound speed are presented onFigure 4. For each of six tanksthe sound speed data has been measured in 6 - 8 entry points. Four tanks contained light oil and two tanks were filled with medium oil. Since the measurements were performed at different temperatures (16 - 21)˚C, the experimental data is recalculated to the same temperature of 20˚C using the formula (3) for oil with API^{0} 40.

It is seen from this data that formula (3) is a quite satisfactory approximation for the evaluation of sound speed, but in practical conditions it is better to measure sound speed directly in the actual tank environment, because possible errors of evaluated inclined distances can be meaningful.

2.3. Sound Attenuation and Absorption in Crude

The general weakening of the acoustic energy ofin sound waves propagating in crude (α_Σ-attenuation) is determined not only by classical absorption but also by scattering of sound by various irregularities that are always present in crude oil. Such irregularities are dissolved gas bubbles, water droplets and solid particles of sand and other minerals. Therefore general attenuation is [2]

$$\alpha_\Sigma = \alpha_{abs} + \alpha_{wio} + \alpha_{gas} + \alpha_{solids} \tag{4}$$

where

α_{abs}—absorption coefficient, which depends on $f^2\alpha_g$, α_w, α_p,—coefficients of sound attenuation due to scattering from gas bubbles, water droplets and solid particles, respectively.

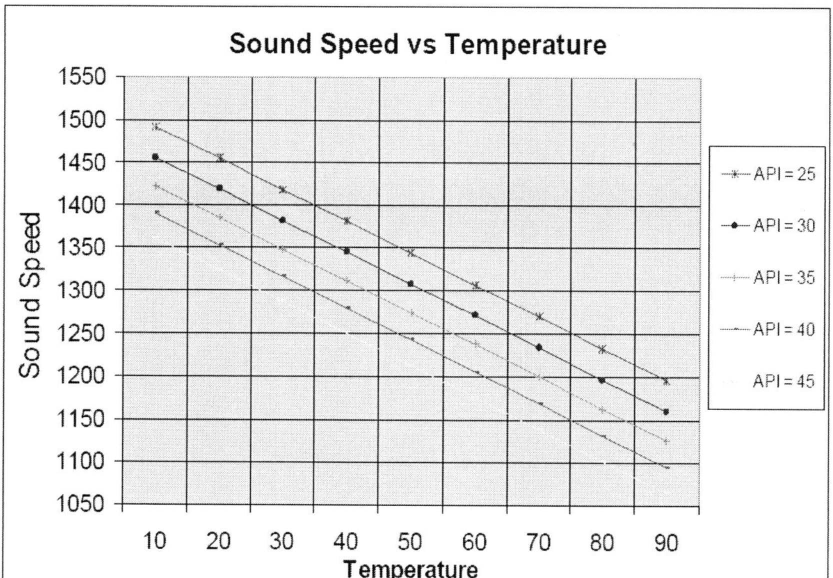

Figure 2. Sound speed in crude oil versus temperature for different API.

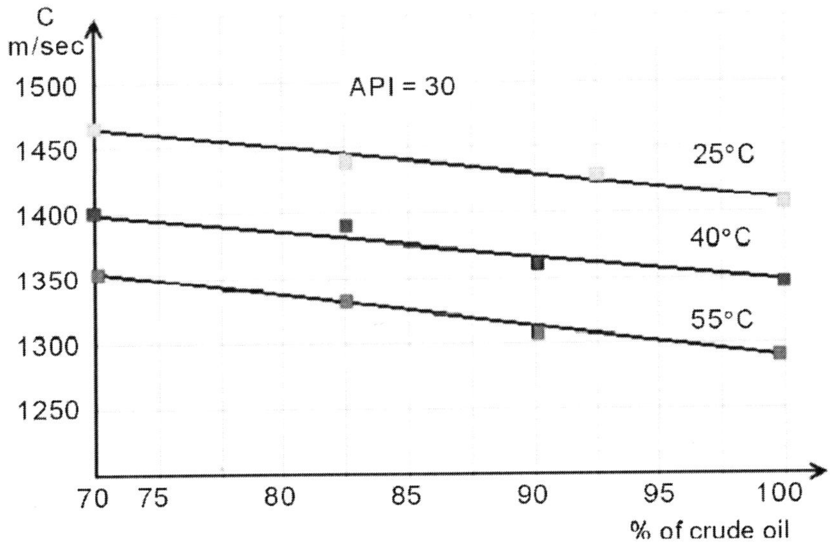

Figure 3. The speed of sound in the crude oil versus % of water content.

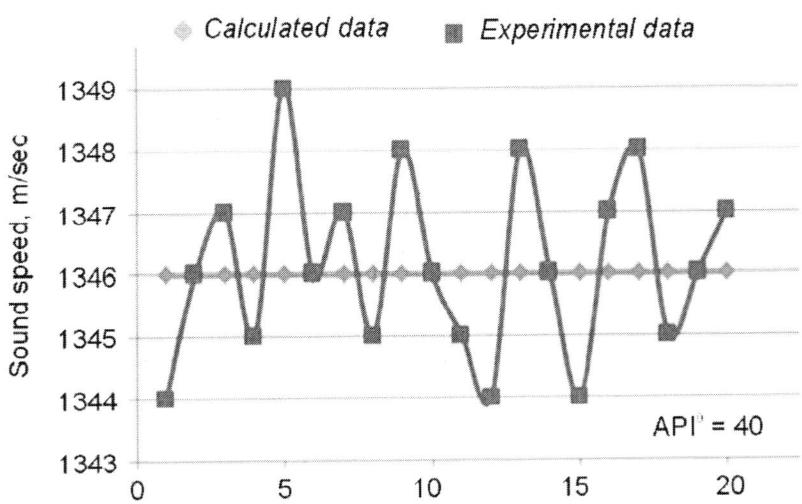

Figure 4. Experimental data of sound speed versus calculated value at T = 20°C for API 40.

It is difficult to quantify the level of scattering by the water droplets because it depends on many factors—the droplet's size, pressure, temperature, frequency and type of crude oil. The presence of dissolved gas also leads to weakening of sound, and can be significant, if the operating frequency is the resonant frequency of the gas bubbles. The level of scattering due to bubbles also depends on their concentration, bubble size, temperature, pressure, frequency and type of oil. If it is assumed that in a crude tank, the relative concentration of the gas bubbles is not very big ($\tau_N < 10^{-5}$), where $\tau_N = \tau_k N$, then in accordance with [7] additional sound attenuation in oil containing N bubbles of radius $r_k$in 1 cm^{3}radius, is equal to:

$$\beta_{\text{bubble}} = \frac{3.12\times10^{-6}\tau_N f_0^2}{\left(1-\frac{f^2}{f_0^2}\right)^2 + 1.21\times10^{-10} f^2} \tag{5}$$

where $\tau_N = \tau_k N = \frac{4}{3}\pi r_k^3 N$, and resonance frequency of gas bubble is determined from relation $f_0 \approx \frac{326}{r_k}$

For the frequency range 100 - 200 kHz the largest contribution comes from bubbles with a radius of 3.26 (2^{-10} - 3^{-10}) mm. For example, if a 1 cm^3 volume contains one resonant gas bubble at 150 kHz, the additional attenuation will be about 0.015 dB/m, which is quite significant. It should also be noted that small bubbles can "live" in the viscous oil for a very long time since their ascent rate is very low. For example, the ascent rate of a bubble with diameter 50 micron in water is less than 0.0005 m/sec, and in oil with a kinematic viscosity of 5 - 7 cSt this rate can be 100 - 1000 times less. The presence of solid particles suspended in the crude oil also leads to an additional attenuation due to the friction, scattering, and internal losses. We have no experimental data on the concentration of particles of various impurities in the crude oil and their size distribution. Therefore, for a rough estimation of the attenuation one can use the empirical formula for water and little concentrated suspensions [8]

$$\alpha_p = 0.167 P_p A_p \tag{6}$$

where $P_{\text{part}} = \frac{4}{3}\pi r_{\text{part}}^3 N \rho_{\text{part}}$ The dependence of the A_p on frequency and size of the particles is shown in Figure 5.

For example, for averaged particle with diameter of 30 microns, concentration of 0.025 to 0.7 kg/m^3 and frequency 150 kHz the additional attenuation can be from (0.007 to 0.05) dB/m. Note that these irregularities are always present in the crude oil. Figure 6 shows the results of laboratory analysis of crude oil samples taken from 6 different storage tanks (different oil fields).

From these data it follows that the real attenuation of sound waves in crude will be always more than pure absorption and this feature should be considered in the design of acoustic tools for correct selection of their operating frequency and source level. Figure 7 confirms this thesis.

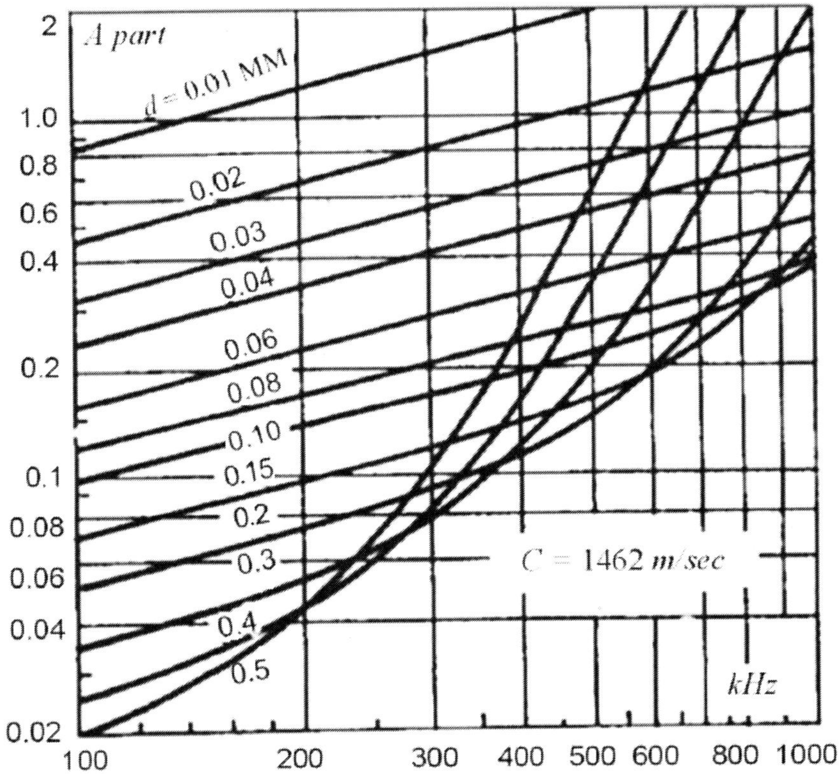

Figure 5. The frequency dependence of the coefficient Ap for particles of different size.

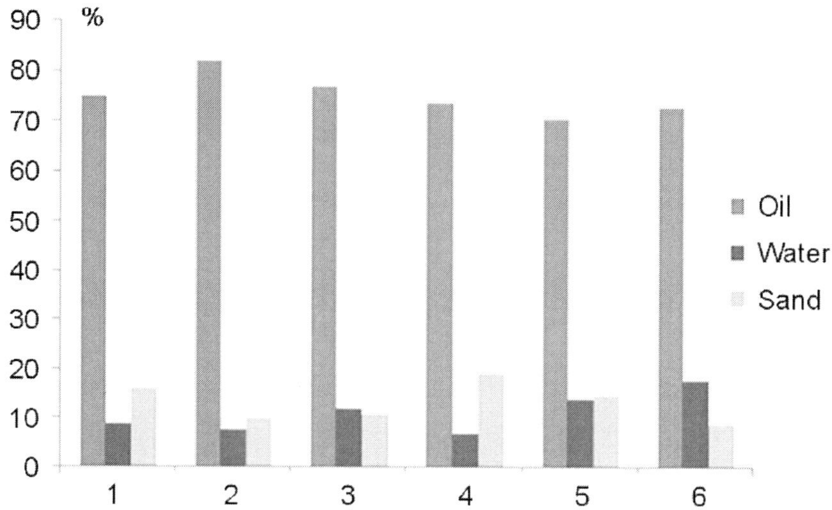

Figure 6. Content of crude in 6 different storage tanks.

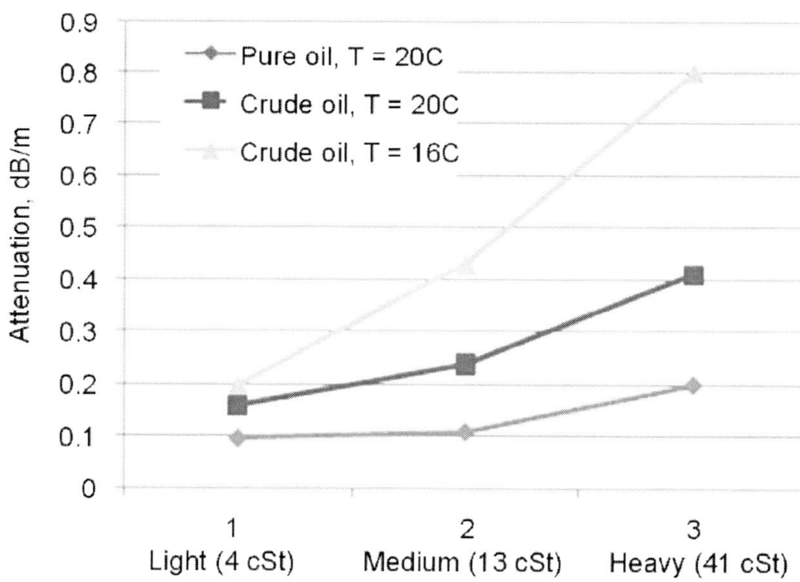

Figure 7. Attenuation on F = 150 kHz in pure oil and crude oil (experimental data).

2.4. Parameters of Sludge

Knowledge of sludge parameters is necessary for the correct selection of carrier frequency and evaluation of energy balance for the acoustic profiling tool, since the coefficient of backscattering from sludge is a very critical parameter depending on angle of illumination and differences in the acoustical impedances of oil and sludge. Oil sludge compositions are extremely varied and complex systems consisting of oil, water and mineral fraction (sand, clay, silt, etc., the ratio of which can vary over very a broad range. On average sludge can contain (by weight) oil (10 ... 60%), water (30 ... 85) % and solids (up to 45%). A more detailed analysis shows that the organic part of sludge, except for free oil hydrocarbons, contains asphaltenes, paraffin and nonorganic inclusions such as iron oxides, sulfides and so forth. Figure 8 presents the content of sludge from different storage tanks [6] .

Viscosity and density for these samples are presented on Figure 9.

We specially considered a large number of storage tanks to emphasize that the sludge composition always varies between tanks. It is seen that the density of sludge varies from 950 to 985 kg/m^{3}and its viscosity can reach 1750 MPa*sec. The solids content is up to 21%, water is up to 40%, Hydrocarbon part can contain 21% of paraffin, 19% of resins and 18% of asphaltenes. It should be noted that due to complex chemical reactions in oil tanks the sludge parameters may change over time, and if the oil tank wasis not in operation for a long time the sludge viscosity may increase so that its structure becomes almost solid and can include hard water scale deposits. This causes large differences in the acoustic impedances of crude and sludge, which of course, makes it easier for acoustic detection, but significantly hinders operations to remove such sludge. However, we shall note that in the crude oil tanks, the boundary between the crude oil and sludge is not sharp. Laboratory samples of sludge from tanks indicate that the sludge is composed of three layers: the first layer—is an actually crude itself but with some percentage of different suspensions. It is practically impossible to detect this layer on the frequencies which are used for acoustic sludge profiling. The second layer is the "transition layer". It is also sludge, but its viscosity and density are less than the third sludge layer. The thickness of the transition layer can range from a few millimeters to ten centimeters or more. The acoustic impedances and backscatter coefficients of the transition layer are not known. But the very important issue is that the presence of the transition layer, if this layer is detected, can be the source of additional errors of in estimation of the sludge volume. The existence of the transition layer is illustrated in Figure 10, which shows the results of direct measurements of the density of the crude oil. It is evident that there is a transition layer with a thickness of more than 10 cm.

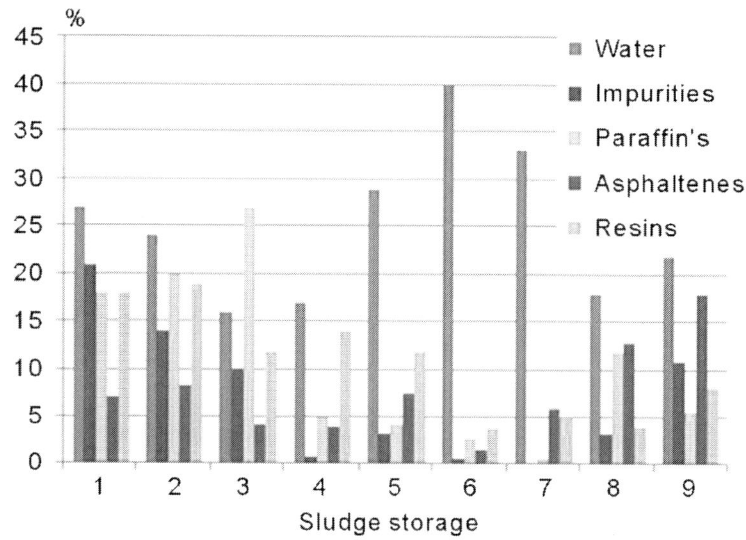

Figure 8. Content of sludge in 9 sludge storage tanks.

The problem in detecting the transition layer is closely related to the values of acoustic impedances for sludge and crude oil. We can directly measure the acoustic impedance of crude oil using measurements of sound speed and density. The problem is how to measure sound speed in the sludge? Unfortunately such data is not available in published literature. The simplest way is direct measurement. However, the realization of such measurements needs to have an echo-sounder with a low operating frequency, since the absorption in sludge is very high due to the very high viscosity. Then knowing the height of the tank, the level of oil, depth of immersion of tool, sound speed in oil and density of the sludge it is possible to measure sound speed in sludge using measured time of flight of forecho-signals reflected from the metal bottom and sludge surface. However, the need to have an additional lower-frequency measuring channel significantly complicates the equipment and increases its cost.

Therefore, we can proceed as follows. We can estimate the sound speed in sludge, using data on the density and viscosity of the sludge and comparing this with known experimental data for the refined petrochemical products of crude oil, which have similar viscosity and density to the analogous parameters of the sludge. Such products range between diesel oil and lubricating oils. Because the mentioned parameters are similar, it does not matter that chemically these substances are different. Experimental data for some petroleum products are presented in Table 1

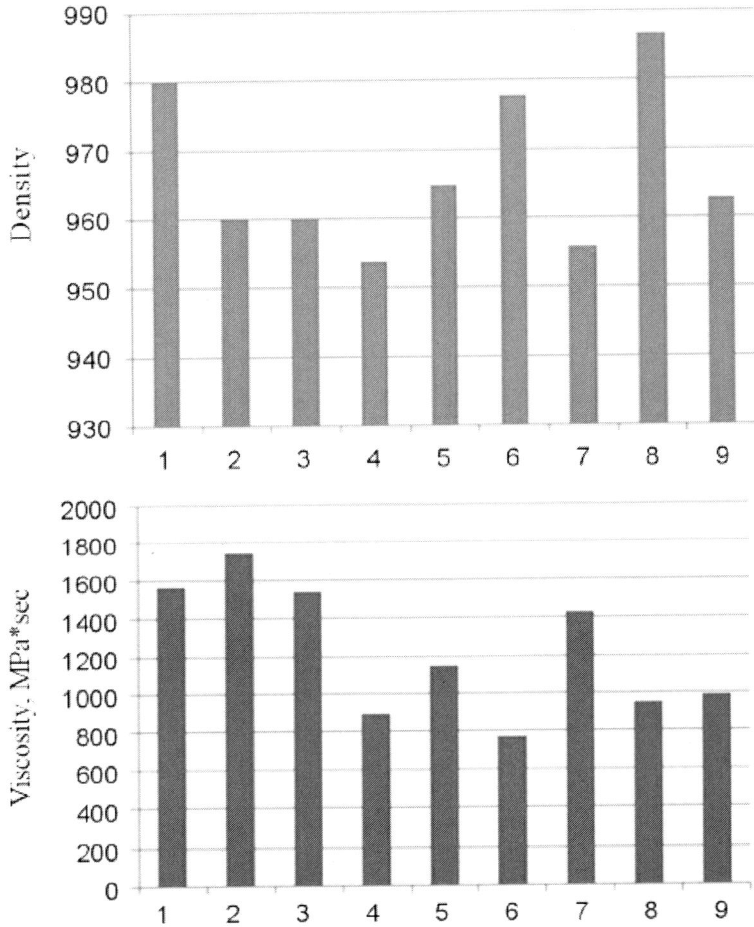

Figure 9. Density and viscosity of sludge in different sludge storages.

Table 1. Parameters of petrochemical products.

Oil product	Density, (20°C) g/cm^3	Sound speed m/sec
Oil	845	1366
2nd product	881	1450
3rd product	893	1474
4th product	907	1490
5th product	923	1509
Fuel oil	938	1513
Tar	988	1581

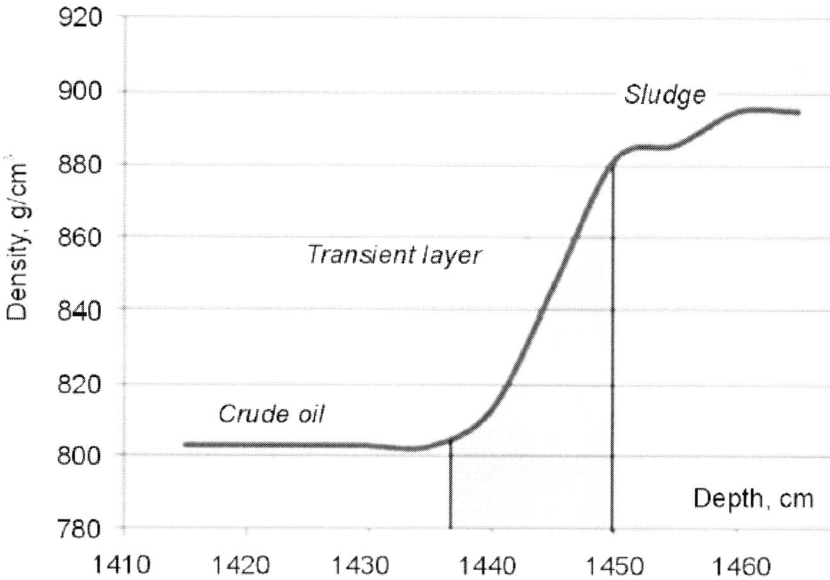

Figure 10. Transient layer.

Experimental data for sound speed versus viscosity for these products and an approximation curve are presented in Figure 11.

The approximation curve describes quite well the behavior of sound speed in high-viscosity oil products and it can be used for estimation of sound speed in sludge. Now, using the obtained data, we can estimate the acoustic impedances of oil and sludge, and calculate the coefficients of reflections from sludge, which in turn will allow the estimation of the levels of scattering echo signals and correct calculating the energy balance of the acoustic profiler.

2.5. Temperature Gradients

The temperature distribution inside the tank may not be uniform. Typically, vertical temperature gradients are the most significant, although in some conditions horizontal gradients can arise. The presence of vertical temperature gradients causes the refraction of sound waves, which in turn can lead to errors of measured slant ranges and, ultimately, errors of volume. Figure 12 shows measurements of the density and temperature of oil versus depth, for a small oil tank of about volume 3000 m^3 and a height of 11 meters.

Calculated ray traces for this temperature profile are presented on Figure 13. It is clearly seen that the refraction of sound rays can be significant. Compare the position of two reflecting points (A and A¢) where the inclined distances differ on 25 meters! Therefore, when there is a significant vertical temperature profile, the resulting refraction of acoustic waves must be taken into account when calculating the surface and its shape corrected. It can be seen that for a given vertical temperature profile, refraction of rays can be substantial, and the position of the reflecting surface point can be evaluated properly and with great accuracy.

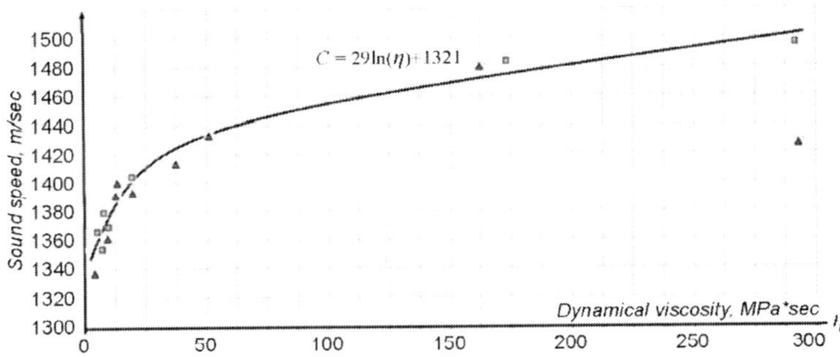

Figure 11. Experimental data on sound speed in petroleum products and approximation curve. (Data provided courtesy of G.M. Volk).

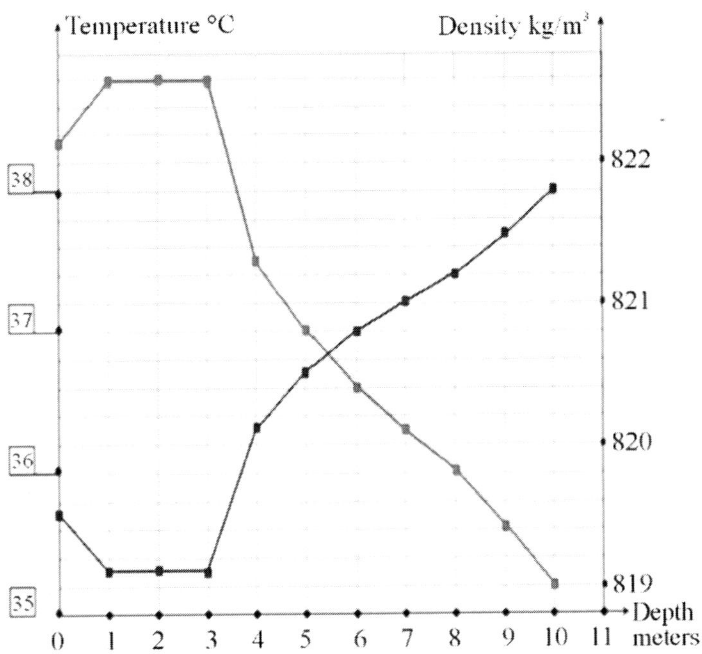

Figure 12. The vertical profile of temperature and density in 8 hours after filling the tank with heated oil.

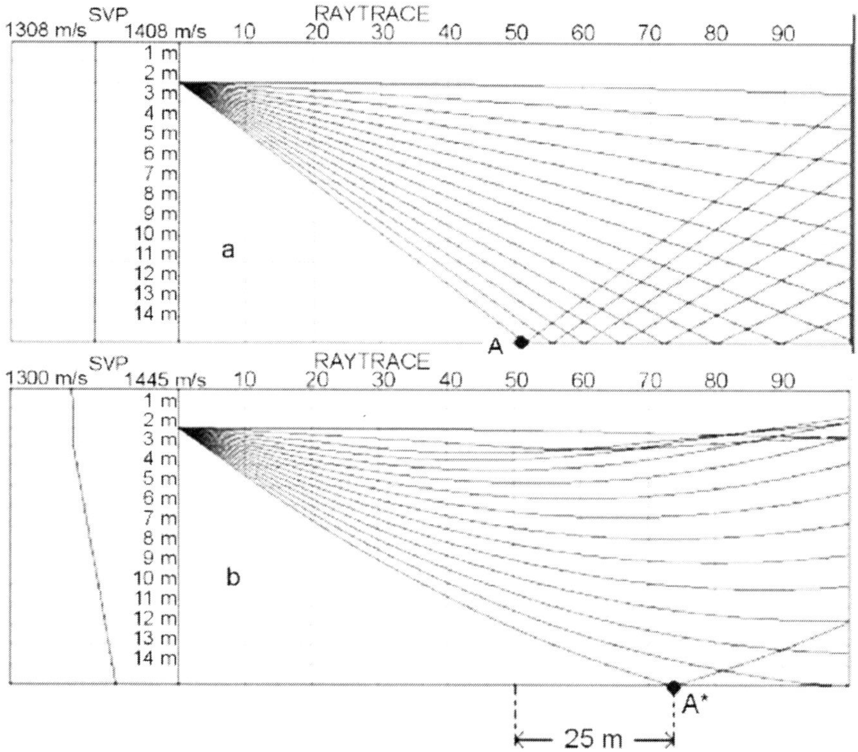

Figure 13. Ray traces for uniform and vertical temperature profiles.

3. GEOMETRY OF MEASUREMENTS

Typically the height of large storage tanks does not exceed 15 - 18 meters, while their diameter may range from 35 to 120 meters and the oil level can be from 10 to 17 meters. Given that the number of possible entry points is limited, the problem of where to install the array arises. It is desirable to provide maximum coverage from one entry point; therefore the installation of the array parallel to the bottom of the tank is not appropriate since the illuminated area becomes very small. Therefore generally acoustic profilers use an inclined array at a certain vertical angle. However, the vertical inclination of the array creates two negative effects. Inclined illumination causes a weak level of backscattering signals from the sludge surface and the character of backscattering becomes close to isotropic, which in turn reduces the effectiveness of the correlation processing of complex broadband signals. The transmitter power and resulting source level is limited by the heat dissipation capability of the immersed array. The only way to increase the signal/noise ratio is to use complex signals with correlation processing. But the isotropic character of scattering leads to decorrelation of complex signals and gain of correlation processing becomes small compared with the radiation of a simple tone burst. There is no reliable data on backscattering levels of echo-signals from sludge. For rough estimations we will use published data for different bottom types in fresh water [7] , Figure 14.

Because the surface of sludge is more similar to irregular silt the design of the acoustic profiler should mostly focus on plots 1 and 2, therefore the averaged backscattering level can be estimated as about -(25 - 30) db.

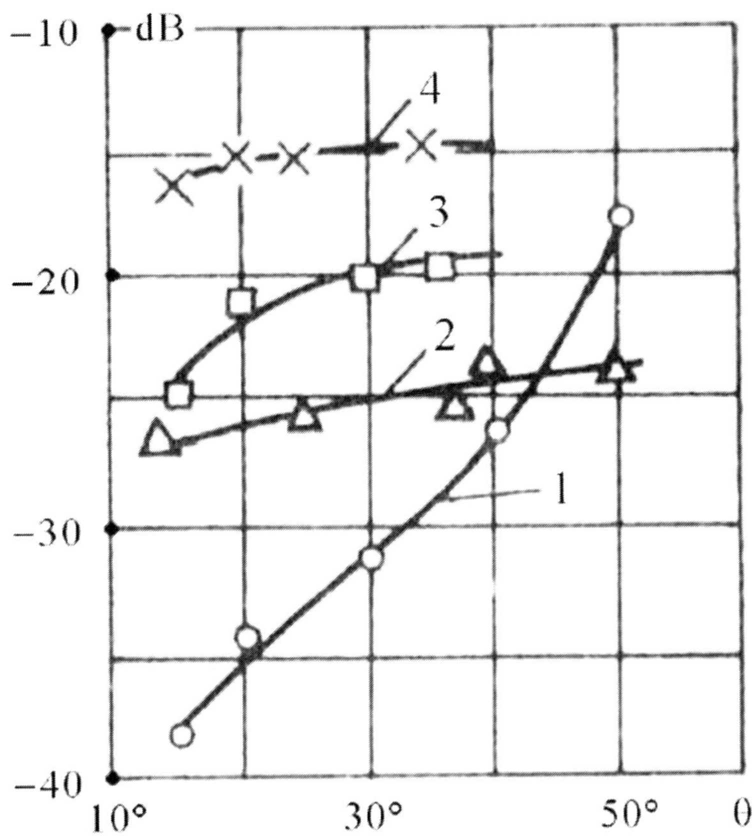

Figure 14. Levels of backscattering for different types of the bottom in fresh water versus sliding angle. 1. Fine sand. 2. Irregular silt 3. Large sand, gravel 4. Stones (10 × 10 × 10 cm in average).

4. ACCURACY OF MEASUREMENTS

What could be the real measurement error of the sludge volume? Despite the fact that the technical (system) accuracy of the profiling tool can be very high, the actual error of a measured sludge volume may not reflect this. The operating frequencies used in these systems generally lies in the range from 120 kHz to 200 kHz and slant ranges, which determine the local heights of sludge, may be estimated with a very high accuracy using short pulses or complex signals. The range resolution for a simple pulse signal is equal to $\Delta R = \frac{C\tau}{2}$ or $\Delta R^{*} = \frac{C}{2\Delta F}$ for complex signal and may be adjusted from 0.5 cm to 2.0 cm, where C—the sound speed in oil, τ—pulse length, ΔF—bandwidth. The mean-square error of measurements of slant range depends on output signal/noise ratio, $\delta R \approx \left(\frac{S}{N}\right)^{0.5}$. Therefore the relative error $\theta = \frac{\Delta R}{\delta R}$ can be very small at high output signal/noise ratio, S/N.

However, this small relative error does not guarantee the same error of measured volume. There are two types of errors—systematical and random. The nature of systematical errors is the errors of positioning of the tool, incorrect information on the co-ordinates of entry points and so on. We will not consider them. The main source of errors is errors in the measured sound speed and reverberation effects.

4.1. Refraction Errors

Let us review in short the principle of acoustic profiling to understand the influence of refraction errors, Figure 15.

Regardless of the type of array, it creates a narrow beam (or beams) which can scan in the vertical and horizontal planes. The width of beam determines the angle resolution and the covered partial area of the sludge surface. The principles of beamforming can be different—2D full sampled arrays, Mills cross, interferometry arrays and so on. Let us propose that the sludge surface has one reflecting point and the distribution of temperature in tank is uniform. Our goal is to estimate the height of this point. First of all we measure the slant distance to this point, knowing the time of flight τ of the echo signal.

$$R_1 = \frac{C_{oil}\tau_1}{2} \qquad (7)$$

Because we know the vertical steering angle ϕ_1 we can calculate h_o as $h_o = R_1 \mathrm{Sin}\,\phi_1$ and the required height of sludge h_1,

$$h_1 = (L - h_s) - \frac{C_{oil}\tau}{2}\mathrm{Sin}\varphi_1 \qquad (8)$$

Acting similarly for all scanning horizontal and vertical angles we can define all theheights of a surface h(x,y). The result of such measurements is the cloud of points h(x,y) which describes the sludge surface in a discrete form, Figure 16. Interpolation procedures are then applied to get the final image of thesurface.

Note that thepartial volumes are presented as "cylinders", though in reality they will have another form. The volume of sludge is the sum of all partial volumes but there is an unknown parameter—sound speed C_{oil}..

Strictly speaking, system parameters such as phase shifts or time of delays which are used in beamforming procedures also depend on the speed of a sound. In particular, the volume of cylinder is equal to:

$$V_1 = \pi r^2 h_1 = \pi r^2 \left[(L - h_s) - \frac{C_{oil}\tau}{2} \mathrm{Sin}\varphi_1 \right] \tag{9}$$

Total volume will be the sum of (9)

$$V_{\mathrm{total}} = \sum_{i=1}^{N} V_i \tag{10}$$

It follows from (8) that errors of volume ΔV are proportional to errors of sound speed ΔC_{oil},

$$\delta V = \left| \frac{\Delta V}{V} \right| = \mathrm{const} \left| \frac{\Delta C_{oil}}{C_{oil}} \right| \tag{11}$$

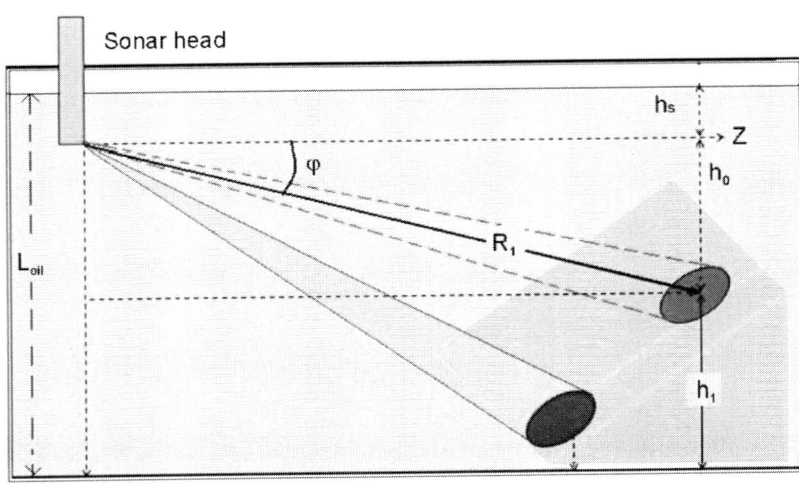

Figure 15. Principle of sludge profiling.

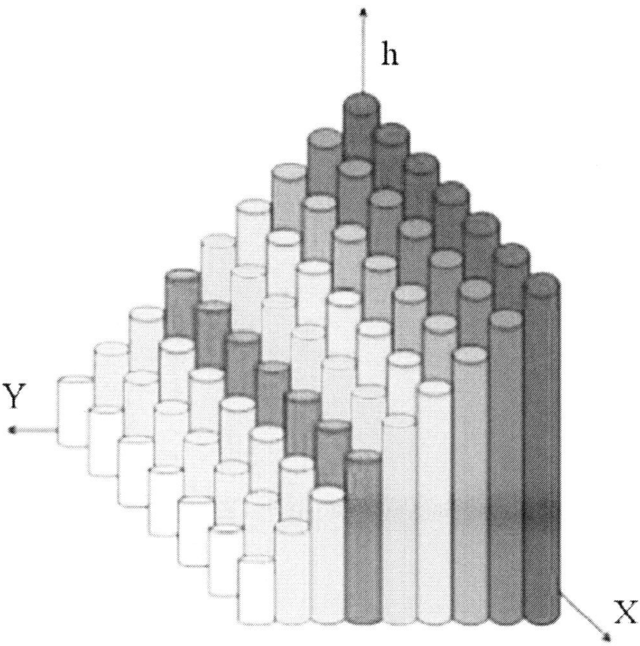

Figure 16. Discrete 3D sludge topography (no interpolation), values of heights are marked by conditional colors.

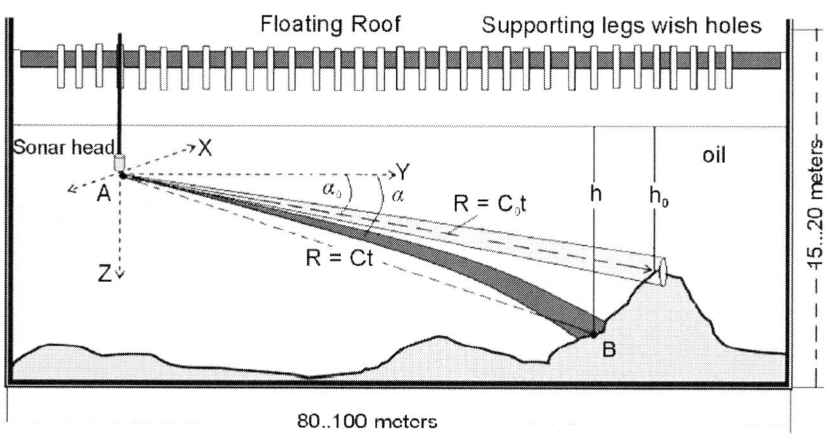

Figure 17. Errors of measured height in a vertical plane.

Note ΔC can be positive or negative. However, sound velocity also depends on the temperature (3). Therefore, when measuring the sound speed we must also measure the temperature.

Roughly the accuracy of sound speed in real conditions can be +/− 2% - 3% and the accuracy of measured sludge volume directly depends on this accuracy.

Now let us consider themore complex case when oil has a vertical temperature gradient like thedependence in Figure 7.

Because of thenon-uniform distribution of sound speed with depth thesound rays will refract. The presence of refraction can affect theaccuracy of themeasurements of heights and consequently, the error of volume, Figure 17.

Let us suppose that we formed a vertical beam with angle α_0. In some time we receive a reflected signal from distance R = C_0t. C_0—is the speed of sound we have measured at point A.

Because of refraction the vertical beam generated with vertical angle α_0 will be bent and result in the distance R = Ct, and ($C \neq C_0$!).We will get the reflection from point B and the image of this point is true. In other words, angle α is a true angle and angle α_0 is the apparent angle. Therefore the height h_0 is the apparent height. Its value is only true for a homogeneous environment. Since point B has a different height from point A, it means, that there is an error in the measurement of height h_0.. An inhomogeneous environment will cause errors in the vertical angle that will result in errors in the estimation of heights.

The true height is equal to:

$$h = C(z)t\text{Sin}\alpha'' \tag{12a}$$

and the apparent height is equal to

$$h_0 = C_0 t\text{Sin}\alpha_0 \tag{12b}$$

where C—some average sound speed on the line AO C_0—sound speed in a point A t—double time of flight from array up to reflecting point.

The true height h can be calculated if we can calculate time of flight along the refracted ray knowing sound speed profile and angle α_0 [7]

$$t = \pm\frac{1}{C_0}\int_{h_0}^{h}\frac{n^2(z)\text{dz}}{\sqrt{n^2(z)-\text{Cos}^2\alpha_0}} \tag{13}$$

In Equation (13) all parameters are known except the true height.

Because of complex function of n(z) for real environments integral (13) can be calculated by numerical methods. There is only one simple case where we can calculate (13) analytically.

If

$$n^2(z) = 1 + 2aZ \tag{14}$$

and $\mathrm{Cos}^2\alpha_0 \approx 1$ and $\mathrm{Sin}^2\alpha_0 \approx \alpha_0^2$

$$h = h_0 + \frac{a(C_0 t)^2}{2} \tag{15}$$

Therefore the correction factor is equal to

$$\Delta h = h - h_0 = \frac{a(C_0 t)^2}{2} \tag{16}$$

Note that the coefficient a can be positive or negative. Equation (16) is only true for small vertical angles. For arbitrary angles of vertical scanning

$$h = \frac{1}{K}\left[\left(\frac{Kh_0}{2} + 1\right)^2 - 1\right] \tag{17}$$

where

$$K = 2\frac{G_C}{C_0}\left(\mathrm{ctg}^2\alpha_0 - 1\right) \tag{18}$$

The results of calculations using (16) and the vertical profile in Figure 13 are presented in Figure 18. Note that this data is only to illustrate the influence of refraction on the accuracy of measurements.

Note that the height of the sludge surface is equal to

$$H_{SL} = L_{oil} - (h_0 \pm \Delta h).$$

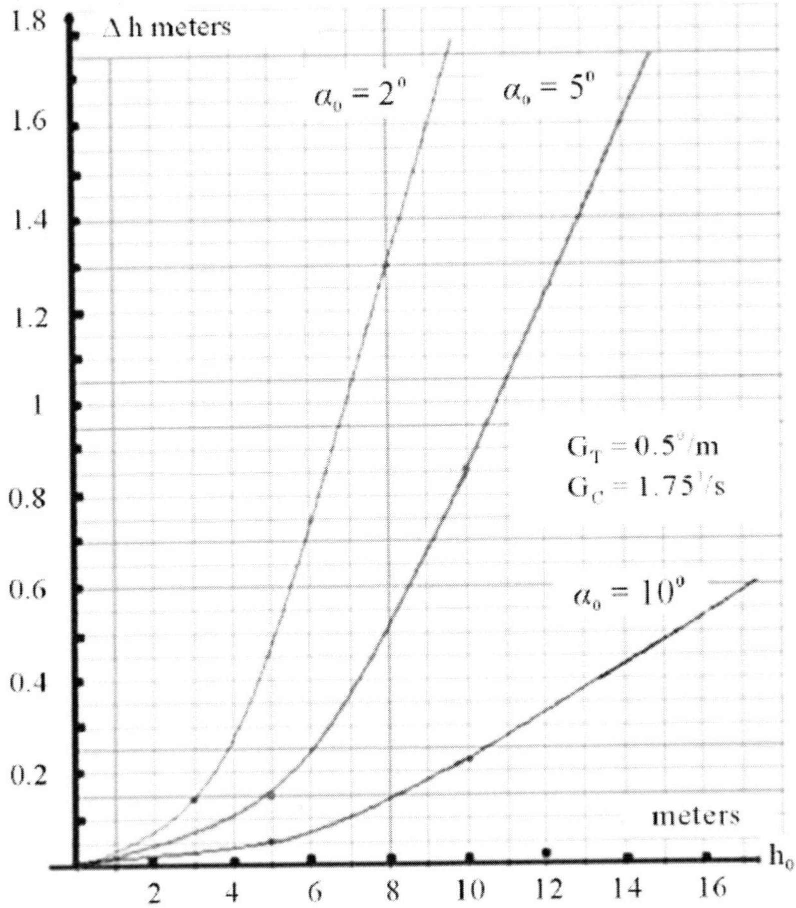

Figure 18. Errors on height versus seemed measured height at different angles of vertical scanning.

From Figure 18 we can conclude that non-uniform distribution of temperature in the oil can lead to significant errors in measuring the form of a surface because of refraction. The error values depend on the vertical scanning angles sector and the values of gradients of temperature and sound speed.

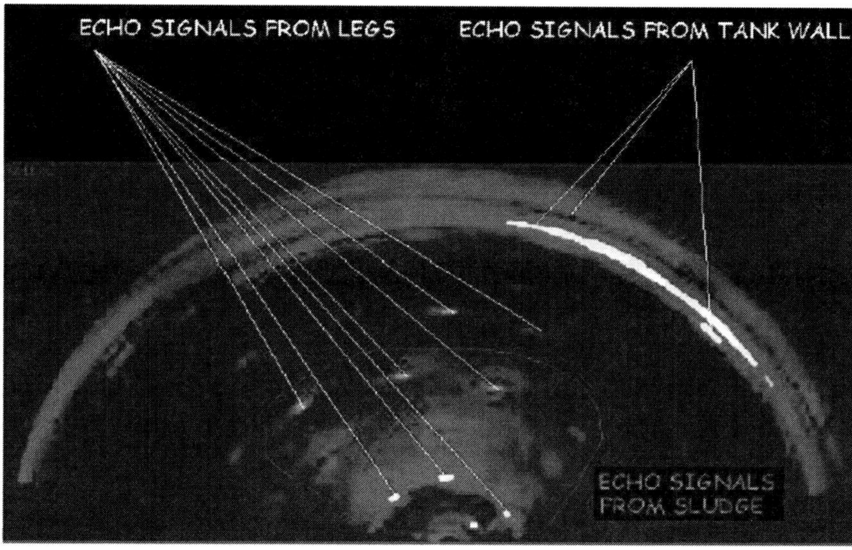

Figure 19. 2D distribution of intensities of echo-signals in tank.

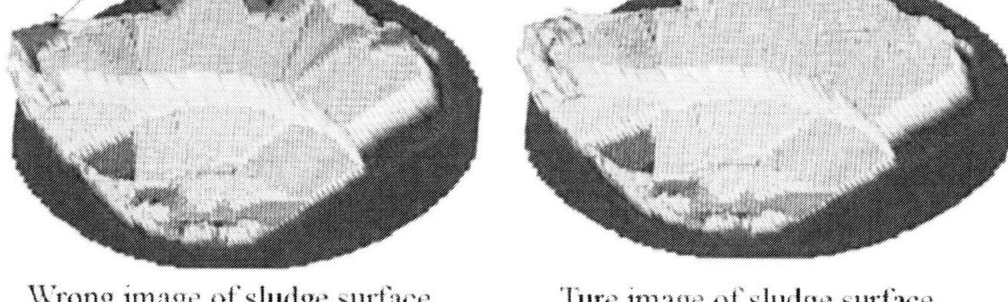

Figure 20. Wrong and true image of sludge surface.

4.2. Reverberation Errors

The second main source of errors is the reverberation of spurious signals received and generated by the main and side lobes of the transmitting and receiving arrays. These false signals appear as a result of the reflection of the pulses from metallic supports the roof, tank walls and various structures located on the bottom (heaters, pipes, etc.). Despite the fact that the arrays typically use apodization to decrease the side field, the levels of spurious side echo signals reflected from the metal structures can still exceed the useful weak echoes from sludge and, in addition, they can appear in unpredictable places. If such signals fall in the area of the sludge surface, they create additional "peaks" or "false" reflection points on the surface, which lead to an incorrect evaluation of heights and incorrect estimation of the volume. It usually leads to an over estimation of the sludge volume Figure 19 shows an example of a two-dimensional distribution of intensity echoes (in in a tank with a diameter of 60 meters).

The sludge surface looks like a cloud of non-uniform brightness; it is outlined by a closed curved red line. The strong echoes like "dashes" visible on the surface of the sludge and beyond it are generated by the reflections from the roof supports. Some of them are received by the main lobes of the array and some of them are generated by the side lobes. The presence of these signals leads to incorrect reproduction of the surface shape of the sludge; refer to the left image in Figure 20. It can be seen that the sludge surface has heights up to 6 m in some places, while the average height of sludge does not exceed 1.9 m. The presence of these spurious signals leads to an overestimation of sludge volume: 3262 m^3 versus the real volume of 2764 m^3.

This example implies that the development of acoustic sludge profiling systems should strive to minimize the level of side lobes in the arrays and use more sophisticated signal processing algorithms.

Accounting for these cumulative errors and the experience of making field measurements results in an error of the sludge volume in the range 3% to 5%. This value of the error is typically indicated by developers of similar equipment.

5. CONCLUSIONS

1) Considered features of sound propagation in the crude oil, located in large closed metal containers and their specific design can significantly affect both the structure of the arrays, and the signal processing algorithms, which may significantly differ from the usual methods of acoustic bathymetry in oceanography.

2) The description of the basic laws of propagation of acoustic signals in the crude oil and their dependence on various parameters can serve as a basis for the development of acoustic profiling systems, selection of the optimal frequency, size of the arrays and the required source level.

3) To minimize the possible errors in the calculated image of the sludge surface and its measured volume it is desirable to measure the vertical temperature profile and sound speed in all entry points.

4) In general, the signal processing algorithms should include procedures to compensate for refraction effects and suppression of false echo-signals caused by reverberation effects.

5) During inspection of crude oil tanks, the satisfaction of these conditions allow to obtain an error value of measured sludge volume in the range 3% to 5%.

REFERENCES

1. ASTM D341 (2003) Standard Test Method for Viscosity-Temperature Charts for LiquidPetroleum Products, USA.
2. Kalivoda, R.J. (1995) Understanding the Limits of Ultrasonicsfor Crude Oil Measurement. www.engineerlive.com
3. Wang, Z., Nur, A. and Batzle, M.L. (1988) Acoustic Velocities in Petroleum Oils. SPE Paper 15646, Proceedings 61st SPE Technical Conference.
4. Kolobaev, P.A. (2000) An Experimental Research of Temperature Relation of Speed of Ultrasonics in a Crude Oil and Some Petroleum. The Transactions of the X Session of the Russian Acoustic Society, 2, 94-96.
5. Bardyshev, V.I. and Kolobaev, P.A. (2005) Experimental Researches of Temperature Dependences of Sound Speed in the Water Selected from Crude Oil and Density of This Water. Transactions of the XVI Session of Russian Acoustical Society, 34-35.
6. Minnigalimov, R.Z. (2009) An Investigation of the Dependences of Dielectric Characteristics of Petroleum Slime. High Temperature, 47, 606-608.http://dx.doi.org/10.1134/S0018151X0904021X
7. Kablov, G.P., Sverdlin, G.M. and Yakovlev, A.N. (2000) Short Range Sonar. Sudostroenie. Leningrad, 200.
8. Stashkevich, A.K. (1988) Acoustics of Sea. Sudostroenie. Leningrad, 276.

Chapter 7

Implementation of Acoustic Analogies in OpenFOAM for Computation of Sound Fields

J. Schmalz, W. Kowalczyk

Chair of Mechanics and Robotics, University of Duisburg-Essen, Germany

ABSTRACT

In this work the turbulence based acoustic sources and the corresponding wave propagation of fluctuating flow values in incompressible fluid flows are considered. Lighthill's and Curle's acoustic analogies are implemented in the open source computational fluid dynamics framework Open- FOAM. The main objective of this work is to visualize and localize the dominated sound sources and the resulting values of fluctuating pressure values within the computation domain represent- ing the acoustical near field. This is all done on one mesh and during the iterative computation of the transient fluid flow. Finally the flow field and acoustical results of different simulation cases are presented and the properties of the shown method are discussed.

Keywords: Acoustic Analogies, OpenFOAM, Sound Sources, Sound Propagation, High Performance Computing, Incompressible Flow

1. INTRODUCTION

Products like air intake ducts for combustion engines emit acoustic noise mainly based on turbulences and their interaction with solid surfaces. The reduction of these acoustical emissions of technical products and components are becoming increasingly important. In product development process the acoustic parameters are often considered not before an existing mechanical prototype. Usually any changes of concepts, products or components themselves are hardly feasible. Numerical methods like finite elements for structural mechanics or finite volumes for computational fluid dynamics are commonly used by engineers and researchers. Using the advantages of each numerical methods and already existing tools are extended to compute e.g. the sound pressure level in the acoustical far field at a specific observer point. The resulting data is comparable to the results of common acoustic measurements. Concerning the scale disparity between the fluid dynamic pressure and acoustical pressure and the limitation of computation resources often there are used hybrid approaches for computational aero acoustics. Hybrid computational aero acoustic approaches solve the fluid dynamics by e.g. unsteady RANS (Reynolds Averaged Navier Stokes) simulations on a fine mesh in a small domain and then transform the relevant parameters for computing the sound propagation on a coarse mesh representing the acoustical far field.

Further developments in computing methods and parallel communication software and growing resources in computing hardware especially in High Performance Computing (HPC) widen and improve the possibilities to simulate complex acoustic effects. A popular CFD tool is OpenFOAM (Open Field Operation and Manipulation). It is a free, open source software toolbox for Computational Fluid Dynamics (CFD) applications, based on [1] and originally produced by OpenCFD Ltd. It has a lot of various features to solve complex fluid flows in- cluding for example turbulences and chemical reactions up to solid dynamics and electromagnetic and multi physics applications. Almost every feature runs in parallel as standard up to 1000 cores at present and much more in nearer future [2] .

There are different approaches in OpenFOAM for computational acoustics, especially for aero acoustics as mentioned in [3] and approaches based on Linear Euler Equations described in [3] - [5] . These approaches are hybrid approaches and have in common, that they use different meshes to compute the fundamental turbulent fluid flow and acoustic sources on one mesh and the propagation of fluctuating values on another mesh respectively. Nilsson [3] even uses different discretization methods, the finite volume method within the OpenFOAM framework to compute the turbulent fluid flow and acoustic sources and the finite element method to do the sound propagation into the acoustical far field. Stein [4] and Wang [5] do not compute the acoustic sources within the CFD-simulation in particular, they use the Stochastic Noise Generation Method to obtain the source terms and use the Linearized Euler Equations (LEE) to compute the sound propagation. Furthermore the acoustic investigations might be also done during the post processing of the CFD results explained by [6] . In this work the far field solution of the Curle's Acoustic Analogy is implemented in OpenFOAM. This allows the user to obtain the sound pressure level and its frequency spectrum at a specific observer point only. All mentioned approaches either has in common, that there is a transformation between CFD and CAA (Computational Aero Acoustics) mesh necessary or the acoustic parameter can be determined during post-processing at selective points only.

In this work, the presented approach for computational aero acoustics in OpenFOAM 2.1.1 is mainly based on Curle's Acoustic Analogy and takes the possibility and availability of high performance computing resources into account. Due to HPC resources, the shown method makes it possible to determine acoustic parameters of flow fields and provides the advantage to compute and visualize the fields of fluid dynamics, acoustic sources and the resulting wave propagation of the fluctuating values of the decomposed pressure (sound propagation) in the acoustical near field and on one mesh only.

2. THEORY

2.1. Lighthill's Acoustic Analogy for Incompressible Fluids

The implementation of acoustic source terms and the corresponding wave propagation of the fluctuating pressure values within the computation domain according to Lighthill [7] - [9] is principally based on the continuity Equation (2.1)

$$\frac{\partial \rho}{\partial t} + \nabla \cdot (\rho u_i) = 0 \tag{2.1}$$

with its density ρ and the velocity u_i of the fluid. Lighthill's acoustics analogy is also on the momentum Equation (2.2)

$$\frac{\partial}{\partial t}(\rho u_i) + \nabla \cdot (\rho u_i u_j + P_{ij}) = 0 \tag{2.2}$$

with

$$P_{ij} = p\delta_{ij} - \tau_{ij} \quad . \tag{2.3}$$

In (2.3) the Tensor $P_{ij} = p\delta_{ij} - \tau_{ij}$ describes the compressive stresses in which τ_{ij} represents the viscous stresses, p represents the pressure of the fluid and δ_{ij} defines the Dirac function. According to [10] the density ρ , the pressure p and the velocity u are decomposed into constant and fluctuating parts. The fluctuating parts represent the acoustical values.

$$\rho' = \rho - \rho_0 \tag{2.4}$$

$$p' = p - p_0 \tag{2.5}$$

$$u' = u - u_0 \tag{2.6}$$

The term ρv_i, representing the mass density flux in the continuity equation but the momentum density in the momentum equation. By taking the time derivative on the continuity Equation (2.1) and subtracting the divergence of the momentum Equation (2.2) as documented in [7] and eliminating the term ρv_i, Lighthill's inhomogeneous wave Equation (2.7) is obtained

$$\frac{\partial^2 \rho'}{\partial t^2} - c_0^2 \cdot \rho' = \frac{\partial^2 T_{ij}}{\partial x_i \partial x_j} \tag{2.7}$$

if using the decomposed Quantities (2.4)-(2.6). The speed of sound is described by c and the Tensor T_{ij} can be written as

$$T_{ij} = \rho v_i v_j - P_{ij} + c_0^2 \delta_{ij} \rho' \tag{2.8}$$

including the decomposed fluid parameters. Finally the Tensor P_{ij} (2.3) results in (2.9)

$$T_{ij} = \rho v_i v_j - \tau_{ij} + \delta_{ij} \left(p' - c_0^2 \rho' \right) \tag{2.9}$$

and is called the Lighthill-Tensor.

For an incompressible$(\rho = \text{const})$, isentropic and Newtonian fluid $(\tau = 0)$ at low Mach numbers and high Reynolds numbers assumptions and simplifications of equation (2.9) might be taken as proposed by Lighthill himself [7] and this yields to the incompressible approximation of the Lighthill-Tensor (2.10)

$$T_{ij} \approx \rho_0 v_i v_j \quad . \tag{2.10}$$

Following the restrictions for the fluid the density corresponds approximately the density ρ_0 of the incompressible fluid within the whole flow domain. As defined in [3] the acoustic sources induced by turbulences are implemented in OpenFOAM 2.1.1 shown in Listing 1, where density 0 can be defined by the user as a constant value.

2.2. Curle's Acoustic Analogy for Incompressible Fluids

Lighthill's acoustic analogy was mainly developed for free jet streams. No rigid objects or surfaces are allowed within the computation domain. Unfortunately a large amount of technical application cases need to take into account such surfaces. Curle's acoustic analogy takes rigid and stationary surfaces and objects within the computation domain into account. It is based on the extended Lighthill Equation [9] (2.11) and represents a special case of the Ffowcs-Williams and Hawkings equation [11] .

$$\left(\frac{\partial^2}{\partial t^2}-c^2\Delta\right)\{(\rho-\rho_0)H(f)\}=\frac{\partial^2}{\partial x_i\partial x_j}\{T_{ij}H(f)\}+\frac{\partial}{\partial t}\left(\{\rho(v_i-u_i)+\rho_0 u_i\}\frac{\partial y}{\partial x}\delta(f)\right)$$
$$-\frac{\partial}{\partial x_i}\left(\{\rho v_i(v_j-u_j)+P_{ij}\}\frac{\partial f}{\partial x_j}\delta(f)\right), \quad (2.11)$$

The three terms on the right hand side of Equation (2.11) can be interpreted as

```
acoustic = density0*fvc::div(fvc::div(U*U));
```

Listing 1. Implementaion of Lighthill's acoustic sources in OpenFOAM, according to [3] .

- the volume distribution of quadrupoles $\frac{\partial^2 T_{ij}}{\partial x_i \partial x_j}$ (cp Chapter 2.1) due to the turbulent flow,
- as surface distribution of monopoles $\frac{\partial}{\partial t}(Q\delta(f))$ due to the kinematics of the body and
- as surface distribution of dipoles $-\frac{\partial}{\partial x_i}(F_i\delta(f))$ due to the interaction of the flow with the body.

For a rigid object within the computation domain its surface can be described by the auxiliary function $f(x,t)$. This means for the Heaviside function $H(f(x,t))$

$$H(f(x,t))=\begin{cases}1, & \text{for } x\notin V; f(x,t)>0,\\ 0, & \text{for } x\in V; f(x,t)<0.\end{cases} \quad (2.12)$$

and this yields to an expression which disappears within the volume of the object and becomes one inside the fluid region. As mentioned above this work focuses on the Curle's Acoustic Analogy, where the considered surfaces are stationary and terms containing velocity components obtain

$$u_i(x_i, t) = 0 \tag{2.13}$$

$$u_n = u_i n_i = 0 \tag{2.14}$$

$$u_j = 0 \tag{2.15}$$

Additionally for non-porous or rigid control surfaces the following equation is valid:

$$\rho(v_i - u_i) = 0 \tag{3.16}$$

These assumptions eliminates the surface distribution of monopoles in the extended Lighthill Equation (2.11).

The remaining dipole term $P_{ij}\frac{\partial f}{\partial x_j}$ with

$$P_{ij} = p\delta_{ij} - \tau_{ij} \tag{2.17}$$

becomes to

$$p\frac{\partial f}{\partial x_i} = pn_i \tag{2.18}$$

if viscous stresses are neglected. p describes the local surface pressure and n_i stands for local unit outward normal to considered surface.

Finally the acoustic source terms of Curle's Acoustic Analogy are simplified to (2.19)

$$\frac{\partial^2 \rho_0 v_i v_j}{\partial x_i \partial x_j} - \frac{\partial}{\partial x_i}(pn_j) \tag{2.19}$$

which are implemented in OpenFOAM acoustic solver according Listing 2.

2.3. Wave Propagation of Fluctuating Pressure Values

Concerning the decomposition of fluid parameters into mean and fluctuating values as described in chapter 2.1 and according the statements in linear acoustics [12] the pressure-density relation can be written as

$$p' = c_0^2 \rho' \tag{2.20}$$

```
Info<< "Calculating Curle Acoustic Source Field" << endl;
acousticSources =    - fvc::div(pSurfaceSource)              //dipole distribution
                     + density0*fvc::div(fvc::div(U*U));     //quadrupole distribution
```

Listing 2. Acoustic source terms according Curle's Acoustic Analogy were implemented in OpenFOAM 2.1.1.

where the pressure fluctuations are described by p' and the density fluctuations are defined by ρ'. This allows writing the left hand side of a homogeneous wave equation as followed

$$\frac{1}{c_0^2}\frac{\partial^2 p'}{\partial t^2} - \Delta p' = 0 \tag{2.21}$$

The wave operator (2.21) with its second time derivative of the fluctuating pressure values is implemented in OpenFOAM as an additional header file shown in Listing 3 where the fluctuating pressure values are defined by the variable pa.

The header files for computation of acoustic sources and for sound pressure propagation respectively are included in the pisoFoam. C-file within the runtime-loop, but after the PSIO (Pressure-Implicit Split-Operator)- loop itself. This provides the computation of the additional acoustic fields during every time step of the pisoFoam-Solver and based on the latest corrected pressure and velocity fields including the computed turbulences of the chosen turbulence model.

The proposed "sound" pressure wave propagation is strongly dependent on the way the computed pressure is decomposed into mean and fluctuating parts. In the presented work the pressure is decomposed into the computed pressure subtracting by the ambient pressure of the initial fluid flow setup. This strongly results in a propagating pressure wave similar to hydrodynamic pressure waves.

Actually there are no specific boundary condition in the OpenFOAM distribution which can perform acoustic behaviors like non-reflecting or absorbing effects neither for incompressible nor for compressible flows. The boundary condition advective works roughly like a non-reflecting boundary condition for incompressible case set-ups. It is based on the OpenFOAM-class mixed FvPatch Field [13] - [16] defining outlet boundary conditions

$$\frac{\partial \phi}{\partial t} + \nabla(\phi U) = 0 \tag{2.22}$$

and its principle behavior is similar to boundary conditions for compressible fluid flow, which is often used for transonic or supersonic simulations. The advective or waveTransmissive boundary condition are applicable on patches defined as outlet only. For a lot of cases these might be insufficient. To overcome this limitation the in this work the computation domain boundary patches are set up as stated in [17] . According to their and Goldsteins [18] proposal the boundary value method is used for several simulation cases. Therefore in inviscid flows the RHS of Equation (2.7) might be set to zero on patches, if the flow velocity on that patches might be neglected.

In Figure 1 the principal program flow of the novel implemented acoustic solver, named acousticFoam for incompressible transient simulations is shown. Before computing the acoustical sources in the acousticSource.H file it is decided due to the case and meshing set up which acoustic analogy has to be chosen for the computation run. Following the wave propagation of the fluctuating pressure values, representing the sound pressure waves within the fluid region are computed by paEqn.H.

Proving the presented principal concept of implementing a computational aero acoustic approach in OpenFOAM several simulations are done and will be presented in the following chapter.

2.4. Acoustic Test Case

For verifying the quality of a numerical solution the results might be compared with the results of the analytical solution of a specific problem, provided that the analytical solution exists. A typical benchmark test case for a novel implemented computational aero acoustic application is the two dimensional acoustic pulse propagation presented by [19] . The advantage of that case is the existing analytical solution obtained from the linearized Euler equations.

```
fvScalarMatrixpaEqn
(
      (1/(c*c))*fvm::d2dt2(pa)
      -fvm::laplacian(pa)
);
solve(paEqn == acoustic);          // sound pressure resulting on acoustic sources
```

Listing 3. Computation of fluctuating pressure wave propagation in OpenFOAM 2.1.1 related to [12] .

2.4.1. Geometric Setting and Meshing

The two dimensional computation domain is a square as showed in Figure 2, where in the center of the domain an initial Gaussian velocity pulse is set up to initiate the wave propagation of the fluctuating pressure values. The domain extends in $-55 \text{ m} \le x \le 55 \text{ m}$ and $-55 \text{ m} \le y \le 55 \text{ m}$.

Due to the geometric simplicity of the case, the computation domain is meshed by blockMesh, the meshing tool distributed by OpenFOAM. The mesh consists of 40,000 hexahedra elements. The grid spacing of the two dimensional structural mesh is $\Delta x = \Delta y = 0.55 \text{ m}$.

2.4.2. Numerical Model

On the left hand side of the domain the boundary conditions are set as a velocity inlet, however there is no background flow $(U_x = U_y = 0)$ implemented. The outlet is set on the right hand side of the region, whereby the upper and lower boundary conditions are defined as walls. At the initial time $t = 0$ an initial Gaussian velocity distribution (cp. Equation (2.23)) is imposed in the center of the domain. The perturbation consists of two pulses in x- and y-direction respectively.

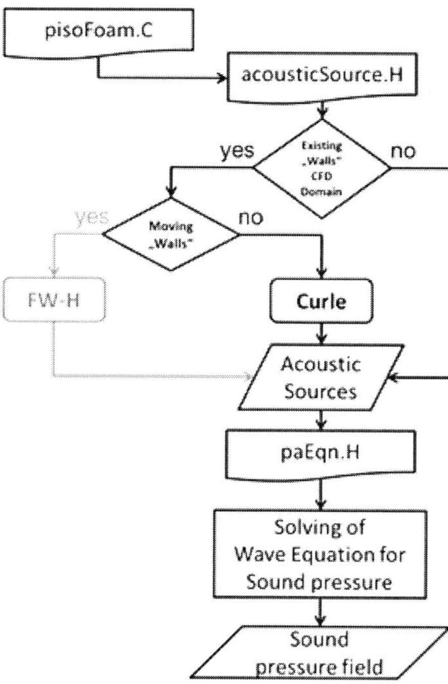

Figure 1. Flow chart of OpenFOAM-Acoustic-Solver acous- ticFoam for incompressible simulations.

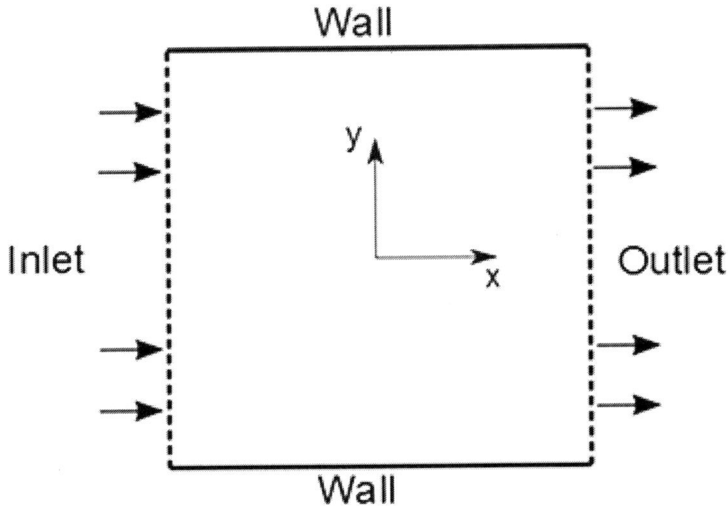

Figure 2. Set up of acoustic test case.

$$U_x = 0.04 y e^{\left[-(\ln 2)\frac{(x-67)^2+y^2}{25}\right]},$$
$$U_y = 0.04(x-67) e^{\left[-(\ln 2)\frac{(x-67)^2+y^2}{25}\right]}. \tag{2.23}$$

The two dimensional initial velocity distribution is shown in Figure 3 where the white line marks a cross section which visualizes in Figure 4 the dimensionless initial Gaussian velocity distribution in x- and y-direction. The described initial set up is chosen to compute the transport of the calculated pressure fluctuations based on the acoustic sources. The used application solver was acousticFoam to compute the acoustic sources and the corresponding wave propagation.

The Gaussian distribution for the initial values of the velocity field was set up at the domain center into x- and y-direction to obtain an almost homogeneous initial velocity distribution. If matching both initial fields and analyze and visualize the velocity field values along the cross section line, mentioned above, this results in Figure 4. Hence the peaks of U_x and U_y cannot be equal along this cross section line.

The described initial set up is chosen to compute the transport of the calculated pressure fluctuations based on the acoustic sources. The used application solver was acousticFoam to compute the acoustic sources and the corresponding wave propagation. In this work for all presented incompressible cases air is used as fluid with its kinematic viscosity $\nu = 1.5787 \times 10^{-5}\,\mathrm{m^2 s^{-1}}$. The density 0 is set to 1.184 kg·m^{-3} at a temperature of 25°C. The speed of sound is declared as a constant with 346.3 m·s^{-1}. The settings are done in an adapted transport Proper ties—file of a common OpenFOAM case structure.

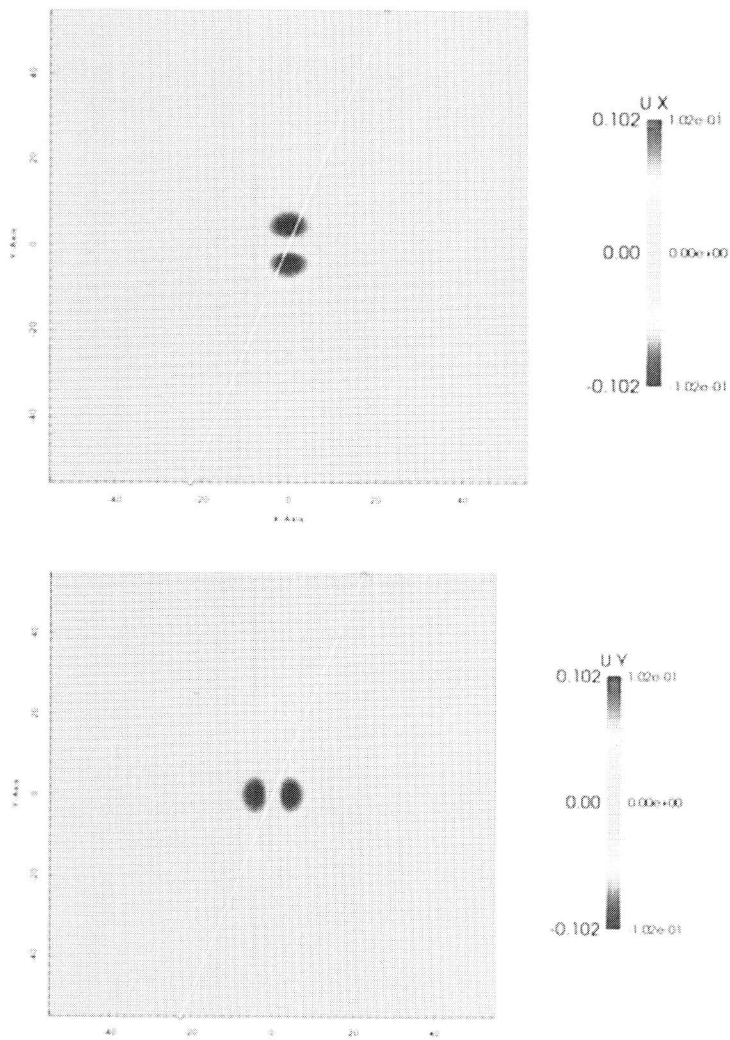

Figure 3. Two dimensional initial velocity distribution in x- and y-direction including the cross section line for further investigations.

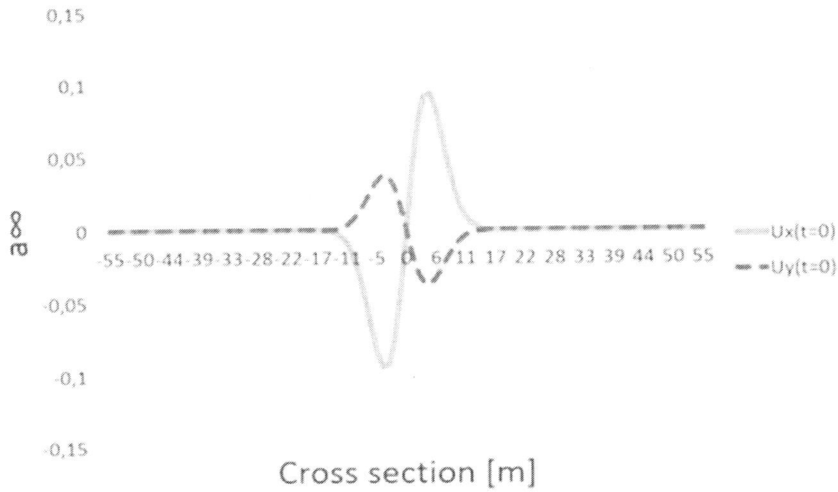

Figure 4. Initial Gaussian velocity distribution (dimensionless) along the cross section.

The Acoustic Test Case was set to laminar where all the turbulence parameters in OpenFOAM are switched off.

The boundary conditions for the acoustic fields, *acousticSources*, *pSurfaceSources* and pa are set each according to Table 1.

To simulate the transient Acoustic Test Case the acousticFoam-Solver based on pisoFOAM-Solver according to chapter 2 is used. The simulation time is set to 0.1 s with a time step of 1.6×10^{-4} s. The simulation run was done in parallel on 4 processors. The parallel execution time is about 1691 s.

Solving the generated linear equation systems the PCG (Preconditioned Conjugated Gradient)-Solver of OpenFOAM is used for pressure and acoustic fields and the PBiCG (Preconditioned Bi-conjugated Gradient)- Solver for asymmetric matrices like the velocity field or fields relevant for turbulence modelling as turbulent kinetic energy k or specific turbulence dissipation ω . To improve the convergence of the solution the preconditioner DIC (Diagonal Incomplete-Cholesky) for symmetric and the DILU (Diagnoal Incomplete-LU) for asymmetric matrices are used. The solver tolerance was set to 1×10^{-6} and the solver relative tolerance relTol was set to 0.

To compute the wave propagation according chapter 2.3 the second time derivative of the sound pressure field has to be calculated numerically. In

OpenFOAM for the second order of time derivative there is the Euler scheme available only.

2.5. Application Case pitzDaily

2.5.1. Geometric Settings and Meshing

The novel implementation of an acoustic solver in OpenFOAM 2.1.1 was tested by running the OpenFOAM distributed tutorial case pitzDaily. This case is a two dimensional set-up of a transient incompressible flow over a backward facing step which is shown in Figure 5 with the corresponding geometrical dimensions given in Table 2.

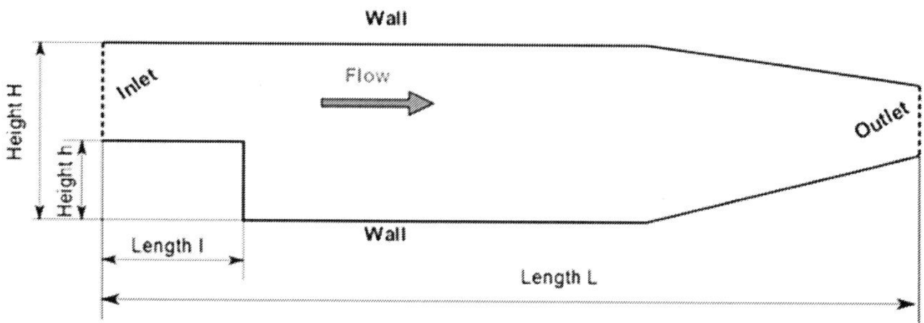

Figure 5. Set up of OpenFOAM-tutorial case pitzDaily.

Table 1. Settings for the boundary conditions of acoustic fields in Acoustic Test Case.

	Inlet	Outlet	Wall	Front and Back
Acoustic Sources	Zero Gradient $\nabla\phi = 0$	Zero Gradient $\nabla\phi = 0$	Zero Gradient $\nabla\phi = 0$	Empty
pSurface Sources	Zero Gradient $\nabla\phi = 0$	Zero Gradient $\nabla\phi = 0$	Zero Gradient $\nabla\phi = 0$	Empty
pa	Fixed Value $pa = 0$	Fixed Value $pa = 0$	Zero Gradient $\nabla\phi = 0$	Empty

Table 2. Geometric dimensions of OpenFOAM-tutorial case pitzDaily.

Length L	310.6 mm
Length l	20.6 mm
Height H	50.8 mm
Height h	25.4 mm

To mesh the computation domain the OpenFOAM-meshing tool blockMesh was chosen. The computation mesh consists of about 12,225 hexahedra elements. This results in an element size of about 0.5 mm concerning the given dimensions in Table 2.

2.5.2. Numerical Model

The physical properties of the fluid are set up as described before in chapter 2.4.2. The velocity u of the fluid flow is set to $u_x = 20\ \mathrm{ms}^{-1}$ in x-direction on the inlet, whereby the pressure p is set to $p = 0$ Pa on the outlet boundary condition. The step itself, the bottom and the top of the computation domain are defined as walls. The height h of the step defines the Reynolds number to Re = 32,000 according to the velocity of the fluid.

To calculate the turbulences the turbulence model kOmegaSST (k-ω-SST) of OpenFOAM is used with the initial values of the turbulent kinetic energy $k = 1.5\ \mathrm{m}^2\mathrm{s}^{-2}$, the specific turbulence dissipation $\omega = 447.2\ \mathrm{s}^{-1}$ and the turbulent viscosity $\nu_t = 0.00271\ \mathrm{m}^2\mathrm{s}^{-1}$ according to the chosen fluid flow velocity. Used wall functions of the turbulence model are kqRWallFunction for turbulent kinetic energy, nutUWallFunction for turbulent viscosity and omegaWalFunction for specific dissipation.

The setup of the boundary conditions of the acoustic fields like *acousticSources*, *pSurfaceSources* and pa is identical to the settings of the Acoustic Test Case above.

The transient pitzDaily—case is computed by the developed application solver acousticFoam with the already mentioned solver settings for linear equation systems. The simulation time is set to 0.1 s with a time step of 1 $\times 10^{-6}$ s. The small geometric dimensions and the relative short simulation time make it possible to run this case in serial on one processor. The serial execution time is about 670 s.

2.6. Application Case Quadratic Brace 2D

2.6.1. Geometric Settings and Meshing

In the quadratic brace 2D simulation case as presented in Figure 6 transient fluid flow around a quadratic brace is simulated. The quadratic brace is positioned in the middle of the domain which is bounded by the inlet and outlet on the left and right hand side respectively and two walls at top and bottom of the domain. The solution of the initial pisoFoam-Simulation represents a virtually steady state solution of a von Kármán vortex sheet behind the obstacle. This solution is used as the initial condition for the aero acoustic simulation with acousticFoam.

The meshing was done with blockMesh. Concerning the dimensions of Table 3 the mesh consists of 2.75 Miohexahedra elements with anelement size of 0.85 mm.

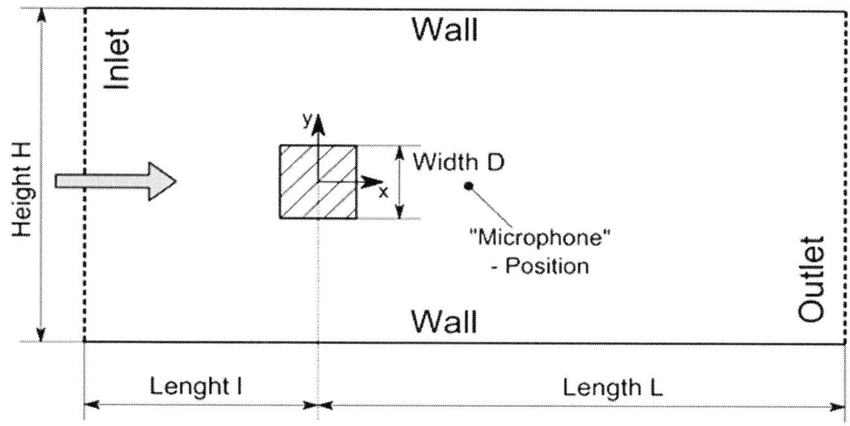

Figure 6. Set up of test case quadratic brace 2 Daccording [20] .

Table 3. Geometric dimensions of test case quadratic brace2D.

Length L	300 mm
Length l	50 mm
Height H	60 mm
Width D	100 mm

2.6.2. Numerical Model

In the quadratic brace 2D simulation case the velocity u of the fluid flow is set to $u_x = 0.535\ \mathrm{ms}^{-1}$ in x-di- rection on the inlet, whereby the pressure p is set to p = 0 Pa on the outlet boundary condition. The brace itself, the bottom and the top of the computation domain are defined as walls. Due to the inlet height H and the velocity of the fluid the case is considered as turbulent with the Reynolds number of Re = 19,000.

Turbulences are calculated by the turbulence model kEpsilon (k-ε) of OpenFOAM with the initial values of the turbulent kinetic energy $k = 1.7 \times 10^{-4}\ \mathrm{m}^2\mathrm{s}^{-2}$, the turbulence dissipation $\varepsilon = 3.6 \times 10^{-5}\ \mathrm{m}^2\mathrm{s}^{-3}$ according to the chosen fluid flow velocity. Used wall functions of the turbulence model are kqRWallFunction for turbulent kinetic energy and epsilonWalFunction for dissipation.

In difference to the previous cases the boundary conditions for the acoustic field pa are set according to Table 4.

To simulate the transient quadratic brace 2D—case acousticFoam with the same solver settings for linear equation systems is used as in both other simulation cases before. The simulation time in this case is limited to 0.1 s with a time step of 1×10^{-6} s. The fine mesh granularity and the high time resolution make it necessary to run this case in parallel as well. To reduce the influence of the initial error in acoustic fields the solution of a steady state case is used as initial values for the transient simulation.

3. RESULTS

3.1. Acoustic Test Case

The numerical solution of acousticFoam is compared with the analytical solution (3.1) presented in [19] . The two dimensional numerical results of the Acoustic Test Case computed by the novel developed acousticFoam- Solver are shown in Figure 7. The figure shows the distribution and the propagation of the fluctuating pressure values after to different time steps. Due to the superposed initial velocity distributions the shape of the distribution of fluctuating pressure values is conspicuous and looks like a doubled quadrupole distribution.

The analytical solution of pressure distribution according to (3.1) for a certain time step

$$p = \frac{1}{2\alpha_1} \int_0^\infty e^{\frac{-\xi^2}{4\alpha_1}} \cos(\xi t) J_0(\xi \eta) \xi \mathrm{d}\xi \qquad (3.1)$$

is compared with the numerical solution of acousticFoam after certain time step along the cross section line. The shape of the numerical solution fits to the analytical solution curve (Figure 8). There are differences between the numerical and analytical absolute values of the pressure values in the peaks due to the numerical errors of the used linear solvers.

Table 4. Settings of the boundary conditions for acoustic field pa in quadratic brace2D.

	Inlet	Outlet	Wall	Brace	Front and Back
pa	Advective $\frac{\partial \phi}{\partial t} + \nabla(\phi U) = 0$	Advective $\frac{\partial \phi}{\partial t} + \nabla(\phi U) = 0$	Fixed Value $\phi = 0$	Zero Gradient $\nabla \phi = 0$	Empty

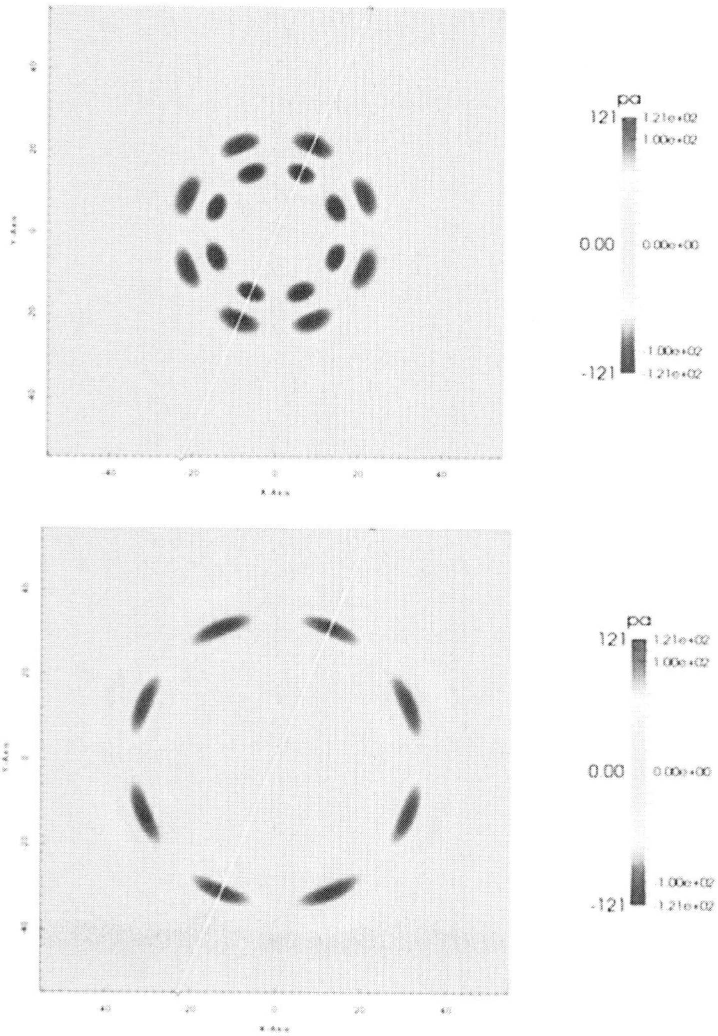

Figure 7. Two dimensional visualization of propagating pressure fluctuations after two different time steps. White line represents cross section for further investigations.

To measure the accuracy of the applied numerical method the error between the simulated solution and the presented analytical solution is calculated. The local error (3.2) at a given time step is the absolute value of the difference between the exact analytical solution and the numerical solution, calculated as following

$$\text{Local Error} = \left| y_{\text{exact}} - y_{\text{num}} \right| \tag{3.2}$$

The Global Error (3.3) is calculated according to the Euclidean norm to

$$\text{Global Error} = \sqrt{\frac{\sum_{i=1}^{n} \left| y_{\text{exact}} - y_{\text{num}} \right|^2}{\sum_{i=1}^{n} \left(y_{\text{exact}} \right)^2}} \tag{3.3}$$

In the presented benchmark test case the global error between the numerical and the analytical solution is computed to 0.38. This global error corresponds to results presented by [21] for compressible two dimensional acoustical wave propagation.

3.2. pitzDaily

The fluid dynamic results of the pitzDaily computation case are well known by OpenFOAM users as a common tutorial case in the OpenFOAM distribution. The first three pictures (a-c) of Figure 9shows the velocity, pressure and turbulent kinetic energy distribution in the computation domain respectively after 20 ms. Figure 9(d) visualize the distribution of the turbulence based acoustic sources travelling downstream towards the outlet boundary condition.

The results of transient simulation are a snapshot of the developing steady fluid flow over the backward facing step. The developing velocity profile moves downstream towards the outlet and the beginning of the vortex on the ground behind the step is captured in Figure 9(a) and Figure 9(c). The detached vortices next to the trailing edge of the step are the basis for the computed acoustic sources by the application solver acousticFoam. These acoustic source are travelling downstream with the fluid flow. Due to the wall boundary condition at the top of the computation domain there are additional vortices developing over the simulation time.

3.3. Quadratic Brace 2D

The quadratic brace 2D-simulation case was run in parallel on 8 processors. For domain decomposition the scotch method was chosen. The computation domain was decomposed in 8 subdomains according to Figure 10.

The solution of the initial pisoFoam-Simulation according to [20] was used as the initial field for the transient simulation with acousticFoam. The vortex street behind the obstacle is visible in Figure 11where the fluid dynamic results of quadratic brace 2D are shown.

In Figure 11(b) the stagnation point in front of the obstacle is very good visualized by the pressure distribu- tion within the computation domain. In the other three items of Figure 11 the periodical flow separation is ob- vious. The detached vortices in the flow behind the quadratic brace are well resolved by the kEpsilon turbulence model of OpenFOAM and are represented by the results of the turbulent kinetic energy field and turbulent viscosity field.

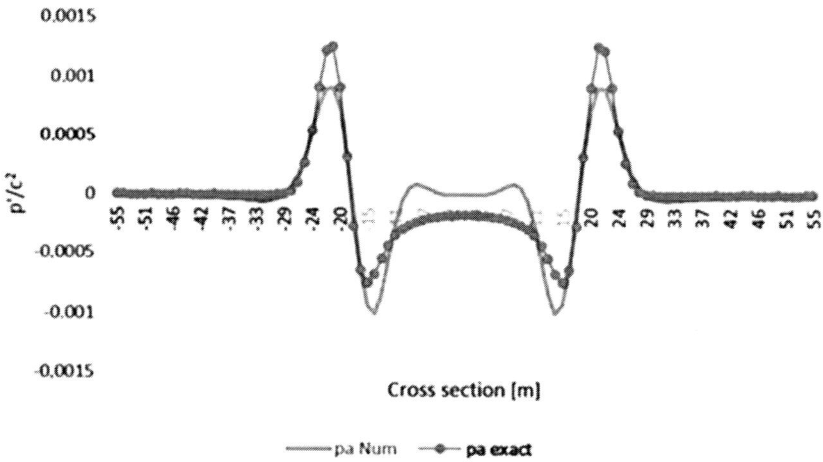

Figure 8. Wave propagation of fluctuating pressure value after certain time steps; comparison of analytical vs. numerical solution along the cross section line.

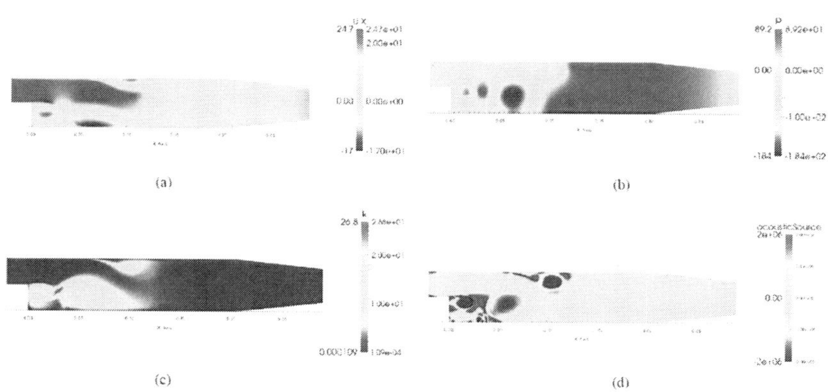

Figure 9. Visualization of CFD and CAA results computed by acousticFoam for incompressible flow fields; (a) velocity in x-direction [m·s−1]; (b) incompressible pressure distribution [m2·s−2]; (c) turbulent kinetic energy [m2·s−2] and (d) acoustic sources [kg·m−3·s−2].

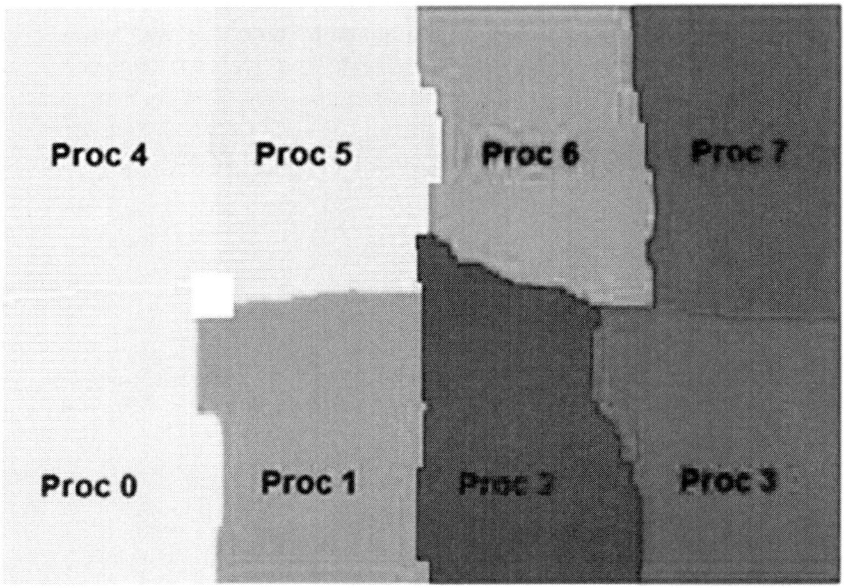

Figure 10. Quadratic brace 2D domain decomposition for parallel run on 8 processors.

In Figure 12 the additionally computed acoustic fields by acousticFoamare shown. The acoustic sources according Curle's Acoustic Analogy are presented in a different scaling within the same time step to provide a better visualization of the acoustic sources travelling downstream through the computation domain.

The propagation of the fluctuation pressure values in Figure 12(c) do not start from the computed acoustic sources around the obstacle only. There are disturbances recognizable in the acoustic source field and the fluctuating pressure field which start travelling from the outlet patches upstream into the computation domain. This disturbances are caused by numerical errors of the outlet boundary condition which is not perfectly suitable for acoustic simulations.

For investigations of acoustic parameters like "sound pressure", "sound pressure level" or their frequency spectra a point located at x y (0.06 0) within the computation domainis considered as a specific probe location. This probe location represents a microphone positioned at a distance of 1.5 x (Width D) behind the obstacle along the x-axis. At this probe location the values of the pressure fluctuations are written out every time step of the computation run. During the post processing the time signal of the pressure fluctuations are analyzed by using the GNU Octave library.

The time signal in Figure 13(a) represents the recorded values of the "microphone" at the probe location within the computation domain. The simulated time was set to 0.15 s. The values of the fluctuating pressure decrease with the time of simulation. The high values at the beginning of the simulation are caused by the influ ence of the numerical boundary conditions, which are not totally suitable for acoustic simulations. This effect is obvious in Figure 12(c) too. These disturbances at the beginning are smoothed over the simulation time. The original time signal is analysed by applying a Fourier Transformation on it and which results in the frequency spectra shown in Figure 13(b). There are distinctive frequency peaks around 400 Hz, 510 Hz and 800 Hz. These multiple peaks are rather indicators for a tonal sound than for a random noise. The logarithmic measure of the fluctuating pressure values represents the "sound pressure level" relative to the reference value of 20 μPa. Ithas a mean level of about 70 dB which might be a realistic value for the simulated application.

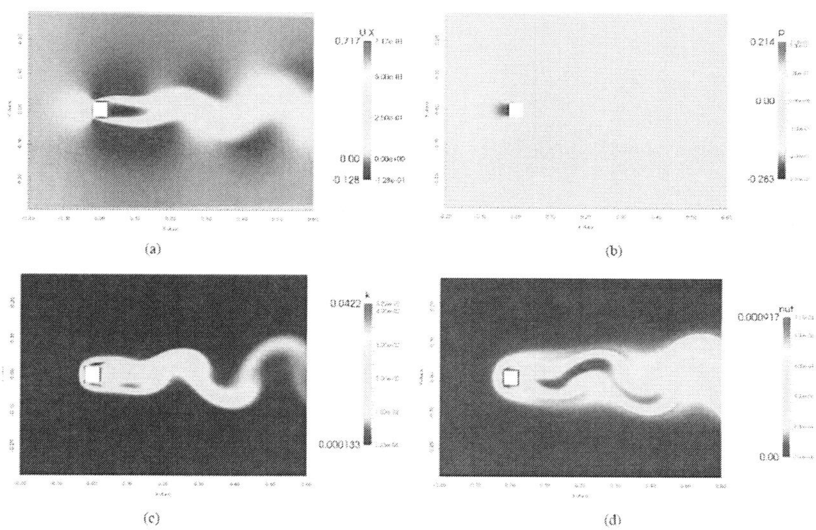

Figure 11. Initial field of quadratic brace 2D—test case according to [20] ; (a) velocity in x-direction [m·s−1]; (b) incompressible pressure distribution [m2·s−2]; (c) turbulent kinetic energy [m2·s−2] and (d) turbulent viscosity [m2·s−1].

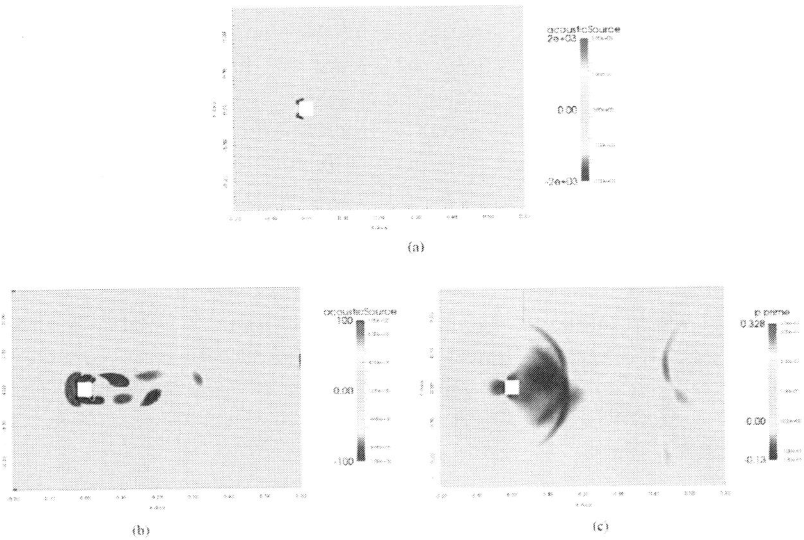

Figure 12. Acoustic fields of quadratic brace 2D after 0.5 ms simulated time; (a), (b) acoustic sources [kg·m−3·s−2] and (c) pressure fluctuation values [Pa].

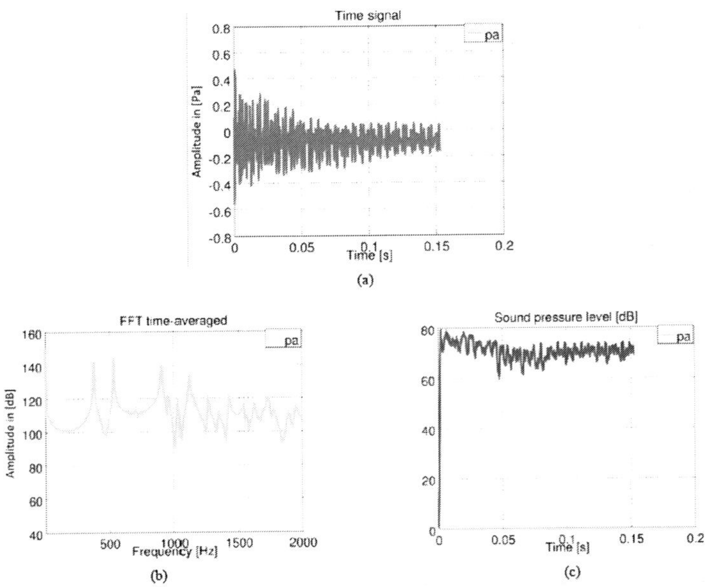

Figure 13. Acoustical results of quadratic brace 2D at specific probe location; (a) time signal of "sound pressure" (pressure fluctuations); (b) the time averaged frequency spectra and (c) the filtered "sound pressure level" over time.

4. CONCLUSIONS

The main target of the presented work was to realize acoustic simulations within the open source computational framework OpenFOAM. The Acoustic Analogies according to Lighthill and Curle were implemented in an application solver of OpenFOAM. These modified application solver, called acousticFoam, is developed to compute and solve incompressible fluid dynamic simulation cases regarding turbulence inducted noise. The solver also takes existing walls and objects in the computation domain into account. Using acousticFoam, acoustic sources due to turbulences can be computed and visualized. Based on this computed acoustic sources the propagation of fluctuating values can be also determined and visualized within the "acoustical" near field. The novel implemented method to compute acoustical fields has been verified and validated on the basis of common computational aero acoustical and computational fluid dynamic benchmark cases. The presented work also includes a parallelized computation case. In this simulation case the time signal on a specific measurement point, represented as a probe location within the computed fluid flow is recorded during simulation time. This time signal is used for further acoustic analysis such as frequency analysis or calculating the sound pressure level over time at this specific observer point.

The presented results clarify the principal applicability of the shown method for computing the sound sources of incompressible turbulent flows. The estimation of sound propagation within OpenFOAM is also feasible. The correctness of the presented results are highly sensitive to the used linear solver, chosen mesh granularity and taken simulation time step. The main error concerning the propagation of the fluctuating pressure values occurs mainly because of the second order time derivative in the wave operator on the RHS of Equation (2.10). Up to now there are no boundary conditions concerning any special acoustical behavior like absorbing or non-reflec- tive characteristics in the OpenFOAM distributions implemented yet. Despite to the numerical and methodical limitations similar to [5] OpenFOAM provides, regarding the presented results of the novel developed application solver, the principal ability for computational aero acoustic applications of external flow fields, especially for computing the turbulence based sound sources in the acoustical near field of an incompressible flow field. Due to the disparity of acoustical and fluid dynamic scales which enforces a small time step and a fine granularity of the mesh, the principal approach of implementing acoustic analogies have to take HPC resources into account. Growing and better disposal resources in High Performance Computing and novel HPC applications generate new potentials for computational aero acoustic applications.

Possible applications of the shown method might be the acoustical investigations of air intake systems for combustion engines, HVAC-systems for heating, ventilation and air condition of vehicles or aircrafts, or any other product or system which causes turbulence based acoustic sources. This investigations might be done in an early stage of development process. Hence the results might be used for the optimization of the investigated geometry concerning the aero acoustic emissions and fluid dynamic properties of technical parts or entire systems.

Further development works have to aim at the enhancement of the presented aero acoustic solver in OpenFOAM. One of the most important development steps might be detailed investigations on acoustically suitable boundary conditions with non-reflecting or absorbing characteristics in OpenFOAM. The implementation of the Ffowcs-Williams-Hawkings acoustic analogy will also enlarge the principal applicability of the novel application solver on more realistic simulation cases. Using a compressible application solver as a basis for an improved novel aero acoustic solver will improve it mainly regarding its wave propagation ability. The implementation of acoustic analogies in a compressible application solver of the OpenFOAM-framework will also emphasize the plausibility of the shown method.

ACKNOWLEDGEMENTS

The authors want to acknowledge the Project "FETоL—Fault Tolerant Framework for peta-scale MPI-Solvers" supported by the German Federal Ministry of Education and Research.

REFERENCES

1. Weller, H.G., Tabor, G., Jasak, H. and Fureby, C. (1998) A Tensorial Approach to Computational Continuum Mechanics Using Object-Oriented Techniques. Computers in Physics, 12, 620-631.
2. Lysenko, D.A., Ertesvåg, I.S. and Rian, K.E. (2012) Modeling of Turbulent Separated Flows Using OpenFOAM. Computers & Fluids, 80, 408-422.
3. Nilsson, J. (2010) Implementation of Acoustical Analogies in OpenFOAM and CALFEM. Lund University. Master- Thesis.
4. Stein, A. and Pelz, P.F. (2012) Möglichkeiten und Grenzen numerischer Strömungsakustik mit OpenFOAM, DAGA 2012.
5. Wang, Q., Pelz, P.F. and Matyschok, B. (2010) Numerische Simulation von turbulenzbedingtem Schall mit Open- FOAM, TU Darmstadt.
6. Kraposhin, M.V. and Strizhak, S.V. (2013) How to Implement Simple Acoustic Analogy in OpenFOAM. 8th Interna- tional OpenFOAM Workshop 2013, Jeju, Korea.
7. Lighthill, M.J. (1952) On Sound Generated Aerodynamically I. General Theory. Proceedings of the Royal Society A, 211, 564-587.
8. Költzsch, P. (2008) Flow Acoustics. In: Mechel, F.P., Ed., Formulas of Acoustics, 2nd Edition, Springer-Verlag Berlin Heidelberg, 945-1016.

9. Curle, N. (1955) The Influence of Solid Boundaries upon Aerodynamic Sound. Proceedings of the Royal Society A, 231, 505-514.
10. Ehrenfried, K. (2004) Strömungsakustik: Skript Zur Vorlesung, Mensch & Buch Verlag, Berliner Hochschulskripte.
11. Ffowcs Williams, J.E. and Hawkings, D.L. (1969) Sound Generation by Turbulence and Surfaces in Arbitrary Motion. Philosophical Transactions of the Royal Society A, 264, 321-342.
12. Oshima, T. and Imano, M. (2008) A Full Finite-Volume Time-Domain Approach towards General-Purpose Code Development for Sound Propagation Prediction with Unstructured Mesh. Proceedings of Inter-Noise 2008, Shanghai, 26-29 October 2008, 15 p.
13. Poinsot, T.J. and Lelef, S. (1992) Boundary Conditions for Direct Simulations of Compressible Viscous Flows. Journal of Computational Physics, 101, 104-129.
14. Colonius, T., Lele, S.K. and Moin, P. (1993) Boundary Conditions for Direct Computation of Aerodynamic Sound Generation, AIAA Journal, 31, 1574-1582.
15. Andreini, Bianchini, Facchin, Giusti, Bellini, Chiti, Grazzini (2011) Large Eddy Simulation for Train Aerodynamic Noise Predictions. Proceedings of the WCRR 2011, Lille, 22-26 May 2011.
16. Jangi, M., Tilley, N. and Merci, B. (2009) Numerical Simulations of Some Possible Fire Scenarios in a Closed Car Park with RANS and LES. Proceedings of the IAFSS Advanced Research Workshop, Santander, 15-17 October 2009, 233-242.
17. Boersma, B.J. (2004) Numerical Simulation of the Noise Generated by a Low Mach Number, Low Reynolds Number Jet. Fluid Dynamics Research, 35, 425-447.
18. Goldstein, M.E. (1976) Aeroacoustics. McGraw-Hill International Book Co., New York.
19. Tam, C.K.W. and Webb, J.C. (1993) Dispersion-Relation-Preserving Finite Difference Schemes for Computational Acoustics. Journal of Computational Physics, 107, 262-281.
20. Schwarze, R. (2013) CFD-Modellierung-Grundlagen Und Anwendungen Bei Strömungsprozessen. Springer Vieweg, Berlin.
21. Tóth, P., Fritzsch, A. and Lohász, M. (2008) Application of Computational Fluid Dynamics Softwares for 2D Acoustical Wave Propagation. Gépészet, 29-30.

Chapter 8

Acoustic Emission Activity for Characterizing Fracture of Marble under Bending

Eleni Tsangouri [1], Dimitrios G. Aggelis [1,†], Theodore E. Matikas [2,†] and Anastasios C. Mpalaskas [2,*]

[1] *Department of Mechanics of Materials and Constructions, Vrije Universiteit Brussel, Pleinlaan 2, Brussels 1050, Belgium*

[2] *Department Materials Science and Engineering, University of Ioannina, Ioannina 45110, Greece*

ABSTRACT

The present paper occupies with the acoustic emission (AE) monitoring of fracture of marble. The specimens belong to two different material types and were tested in three-point bending after being ultrasonically interrogated. Consequently, they were repaired by means of suitable epoxy agent and mechanically re-loaded. Apart from the well-known correlation of pulse velocity to strength, which holds for the materials of this study as well, AE provides some unique insight in the fracture of the media. Parameters like the frequency content of the waveforms, and their duration among others show a transition in relation to the load. According to their strength class, the specimens exhibit distinct AE characteristics even at low load, enabling to judge their final strength class after having sustained just a small percentage of their ultimate capacity. More importantly, the AE activity during reloading indicates the quality of repair; specimens with good restoration of strength, exhibited similar AE activity to the intact specimens, while specimens with lower repaired capacity exhibited random behavior. This work discusses the passive monitoring of marble fracture and shows that AE parameters that have been used to successfully characterize cementitious materials, provide good results in monolithic materials like marble as well. It is suggested that AE monitoring during a proof loading can provide good information on the potential strength class, which is especially useful for repaired specimens, where the pulse velocity cannot be easily used.

Keywords: marble; acoustic emission; ultrasound; bending; epoxy

1. INTRODUCTION

Acoustic emission (AE) is a technique used for many decades in structural health monitoring (SHM) applications. It uses sensors to record the elastic motion that follows crack nucleation and propagation events within the materials and transform them into electric waveforms. Their intensity (energy), number and other parameters reveal information on the crack density and propagation rate, the location and the mode of fracture [1,2,3,4]. The use of the technique in a typical laboratory experiment is shown in Figure 1. In typical rock materials like marble and granite, acoustic wave techniques have been used for quite some time. The cumulative AE activity was studied in relation to the fracture process of pre-cracked marble under compression. It was shown that specimens with "zigzag" cracks failed in a more unpredictable way since most of the activity was registered at the moments of macroscopic failure [5]. The evolution of AE activity was very indicative of the strain at creep experiments of granite [6]. The AE activity and energy release during compression was used to indicate the damage level showing that in marble, main damage comes with the peak stress, while earlier only micro-cracking is formed compared to granite and rock salt [7]. The frequency content of AE has shown to increase with the load, being essentially a precursor of failure in confining pressure experiments, while close to the peak load a strong drop was noticed [8]. The Kaiser effect (absence of detectable AE before the previous maximum load is overpassed) in Brazilian and bending tests was investigated to reveal the previous tensile stress imposed in marble [9]. Energy rate, hit counts, amplitude and location of AE events were used for characterization of crack propagation during the indentation test [10] and compression test in rock materials [11].

AE monitoring of rock material under compression showed mainly two time zones of high "avalanche-like" activity, interrupted by a long period of acoustic silence [12]. The effect of temperature on brittleness of marble was studied with AE corresponding very well to the "rupture evolution law" [13]. Complementary to AE, another elastic wave method, ultrasonic pulse velocity (UPV) has also been used for characterization of marble condition and mechanical performance. Specifically UPV has been used to derive elastic constants of marble [14] and has been positively correlated to the mean grain size of marble [15], while the anisotropy has also been highlighted due to the noticeable difference between the velocities at the three different axes [16]. The influence of water saturation in granite has also been examined by ultrasonic measurements. The effect of microstructure is revealed by the systematic increase of velocity for higher ultrasonic frequencies (150 kHz compared to 54 kHz) [17].

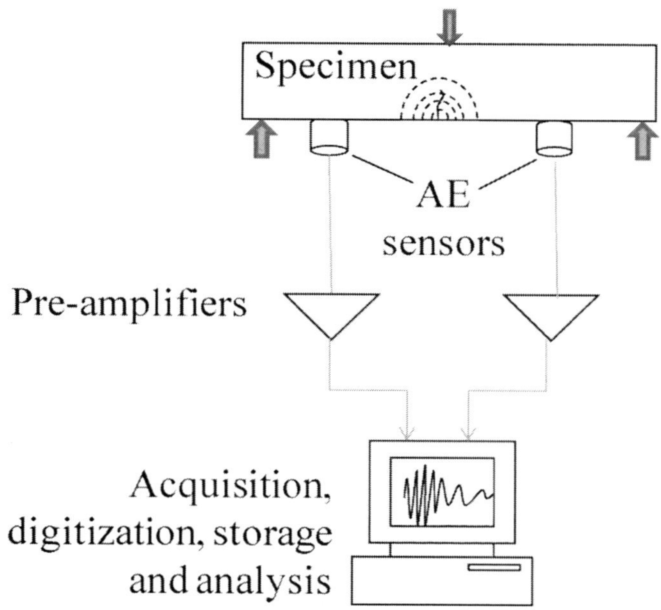

Figure 1. Typical acoustic emission (AE) setup.

The present study is a continuation of preliminary tests on granite and marble. Previously, different fracture modes were targeted showing that AE is quite sensitive to the loading conditions [18,19]. Specimens loaded in pure bending emitted higher frequencies and signals of shorter duration relatively to shear-loaded ones. After encouraging results were found, in this case, a large population of specimens of two marble types was tested in bending with concurrent AE recoding to increase the statistical adequacy in an effort to check possible correlations between passively monitoring parameters and mechanical properties. The AE behavior proved very consistent exhibiting three distinct stages with different AE rate. Specific parameters can be used from the early loading to characterize the capacity of the material, like average frequency and RA value, which have been adopted from concrete practice. The fractured specimens were consequently repaired and reloaded to failure. The AE activity of repaired specimens showed capability to characterize the repair efficiency since the specimens with high reloading capacity exhibited similar AE stages with the intact material, while the ones with low strength restoration exhibited a more random behavior driven by the weak bond between the epoxy and the marble. To the authors knowledge, this is the first effort to relate low load AE (approximately one-third of the ultimate load) to the final load bearing capacity of marble both in the intact and the repaired state.

2. EXPERIMENTAL SECTION

2.1. Materials and Mechanical Details

Two types of marble were used, one named California Honey from Morocco with an average measured density of 2280 kg/m^3 and the other named Crema Mocha with an average measured density of 2508 kg/m^3 from Spain [20]. For convenience the first marble type will be addressed as MA and the next as MB. The specimens were prismatic of size 40 × 30 × 160 mm. Thirty-three specimens of the first type of marble and forty specimens of the second type of marble were subjected to three-point bending according to EN 13892-2:2002, see Figure 2a,b. The load was applied at a constant rate of 50 N/s and the loading was automatically terminated at the moment of instant load drop. At the three-point bending test, the crack starts from the central point of the bottom side due to the tensile stresses (Figure 2b).

(a)

(b)

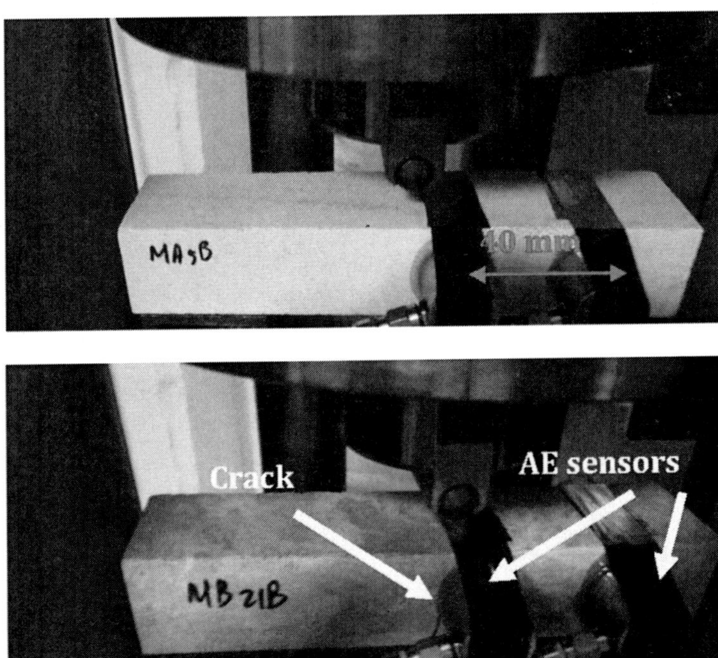

Figure 2. (**a**) Three point bending of (**a**) MA and (**b**) MB marble with concurrent AE monitoring by two sensors.

After bending fracture has occurred all the specimens were repaired in the crack surface with a two-component solvent-free bonding system based on epoxy resin and hardener named "Epoxol" in its commercial product name. It is a specially formulated, high viscosity, polyester marble adhesive-putty, used to bond and fill for repair purposes with hardening time of approximately 5–6 h [21]. The bonding was carefully established so that the original geometry of the specimen was not altered. Figure 3a,b show the populations of healthy and repaired specimens for MA marble, respectively.

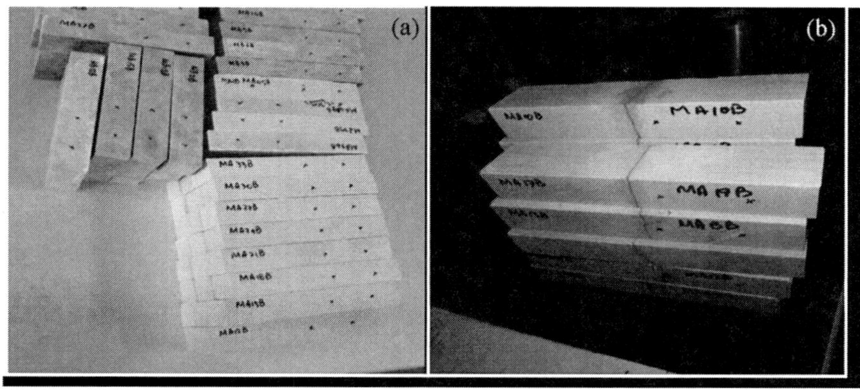

Figure 3. The population of (**a**) healthy and (**b**) repaired specimens for MA marble.

2.2. Ultrasonic Inspection

The whole population of specimens was examined by ultrasound before the fracture test. Two AE transducers (pulser and receiver) were placed on the longitudinal axis of the specimens as seen in Figure 4, having a wave path of 160 mm. A layer of grease between the transducers and the specimen enhanced acoustic coupling. The transducers were of 150 kHz resonance (R15, Mistras Group). The excitation came by a pulse generator with an electric cycle of 150 kHz directly fed to the pulser as well as the acquisition board to act as reference. The pulse velocity (UPV) was calculated by the wave path over the time delay to the first detectable disturbance of the received waveform. The signal was clear and there was no need for extra processing like stacking or filtering.

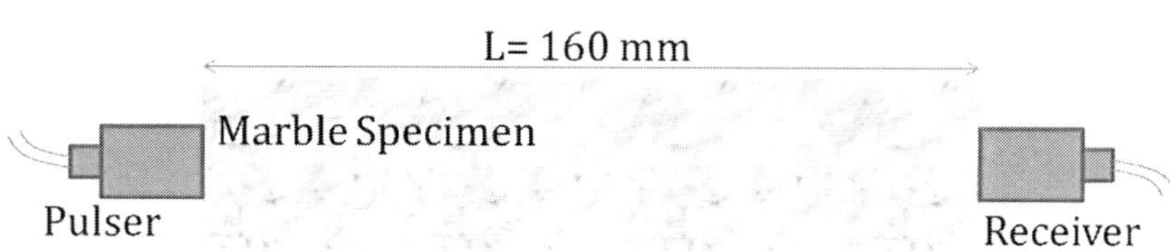

Figure 4. Schematic representation of ultrasound testing.

2.3. Acoustic Emission

AE monitoring took place by means of two piezoelectric sensors, (R15, Mistras). Their resonance comes at 150 kHz and their positions on the specimens are shown in Figure 2. The center of one sensor was placed 15 mm away from the point of the expected crack, which was secured by a notch. The second sensor was placed 40 mm away. Both sensors were positioned at the same side of the crack in order to be able to characterize the distortion of the signals as they propagate for the additional distance of 40 mm. Nevertheless, the effect of distortion will not be discussed herein and will be treated separately in another study. Acoustic coupling was improved by silicon grease between the sensors face and the specimens' surface. AE activity was captured by a two-channel PCI-2 Mistras board with sampling rate of 5 MHz. The threshold was 40 dB, as well as the pre-amplification. A schematic representation of a waveform is seen in Figure 5. Some of the main features are the maximum amplitude, AMP (usually in dB), and the duration, DUR (period between the first and the last threshold crossing). The "rise time" (RT) (which is the time between the first threshold crossing and the point of peak amplitude in μs) is related to the fracture mode of the crack and so is the inverse of the slope of the initial part of the signal (RA value, RT/A in μs/V). The energy (ENE) is defined as the area under the rectified signal envelope. Frequency content can be measured by AF (average frequency), which is the total number of threshold crossings divided by the duration, while there are other indices based on the spectrum of the FFT.

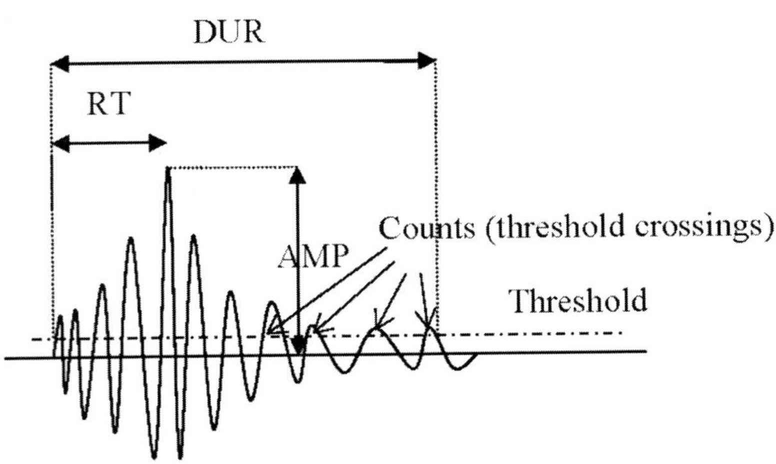

Figure 5. Typical AE waveform and its basic characteristics.

3. RESULTS AND DISCUSSION

The mechanical results will be discussed along the monitoring ones in the corresponding ultrasonic and AE paragraphs.

3.1. Ultrasonics

The ultrasonic velocity results are summarized in Figure 6 in correlation to the maximum load of the specimens. Clearly, the two different types of marble formed two clusters. Type MB, which exhibited higher maximum load, also showed much higher pulse velocity. The differences are quite strong, as there is no overlap between the clusters, while there is also a weak (but not negligible) correlation within each cluster. In any case, the sensitivity of pulse velocity to the stiffness and indirectly to the strength of the engineering materials is well known [22,23].

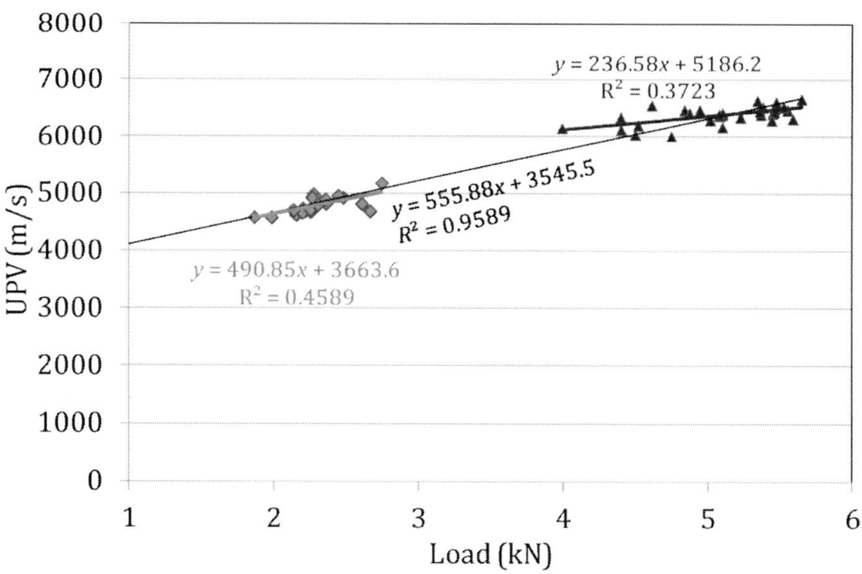

Figure 6. Correlation between pulse velocity and maximum load for the marble specimens.

3.2. Acoustic Emission

Figure 7a,b show typical examples of cumulative AE activity as received by both sensors for marble type MA and MB, respectively. In all of the cases, three distinct stages of activity can be defined. At the start of the experiment a quite high rate of events is recorded (Stage A) followed by a long time window of moderate activity (Stage B). At the end and quite close to the macroscopic fracture of the material, the rate increases again (Stage C). Similar AE activity patterns have been seen recently in similar specimens during compression. The initial saturation was attributed to the depletion of a "weakpoint reservoir in the deforming material", while the increase at the final stage was attributed to clustering of the individual defects due to decreased distance among the sources [12]. It seems that after application of load, formation of micro-cracks is quite fast. However, coalescence of them into larger ones that lead to final fracture takes more time and accumulated energy. This energy is released at the end with another "avalanche" like activity. Although the number of AE hits of Stage C are just a small fraction of the total activity, their energy is much higher.

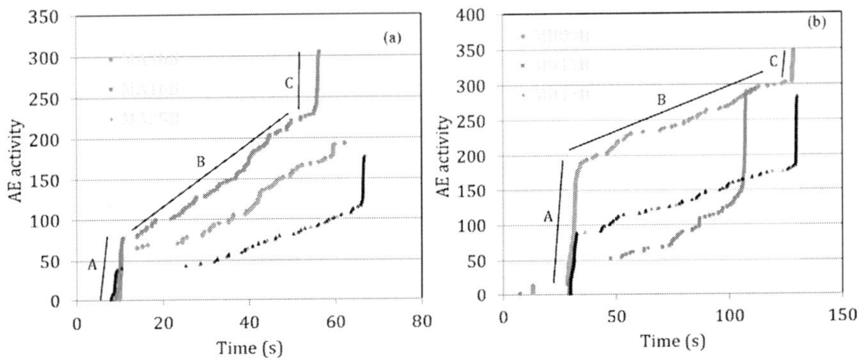

Figure 7. Cumulative AE activity history for marble type (**a**) MA and (**b**) MB.

The behavior of the different marble types was quite repeatable. Apart from the three indicative cases included in the figures, nearly all specimens exhibited the three stages of high, moderate and high activity again.

However, in addition to the total number of hits, high importance is carried by the AE waveform parameters. To give an idea of the transient changes of AE characteristics during loading, Figure 8 shows the RA values for two representative specimens of type MA marble (Figure 8a) and other two from type MB marble (Figure 8b). The data corresponding to the last Stage C are included in a transparent pattern since it is intended to focus on the behavior prior the macroscopic failure. It is obvious that the RA initial level is at much higher levels for type MA than type MB, since values up to 50 ms/V are quite frequent, while for MB specimens values are almost restricted below 30 ms/V. As loading continues high values become scarce resulting in a decreasing trend for RA for both types. At the end of Stage B, and just before the macroscopic failure, the RA values of both types of specimens seem to deepen below 10 ms/V. Concerning the whole population of specimens (more than 30), the average values are shown in Table 1.

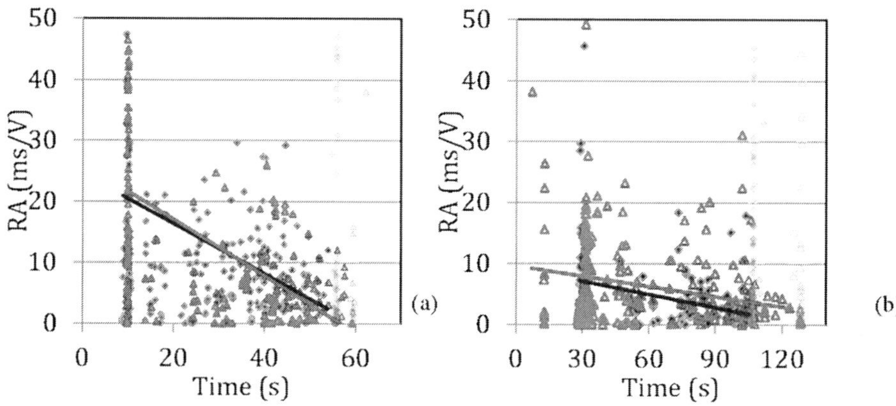

Figure 8. RA values for two specimens of marble type (**a**) MA and (**b**) MB.

Table 1. Results of the bending tests for two types of marble.

Type	Maximum Load (kN)	UPV (m/s)	Fracture Stage	RT (μs)	ENE (-)	DUR (μs)	AMP (dB)	AF (kHz)	RA (μs/V)
MA	2.30	4798	A	455	10.9	1321.0	47.0	25.8	20305
			B	167	7.9	820.8	46.8	30.6	8222
			C	380	8.7	1158.7	46.8	31.0	15829
MB	5.08	6383	A	231	8.6	832.6	48.3	35.8	8452
			B	95	5.8	508.3	47.9	41.0	4344
			C	964	29.4	2904.5	50.7	38.2	22355

From the above table, a general conclusion is that the AE features differ for the successive stages. As an example, the rise time and duration of the second stages (Stage B) is clearly lower than the first (Stage A), while the final stage (Stage C) produces again higher values. Inversely, the frequency content as measured by AF increases for Stage B relative to Stage A. Duration of AE signals is related to the intensity of the source and it shows again the same transition as the total AE activity. It would be premature to interpret these fluctuations of AE parameters in terms of shift between shear and tensile phases, as it should be combined with the knowledge of the microstructure. However, this lowering of RA in the intermediate stage, follows the rate of damage accumulation, which is slower in Stage B, as revealed by the low AE activity rate denoting the creation of micro-cracks, the coalescence of which leads to the final fracture at Stage C.

Apart from the differences between the successive stages, there are considerable differences in between the different marble types. Type MB exhibits for the first two fracturing stages, AF higher by as much as 10 kHz than marble MA. This difference is quite high considering that it was recorded by resonant sensors. Correspondingly, it exhibits lower duration, RT and RA values. In addition, while MA exhibits lower RA values for the last stage, type MB exhibits a strong increase. These different trends are attributed to the different fracturing behavior of the two media since the specimen geometry and the whole experimental lay-out was constant. As aforementioned changes of RA are due to the fracture mode. Although the macroscopic stress field is similar, the microstructure of the different marbles may induce different motion of the crack tips strongly influencing the AE parameters received.

It has been seen that high RA values correspond to material closer to fracture while the inverse is measured for frequencies. The frequencies drop just before the final failure. It comes as no surprise therefore, that the material type with higher frequency and lower RA exhibited higher density and load bearing capacity. What should be stressed is that these differences are obvious from the initial fracture stage; MB exhibits 39% higher frequency and 58% lower RA allowing predictions as to the final capacity of the beams.

By separating the activity at the different stages we can study the AE activity at low, moderate and high load. Figure 9a shows the graph of AF *vs.* RA for the first stage of fracturing (Stage A) as explained in the previous section, while the corresponding graph of the intermediate Stage B is given in Figure 9b. The whole group of points extends from frequencies of approximately 20 kHz to 45 kHz and from RA values of 3 to 30 ms/V. However, it is important to notice that the type MB occupies most of the spots of higher frequency and low RA, while the two populations (MA and MB) could even be separated by a single line (see dashed diagonal line) having only three overlapping points. For the next Stage B of moderate activity, the whole population shifts to higher frequencies and lower RA values (see Figure 9b) but there is still a substantial difference of the centers of the two populations. Marble MB again exhibits higher frequencies and lower RA values and the separation line attains a much higher slope. The separation lines are certainly not unique and successful clustering could achieved by means of pattern recognition algorithms. However, this is left for the next stage, when larger populations will be examined.

Based on the AE activity of the preliminary loading stage it can be concluded that if the ratio of the average AF over RA is more than 1.67, the ultimate load will be higher than 4 kN (denoting type MB) while if it is lower the load capacity will be less than 3 kN (type MA). The importance of such a correlation lies in the fact that the strength capacity of the material can be assessed by a quite reliable way based on AE data resulted from loading the specimen, without however, inflicting any serious damage: the initial activity (Stage A) is recorded while the load is less than 10% of the ultimate.

Results are quite similar for the intermediate stage where the load is between 20% and 90%. The analogy between AF and RA now changes to 5.2 but again the separation with a straight line is similarly successful.

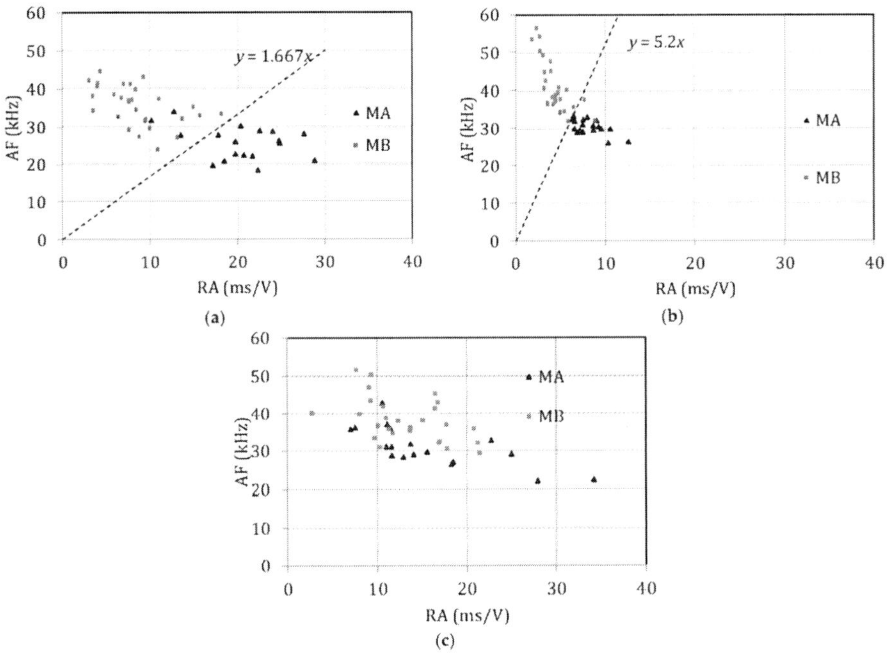

Figure 9. Correlation plot between AF and RA for (**a**) the initial loading Stage A; (**b**) the intermediate loading Stage B; and (**c**) the final Stage C.

For the last stage of macroscopic fracture, the results are more mixed, showing that in macro-fracture the intensity of the phenomena "covers" delicate changes that are imposed by the microstructure (Figure 9c).

3.3. Behavior of Repaired Specimens

As aforementioned, the fractured specimens were repaired by means of adhesive epoxy, suitable for cultural heritage projects. In order to check the strength restoration they were tested again with the same bending setup. Their AE activity was once again monitored. Table 2 shows the average value of maximum load for both repaired marble types. The maximum load is approximately 1.63 kN and 2.46 kN for types MA and MB, respectively, corresponding to 70% and 48% of their original strengths. The marble with higher original strength (MB type) exhibited the higher capacity even after repair, although it suffered larger loss as a percentage of the original strength.

Table 2. Results of the bending tests for two types of repaired marble (reloading).

Marble Type	Maximum Load (kN)	UPV * (m/s)	Fracture Stage	RT (μs)	ENE (-)	DUR (μs)	AMP (dB)	AF (kHz)	RA (ms/V)
MA	1.63	4798	A	146	5.6	629	47.6	38	7.33
			B	61	4.6	416	48.1	43.9	2.93
			C	372	15.2	1153	49.5	36.7	9.45
MB	2.46	6383	A	179	7.2	740	48.8	37.2	5.60
			B	71	4.5	454	47.6	39.9	3.07
			C	344	16.4	1022	50.9	43.9	9.08

* UPV was measured before fracture.

Compared to the nearly uniform behavior of the sound specimens, the AE behavior of the repaired MA type specimens showed mixed trends. Most of the specimens (20 out of 33) followed a similar behavior to the sound (first loading) with the same stages (high initial rate-plateau-high final rate) but the rest of the specimens exhibited a more random behavior where the specific stages could not be defined. Examples are seen in Figure 10a,b for the aforementioned cases. The absence of fracture stages can be attributed to the inefficient interfacial bonding between the marble and the epoxy.

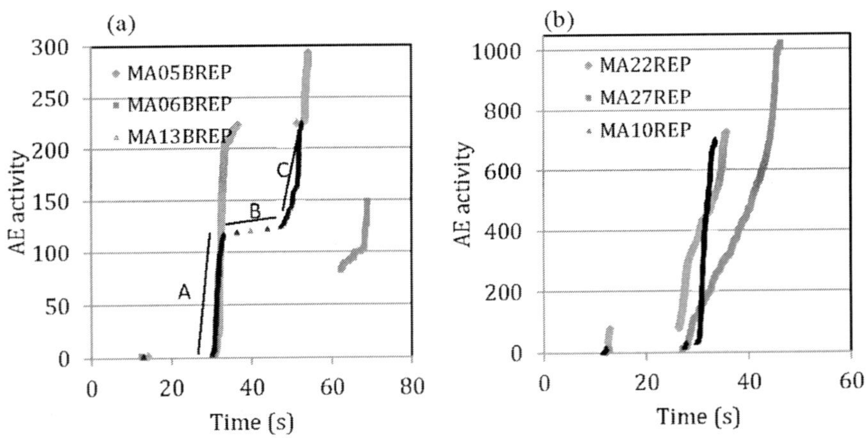

Figure 10. (**a**) Cumulative AE history for repaired MA specimens following three stages. (**b**) Cumulative AE history for repaired MA specimens without similar three stages.

Table 3 shows the average values of ultimate load and reload for the specimens of marble MA that exhibited the three stages at the reloading (20 specimens), the specimens with random behavior (13 specimens) as well as the average of the whole population (33 specimens). It also includes the repair factor (n) as the ratio of ultimate load of the repaired to the initial load of the intact specimens.

Table 3. Results of the bending tests for MA type of repaired marble.

MA	Maximum Load (kN)	UPV * (m/s)	Maximum Load(REP) (kN)	n = Load(REP)/Load (-)
no stages in AE	2.24	4713	1.46	0.66
3 stages in AE	2.33	4846	1.75	0.76
total	2.30	4798	1.63	0.72

* UPV was measured before fracture.

It is worth noticing that the specimens that during reloading followed a behavior similar to the initial loading exhibit substantially higher repaired load. Specifically this group of specimens exhibited an average maximum load of 1.75 kN substantially higher than the group of the random behavior (1.46 kN). This shows that the bonding was of better quality and therefore the structural capacity, the fracture process and as a result the AE behavior was closer to the behavior of the intact material achieving a restoration factor of 76% which is quite high for similar cases [24].

Concerning type MB, nearly all specimens (38 out of 40) behaved similarly to the sound specimens. This has to be related to the bond between the epoxy and the matrix. For the case of MB, the epoxy and its bond to the marble works well and results in the same behavior like the initial monolithic material. This is confirmed by the maximum load after repair.

Additionally, a general observation is that the plateaus (silent periods) are longer for the repaired specimens, while stage B is characterized by lower AE rate than the initial loading. Specifically Stage B exhibits an activity of 3.3 hits/s for marble MA and 2.2. hits/s for marble MB. For the repaired version of the same specimens, the corresponding rates was negligible with only sparse AE hits for Stage B (check again Figure 10a). This lower activity (and the longer plateaus in the beginning of the test) can be attributed to the lower elasticity of epoxy. The stiffness of this layer is much lower than marble, allowing it to deform without cracking for longer period than marble. Therefore, it withstands the imposed deflection with limited AE. Additionally it has been seen that the type of epoxy may influence this behavior. In recent experiments in concrete repaired by flexible healing agent, the AE development was smooth since the bonding took place in small and frequent events. On the other hand, rigid epoxy follows a slightly different trend by fewer but larger debonding events [24].

Excluding the specimens that exhibited random behavior (13 of type MA and two of type MB) one can again separate the AE behavior in the successive stages. Results are included in Table 2 above. It is evident that the parameter values are much closer between the different marble types for each stage. For the first stage received at low load, average frequencies are almost identical and RA values are very close. It should be recalled that in the first loading, marble MB exhibited 39% higher frequency, while the RA value of B was 58% lower than MA (see Table 1). Now these differences are nearly eliminated; frequencies are essentially the same and the average RA of MB is only 23% lower than MA. Differences in rise time and duration are similarly compressed. The AE parameters show indeed that in the repaired specimens, the importance lies primarily in the bonding between epoxy matrix and less in the marble itself. The difference of 23% in RA although may seem substantial, does not allow successful separation of the two classes, due to the overlap. Indeed in AE, the experimental scatter of the parameter values is high and a difference between averages of the order of 20% of even 50% is not enough to separate different populations [25]. However, it can still be seen that type MA with higher RA value in the preliminary stage of loading exhibited lower maximum load. Therefore, there is certainly potential in the study of AE in the repaired specimens as to predict the final behavior. It should be kept in mind that it is possible to apply pattern recognition combining not only two but many features which offer the highest characterization capacity or even apply principal component analysis.

4. DISCUSSION

The reason why AE inspection acquires even more significance is that ultrasound cannot reliably assess the condition of the repaired specimen. As shown in Figure 4, the nominal wave path is 160 mm. In the repaired material, the layer of epoxy resin is approximately 1 mm corresponding to less than 1% of the total. Therefore, the stiffest marble (in this case type MB) would always exhibit much higher velocities and the sensitivity to the epoxy layer and its adhesion to the matrix would be negligible. Despite the fact that ultrasound shows very good correlation to the material properties and the final strength, it would not be possible to characterize by the same test the capacity of the repaired specimens.

It should also be mentioned that although AE gets signals from fracturing of the material it is not necessarily “close to failure”, as micro-cracking activity is induced very early in the loading. There are several other examples in literature where AE starts in a small fraction of the ultimate load [7,11,12,25,26]. Therefore, AE can be used in the concept of “proof loading”. This kind of test is a usual application in different types of structures (e.g., pressure vessels). The structure is loaded up to a safety point and the AE is monitored. According to the received AE the structure can be categorized either as safe for continuous use, as safe for a period of time before another monitoring, or unfit and should be replaced. AE, due to the sensitivity of the sensors records the very early fracturing incidents in the micro-scale before the structural capacity of the member/specimen is compromised. In the specific case, the loading rate was constant. Taking into account the accumulated AE activity for marble MA, the initial AE stage is completed well before 15 or 20 s in a total of 50 to 60 s until fracture (checkFigure 7a). This means that the load until that moment was approximately one-third of the maximum load. For marble MB, the initial activity (first stage) was completed at about 30 s while the duration of the loading until fracture was more than 100 s (Figure 7b). Again this means that the loading for the first AE stage is well below one-third of the maximum. Therefore, it can be supported that the AE data used for correlation with the final load capacity, as presented in Figure 9a correspond to a stage when the material is still macroscopically intact since the load was about 30% of the ultimate. For the repaired specimen, the initial stage of AE represents approximately half the duration of the experiment, so we can say that the "early" AE data were recorded until 50% of the capacity (Figure 10a). After the initial stage, the specimen continued to receive increasing load within a period of acoustic silence before it was fractured 20 to 40 s later. Therefore, it can again be argued that the early loading did not effectively compromise the load bearing capacity of the specimens.

Concerning the localization of the AE sources, the present setting with two sensors enables only linear location. More sensors would not add much to the accuracy of the AE events. This is because the accuracy is compromised by different factors. The specific type of sensors is widely applied in practice due to their high sensitivity but on the other hand they are far from being considered point sensors, as their diameter is approximately 16 mm. This is comparable to the propagation distance between the crack (middle of specimen) and the 1st sensor and also comparable to the specimen dimensions (thickness of 40 mm). Therefore, AE location would enable to separate events that occurred centimeters away but within a single damage zone it would be troublesome to make more accurate statements. As an example, in our study the thickness of the epoxy in the repaired specimens is approximately 1 mm. AE location results would not be reliable enough to check if an event occurred within this mm (crack in epoxy) or 0.5 mm to the right or left (marble crack), as the typical error is certainly higher. In addition to the above mentioned geometry reasons, the heterogeneity of the material should be kept in mind that creates a variation in the pulse velocity and further compromises the accuracy. Therefore, in the specific case, AE location could not be used to clarify the exact point of the fracture and specifically discriminate between fracture in the epoxy or in the marble matrix.

5. CONCLUSIONS

The present paper occupies mainly with the AE behavior of large populations of two marble types under bending and the repair efficiency as tested by a second loading after application of adhesive epoxy. The main conclusions of the work are summarized below:

i. The fracture of the material follows three successive stages of high, low and high AE activity.

ii. The AE parameters at each stage vary showing that the fracture pattern changes throughout the damage process even though the macroscopic failure of both marble types is derived from the same three-point bending setup.

iii. Different marble types exhibit fairly different behavior, even from the initial stage of loading. Based on AE parameters from the early stage (AF and RA), an almost complete separation of the two populations (high and low strength) is possible, allowing the prediction of the strength group that each specimen belongs.

iv. Repaired specimen with adhesive epoxy exhibited a restoration of flexural strength between 50% and 75% depending on the bonding efficiency.

v. The repaired specimens, which followed AE behavior similar to the intact ones (same three distinct fracture stages), exhibited much higher strength restoration (20% higher ultimate load at the test after repair), something attributed to the good adhesion that dictates the specimens to fracture similar to the intact ones.

Study should continue with different fracture modes apart from bending, as well as with other natural stones like granite (different morphology, greater flexural strength and density than marble). In addition, pattern recognition approaches should be applied in order to achieve even better characterization or classification of the different AE populations.

AUTHOR CONTRIBUTIONS

All authors contributed to the work presented in this paper, and to the writing of the final manuscript. E.T. contributed to the data analysis. D.G.A. was responsible for the ultrasonic experiment and the analysis of the data. T.E.M. provided discussion on the testing procedure and the results. A.M. organized the study and was responsible for the mechanical testing and AE monitoring.

REFERENCES

1. Grosse, C.U.; Ohtsu, M. *Acoustic Emission Testing*; Springer: Berlin, Germay, 2008.
2. Zitto, M.E.; Piotrkowski, R.; Gallego, A.; Sagasta, F.; Benavent-Climent, A. Damage assessed by wavelet scale bands and *b*-value in dynamical tests of a reinforced concrete slab monitored with acoustic emission. *Mech. Syst. Signal Process.* **2015**, *60*, 75–89.

3. Shiotani, T.; Oshima, Y.; Goto, M.; Momoki, S. Temporal and spatial evaluation of grout failure process with PC cable breakage by means of acoustic emission. *Constr. Build. Mater.* **2013**, *48*, 1286–1292.
4. Carpinteri, A.; Lacidogna, G.; Accornero, F.; Mpalaskas, A.C.; Matikas, T.E.; Aggelis, D.G. Influence of damage in the acoustic emission parameters. *Cem. Concr. Compos.* **2013**, *44*, 9–16.
5. Li, Y.; Chen, L.; Wang, Y. Experimental research on pre-cracked marble under compression. *Int. J. Solids Struct.* **2005**,*42*, 2505–2516.
6. Chen, L.; Wang, C.P.; Liu, J.F.; Liu, Y.M.; Liu, J.; Su, R.; Wang, J. A damage-mechanism-based creep model considering temperature effect in granite. *Mech. Res. Commun.* **2014**, *56*, 76–82.
7. Zhang, Z.; Zhang, R.; Xie, H.; Liu, J.; Were, P. Differences in the acoustic emission characteristics of rock salt compared with granite and marble during the damage evolution process. *Environ. Earth Sci.* **2015**, *73*, 6987–6999.
8. Zhang, L.; Ren, M.; Ma, S.; Wang, Z.; Wang, J. Acoustic emission and fractal characteristics of marble during unloading failure process. *Chin. J. Rock Mech. Eng.* **2015**, *34*, 2862–2867.
9. Fu, X.; Xie, Q.; Liang, L. Comparison of the Kaiser effect in marble under tensile stresses between the Brazilian and bending tests. *Bull. Eng. Geol. Environ.* **2015**, *74*, 535–543.
10. Yin, L.J.; Gong, Q.M.; Ma, H.S.; Zhao, J.; Zhao, X.B. Use of indentation tests to study the influence of confining stress on rock fragmentation by a TBM cutter. *Int. J. Rock Mech. Min. Sci.* **2014**, *72*, 261–276.
11. Carpinteri, A.; Corrado, M.; Lacidogna, G. Heterogeneous materials in compression: Correlations between absorbed, released and acoustic emission energies. *Eng. Fail. Anal.* **2013**, *33*, 236–250.
12. Chmel, A.; Shcherbakov, I. A comparative acoustic emission study of compression and impact fracture in granite. *Int. J. Rock Mech. Min. Sci.* **2013**, *64*, 56–59.
13. Chen, G.-F.; Yang, S.-Q. Study on failure mechanical behavior of marble after high temperature. *Eng. Mech.* **2014**, *31*, 189–196.
14. Prassianakis, I.N.; Prassianakis, N.I. Ultrasonic testing of non-metallic materials: concrete and marble. *Theor. Appl. Fract. Mech.* **2004**, *42*, 191–198.
15. Sarpun, H.; Kilickaya, M.S.; Tuncel, S. Mean grain size determination in marbles by ultrasonic velocity techniques.*NDT E Int.* **2005**, *38*, 21–25.

16. Sáez-Pérez, M.P.; Rodríguez-Gordillo, J. Structural and compositional anisotropy in Macael marble (Spain) by ultrasonic, XRD and optical microscopy methods. *Constr. Build. Mater.* **2009**, *23*, 2121–2126.
17. Vasconcelos, G.; Lourenco, P.B.; Alves, C.A.S.; Pamplona, J. Ultrasonic evaluation of the physical and mechanical properties of granites. *Ultrasonics* **2008**, *48*, 453–466.
18. Aggelis, D.G.; Mpalaskas, A.C.; Matikas, T.E. Acoustic signature of different fracture modes in marble and cementitious materials under flexural load. *Mech. Res. Commun.* **2013**, *47*, 39–43.
19. Mpalaskas, A.C.; Matikas, T.E.; van Hemelrijck, D.; Iliopoulos, S.; Papakitsos, G.S.; Aggelis, D.G. Acoustic signatures of different damage modes in plain and repaired granite specimens. In Proceedings of the SPIE—The International Society for Optical Engineering, San Diego, CA, USA, 9–11 March 2015.
20. Skandalis marbles. Available online: http://www.skandalis.gr/el/products/marbles.html (accessed on 19 December 2015).
21. Neotex, S.A. Technical Data Sheet. Available online: http://www.neotex.gr/frontoffice/portal.asp?cpage=NODE&cnode=122&clang=1 (accessed on 15 December 2015).
22. Naik, T.R.; Malhotra, V.M.; Popovics, J.S. The Ultrasonic Pulse Velocity Method. In *CRC Handbook of Nondestructive Testing of Concrete*; Malhotra, V.M., Carino, N.J., Eds.; CRC: Boca Raton, FL, USA, 2004.
23. Kourkoulis, S.K.; Prassianakis, I.; Agioutantis, Z.; Exadaktylos, G.E. Reliability assessment of the NDT results for the internal damage of marble specimens. *Int. J. Mater. Prod. Technol.* **2006**, *26*, 35–56.
24. Tsangouri, E. Experimental Assessment of Fracture and Autonomous Healing of Concrete and Polymer Systems. Ph.D. Thesis, MEMC-Vrije Universiteit Brussel, Brussel, Belgium, 2015.
25. Aggelis, D.G.; Verbruggen, S.; Tsangouri, E.; Tysmans, T.; van Hemelrijck, D. Characterization of mechanical performance of concrete beams with external reinforcement by acoustic emission and digital image correlation.*Constr. Build. Mater.* **2013**, *47*, 1037–1045.
26. Lockner, D. The role of acoustic emission in the study of rock fracture. *Int. J. Rock Mech. Min. Sci. Geomech. Abstr.* **1993**,*30*, 883–889.

Chapter 9

Study of the Tensile Damage of High-Strength Aluminum Alloy by Acoustic Emission

Chang Sun [1], Weidong Zhang [1,†,*], Yibo Ai [1,†] and Hongbo Que [2]

[1] *National Center for Materials Service Safety, University of Science and Technology Beijing, Beijing 100083, China*

[2] *Qishuyan Institute Co., Ltd., China South Locomotive & Rolling Stock Corporation Limited, Changzhou 213011, China*

ABSTRACT

The key material of high-speed train gearbox shells is high-strength aluminum alloy. Material damage is inevitable in the process of servicing. It is of great importance to study material damage for in-service gearboxes of high-speed train. Structural health monitoring methods have been widely used to study material damage in recent years. This study focuses on the application of an acoustic emission (AE) method to quantify tensile damage evolution of high-strength aluminum alloy. First, a characteristic parameter was developed to connect AE signals with tensile damage. Second, a tensile damage quantification model was presented based on the relationship between AE counts and tensile behavior to study elastic deformation of tensile damage. Then tensile tests with AE monitoring were employed to collect AE signals and tensile damage data of nine samples. The experimental data were used to quantify tensile damage of high-strength aluminum alloy A356 to demonstrate the effectiveness of the proposed method.

Keywords: acoustic emission; aluminum alloy; tensile damage; quantification model

1. INTRODUCTION

The gearbox is one of the important parts of a high-speed train, which is subjected to damage in its service life. For example, the gearbox bracket area suffers static and dynamic loads during service. The static load is mainly caused by the weight and load of the train, while the dynamic load is related to a lot of factors including the natural wind, shock and vibration generated by operation of the train, the centrifugal force generated by curve movement of the train and so on. Static and dynamic loads bring both tensile damage and fatigue damage to gearbox bracket. Thus, it is of great importance to ensure the service safety of the gearbox. In this paper, high-strength aluminum alloy A356 is the key material of high-speed train gearbox shells. The performance analysis of the service process for this material has not yet been done [1,2]. It is necessary to develop a structural health monitoring method to study the tensile performance of this material.

In recent years, many studies have focused on developing methods to study in-service material performance by structural health monitoring techniques. Acoustic emission (AE), one of the efficient structural health monitoring techniques, which is defined as a phenomenon of rapid release of energy and generation of transient elastic wave from a localized source of the material [3]. Most of the material damage is related to microscopic processes involving some stress relaxation. Monitoring AE signals from a damaging material can give significant information about the microscopic mechanisms involved, because the magnitude of this stress relaxation is dependent upon the particular process [4].

The AE technique is widely used to detect the occurrence and growth of damage and quantify damage of in-service materials and is not limited by the materials. It has been identified that AE monitoring is capable of detecting material damage over the past three decades [5,6,7]. Previous studies have also proved that AE technique is sensitive and reliable in the detection of material damage for in-service structures [8,9]. The key of studying material tensile damage by using the AE technique is to relate AE parameters to tensile processes. Several attempts have been made to find the relationship between AE parameters and material damage during tensile processes. Haneef *et al.* [4] studied the tensile behavior of AISI type 316 stainless steel using AE and infrared thermography techniques. They discovered that AE root mean square voltage increased with an increase in strain rate due to the increase in source activation. The dominant frequency of the AE signals generated during different regions of tensile deformation has also been used to compare the results for different strain rates. They found that the dominant frequency of AE signals increased from elastic region to around 580–590 kHz during work hardening and 710–730 kHz around ultimate tensile strength for different strain rates, but they did not give an exact model to quantify tensile damage of the material. Lugo *et al.* [10] used AE to quantify the microstructural damage evolution under tensile loading for a 7075 aluminum alloy. They proved that the AE activity is related directly with damage progression in this alloy by building a model correlating AE counts to number density, but they focused on capturing ductile material failure. The difference between the elastic stage and plastic stage of tensile processes was ignored. Patrik *et al.* [11] monitored and analyzed the AE signal during tensile tests of pure Mg and Mg alloys of the AZ series in order to study the influence of alloy composition on plastic deformation. The Kaiser effect was used to determine the stability of the microstructure. They found the post-relaxation effect was sensitive to alloy composition and the strain at which the stress relaxation was performed, but they aimed to study the deformation mechanisms of plastic stage. The deformation mechanisms of the elastic stage were ignored. Godin *et al.* [12] used the AE technique to discriminate between the different types of damage occurring in a constrained composite. Two main types of signals were identified, originating from the two expected damage mechanisms *i.e.*, matrix cracking and decohesion. They used the k-means algorithm to split the AE data into two classes by counts, duration, and average frequency. Then k-nearest neighbors (KNN) method was used to classify the AE data. However, in their paper, AE signals were from composite materials, the AE signal of aluminum alloys is different. Cousland *et al.* [13] recorded AE signals during the unidirectional tensile deformation of aluminum alloys 2024 and 2124 to identify the sources of the emission. They concluded that the fracture of brittle inclusions in the primary source of the AE detected during the tensile testing of the alloys in the temper condition T351, but their main work was to find the source of AE signals during tensile fracture. They did not pay attention to quantify elastic deformation. Wen and Morris [14] investigated the effect of different thermal treatment temperatures (from 472 to 783 K) on the characteristics of serrated yielding of three commercial aluminum alloys, AA5052, AA5754, and AA5182 by using the AE technique. They discovered that the acoustic emission appears to be related to the number of Mg atoms actually participating in the dynamic strain aging process, but they concentrate on the yield stage of tensile process.

The AE technique is mainly applied to study plastic deformation during tensile processes. Bohlen *et al.* [15] observed and analyzed AE signals during plastic deformation of an AZ31 sheet in an H24 original condition, as well as after a heat treatment at elevated temperature. The AE count rates show a well-known correlation with the stress-strain curves. Máthis*et al.* [16] investigated mechanisms of plastic deformation of a commercial AM60 magnesium alloy by using AE measurements, TEM, and light microscopy. They found that the deformation behavior of the AM60 alloy exhibits three significant stages. Cakir *et al.* [17] measured the AE response of an implant-quality 205L stainless steel during slow strain rate tensile testing at a constant strain rate of 7.35×10^{-6}. They believed the attenuation of the AE activity beyond necking is attributed to the localization of plastic deformation. Vinogradov *et al.* [18] performed AE measurements during room temperature tensile deformation of high-alloyed cast model steels with different austenite stability to get a better understanding of the kinetics of TRIP/TWIP-assisted plastic deformation. They identified four different microstructure-related major mechanisms of plastic deformation as AE sources. Kocich *et al.* [19] discovered the special character of AE signal during plastic deformation, which can be called white noise with low energy. They used UFG materials to identify the limit of detectability. Therefore, the AE technique is available and effective for studying tensile deformation and quantifying tensile damage of materials. Presently, articles focused on quantifying elastic deformation during tensile process by the AE technique are relatively few. There is a need to develop an AE method to quantify elastic deformation of tensile damage.

In this paper, a method of tensile damage quantification by the AE technique was presented. The proposed method was based on the relationship between AE counts and tensile damage. This research focused on quantifying the elastic deformation of tensile damage. The method presented in this paper was a prognostic method only if data obtained from tensile tests was applied. The application of the developed method in high-strength aluminum alloy A356 of high-speed train gearbox shells was given to demonstrate the effectiveness of the proposed method.

The paper is organized as follows. Experimental procedures are described in Section 2. Section 3 then introduces the theoretical method for tensile damage quantification. The application of the proposed method is also introduced in this section. Conclusions will be discussed and summarized in the last section.

2. EXPERIMENTAL PROCEDURES

An AE instrument was used to detect and record AE signals during tensile tests. Tensile tests with AE monitoring contained two parts of data, which are tensile damage data and AE monitoring data. The purpose of this study is to analyze the relationship between tensile damage and AE signals to quantify tensile damage of the high-strength aluminum alloy A356 by using AE signals.

2.1. Tensile Test

The material used in this study is high-strength aluminum alloy A356, which has reached excellent properties of rigidity, strength, ductility, fatigue, casting, and shrinkage tendency [20]. Its nominal chemical composition is presented in Table 1.

Table 1. Chemical compositions (wt. %) of the improved aluminum alloy type A356.

Composition	Si	Mg	Ti	Sr
wt. %	6.5~7.5	0.20~0.35	0.08~0.2	0.005~0.015

An 810 Material Test System (MTS) was used to carry out tensile tests. The strain rate was 10^{-4}/s, and test environment was 27 °C and 40% RH. There are a total of nine tensile samples, which were numbered S1 to S9. Figure 1 shows the geometry of the sample [21]. During the test, increased longitudinal tensile force was applied to the sample. Samples were broken in the end.

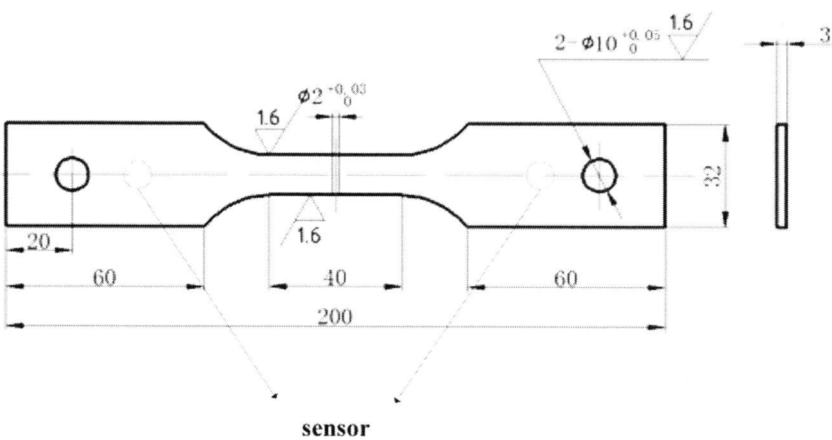

Figure 1. Geometry of tensile sample and arrangement of AE sensors, dimensions in mm [21].

2.2. Acoustic Emission Technology

AE is defined as "a phenomenon of rapid release of energy and generation of transient elastic wave from a localized source of the material" [3]. AE signals are generated from the sudden release of strain energy at the damage sources, which are plastic deformation, crack propagation, wear, friction, and so on [12,22]. AE registration is an effective methodology, which allows "hearing" and registering damage during loading of samples [12].

An AE instrument PCI-2 supplied by American PCA company was performed to record and process AE signals during tensile tests. AE signals were recorded from tensile tests. During elastic deformation of tensile process the AE signals are mainly of continuous type signals. Typical parameters of AE signal are count, energy, amplitude, duration, rise time, and so on. The most commonly used AE parameter is count, which is defined as the number of times that the AE signal amplitude exceeds a predefined subjective threshold value [23]. Figure 2 shows some typical parameters in a burst-type waveform of AE signal.

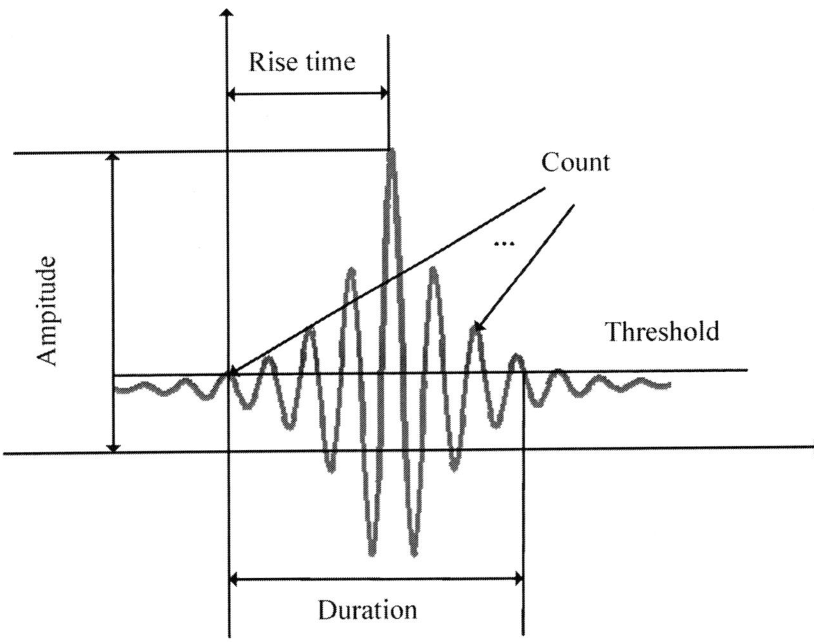

Figure 2. Typical parameters of AE signal.

The sensor used in tensile test was standard resonant sensor of type R15A, a 40dB pre-amplification was employed to amplify signals. For all the nine samples Vaseline was used to mount the sensor to the sample surface to ensure the extreme sensitivity of the sensor. Figure 1 shows the arrangement of sensors. Threshold values were set to remove noise, the fixed threshold (trigger level) value was 45 dB in all tests; the energy threshold value was 1. The sampling frequency is 1 MHz; peak identification time (PDT) was 300 μs; impact identification time (HDT) was 600 μs; hit lockout time (HLT) was 1000 μs; and the crash file length was 2kB.

3. RESULTS AND DISCUSSION

In this section, a characteristic parameter was developed to connect AE signals with tensile processes. Then, a tensile damage quantification model was presented based on the relationship between AE counts and tensile behavior to quantify tensile damage evolution of high-strength aluminum alloy A356. This paper focused on quantifying the elastic deformation of tensile damage.

3.1. Material Tensile Damage

The tensile process of metal can be divided into four stages; elastic stage, yield stage, plastic stage, and fracture stage [20].Figure 3 displays the four stages of tensile process identified on the axial load-axial elongation curve. The four stages can be divided according to different characteristics. At the elastic stage, the material deformation is elastic, the force and elongation is proportional. The elongation will disappear with the force gone, and no residual elongation will occur. When the in-service material is in the elastic stage, it can be considered safe [21], but when the material comes into the yield stage, unrecoverable deformation will be produced. The reliability of the material will be reduced, and the final fracture will come quickly. Thus, the in-service material can be considered unsafe. In this paper, the transition point from elastic stage to yield stage is the failure point of the tensile process [21].

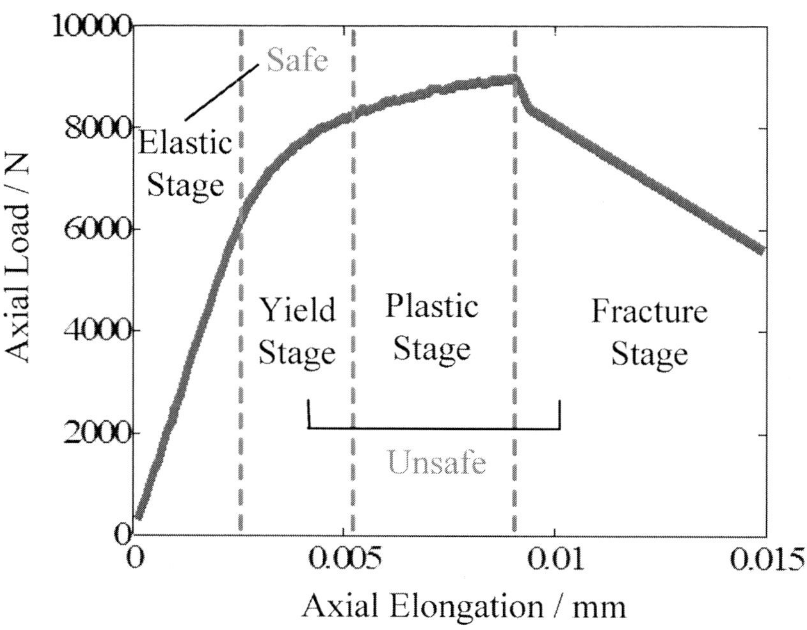

Figure 3. The axial load-axial elongation curve of the gearbox shell material.

If in-service materials reach the failure point of the tensile process, catastrophic failure may occur. It is of great importance to study tensile damage of in-service materials, especially the elastic stage of the tensile process. This paper aims to quantify the elastic deformation of material by the AE technique.

3.2. Characteristic of AE Signal for the Tensile Damage

Figure 4 displays the relationship between AE counts and axial elongation in the same tensile process. It is indicated that AE counts increased significantly at the transition point from elastic stage to yield stage [21]. AE counts can be used to characterize the tensile damage, but the AE counts of each sample at the transition point are different, and the values have a great range. Thus, it is necessary to develop a characteristic parameter involving AE counts to find the transition point between the elastic stage and yield stage of the tensile process. The characteristic parameter is the key to relate tensile damage with AE signals.

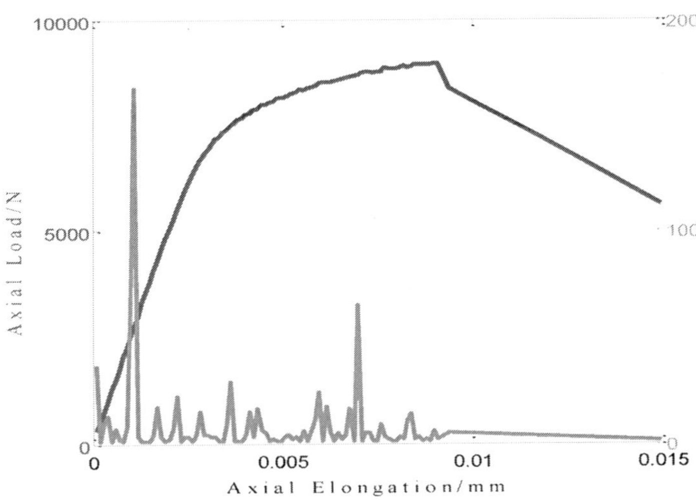

Figure 4. AE counts and axial elongation curve.

In this paper, the characteristic parameter is named *AC*(*t*). It can be calculated by Equations (1) and (2):

$$M_j(t) = \text{median}_m\left(\frac{t_{j+1}-t_j}{c_j} + \frac{t_{j+2}-t_{j+1}}{c_{j+1}} \cdots \frac{t_{j+m}-t_{j+m-1}}{c_{j+m-1}}\right) \qquad j = 1, 2, \cdots, n \tag{1}$$

$$AC_j(t) = M_{j+1}(t) - M_j(t) \quad (2)$$

where c_j is the AE count at t_j time; m is the quantity of values used to calculate the median of them; j is the AE count number. Once the value of $AC(t)$ comes into the preset threshold interval, it means that the tensile process reaches the transition point from elastic stage to yield stage. In other words, the duration of the elastic stage t_f can be obtained.

AE signals were collected and recorded from tensile tests with the AE instrument. The characteristic parameter $AC(t)$ was calculated based on Equations (1) and (2). Herein, m = 10 and the threshold interval was [7.65×10^{-3}, 7.85×10^{-3}], Once the value of $AC(t)$ reached the preset threshold interval, the duration of the elastic stage t_f can be obtained. Table 2 shows the characteristic parameter $AC_f(t)$ at the transition point and duration of the elastic stage t_f of all nine samples.

Table 2. The results of characteristic parameter $AC(t)$ and duration of the elastic staget_f.

Sample	t_f/s	$AC_f(t)$
S1	88.6	0.000780
S2	73.0	0.000769
S3	88.9	0.000779
S4	95.8	0.000782
S5	122.0	0.000773
S6	129.0	0.000781
S7	111.8	0.000777
S8	159.0	0.000775
S9	92.6	0.000776

3.3. Tensile Damage Quantification Model

A tensile damage quantification model was presented to quantify the remaining time of the elastic stage. The actual remaining time of the elastic stage in tensile process $TR_j(t)$ can be described by Equation (3):

$$TR_j(t) = t_f - T_j(t) \qquad j = 1, 2, \cdots, n_1 \tag{3}$$

where $T_j(t)$ is the tensile test running time; j is the AE count number.

The remaining time of elastic stage in tensile process calculated from AE signals $TPF_j(t)$ can be described by the following function:

$$TPF_j(t) = TPM_j + TDM_j = am \cdot \exp(bm \cdot cn_j) + fm \cdot cn_j^2 + gm \cdot cn_j + hm \tag{4}$$

where am, bm, fm, gm and hm are parameters needed to be estimated; cn_j is the normalized cumulative AE counts.

3.4. Model Parameters Estimation

AE signals were collected and recorded from tensile tests with AE instrument. There were nine samples numbered from S1 to S9. Eight samples were selected randomly as training data, while the remaining sample remained as testing data. Training data were used to estimate parameters of the elastic stage remaining time quantification model. Testing data were used to verify the effectiveness of the model. If the testing sample was S9, then the following steps should be applied to get model parameters by using training samples S1–S8.

Step 1. There are a total of eight samples in training data. According to the relationship between cumulative counts and the remaining time, the remaining time $TP_{i,j}(t)$ Can be described as Equation (5):

$$TP_{i,j} = a_i \cdot \exp(b_i \cdot cn_{i,j}) \qquad i = 1, 2, \cdots, 8 \qquad j = 1, 2, \cdots, n_2 \tag{5}$$

where $cn_{i,j}$ is the normalized cumulative counts; i is the sample number; j is the AE count number; a_i and b_i are parameters obtained by performing a linear least squares regression, herein, $TP_{i,j} = TR_{i,j}$. Therefore, eight groups of (a_i, b_i) can be obtained.

Step 2. Set cn = [1,2,...,M], herein, M = 1000. Calculate $TP_{i,j}'$ based on Equation (5), cn, a_i and b_i. Each element in cncorresponds to 8 groups of $TP_{i,j}'$. Calculate the mean value of eight groups of $TP_{i,j}'$. The quantity of mean values is M. Then get the values of parameters am_9 and bm_9 based on calculated mean values and cn by performing a linear least squares regression. Parameters am_9 and bm_9 are taken as the parameter of sample S9. Thus, parameters am_9 and bm_9 in $TPM_{9,j}(t)$ can be obtained:

$$TPM_{9,j} = am_9 \cdot \exp(bm_9 \cdot cn_{9,j}) \qquad j = 1,2,\cdots,n_3 \tag{6}$$

Step 3. Samples S1–S8 are training samples. Select samples S1–S7 to calculate a_k and b_k according to Equation (5) by performing a linear least squares regression, herein, k = 1,2,...,7, $TP_{k,j} = TR_{k,j}$. Calculate $TP_{k,j}'$ based on Equation (5), cn, $a_k$and b_k. Each element in cn corresponds to 7 groups of $TP_{k,j}'$. Calculate the mean value of 7 groups of $TP_{k,j}'$. The quantity of mean values is M. Then get the values of parameters ad_8 and bd_8 based on calculated mean values and cn by performing a linear least squares regression. So parameters ad_8 and bd_8 in can $TPD_{8,j}(t)$ be obtained:

$$TPD_{8,j} = ad_8 \cdot \exp(bd_8 \cdot cn_{8,j}) \qquad j = 1,2,\cdots,n_4 \tag{7}$$

Step 4. The variance between $TPD_{8,j}(t)$ and $TR_{8,j}(t)$ can be given by Equation (8). According to the relationship between cumulative counts and $TD_{8,j}(t)$, $TD_{8,j}(t)$ can be described as Equation (9):

$$TD_{8,j} = TR_{8,j}(t) - TPD_{8,j}(t) \qquad j = 1,2,\cdots,n_4 \tag{8}$$

$$TD_{8,j} = f_8 \cdot cn_{8,j}^2 + g_8 \cdot cn_{8,j} + h_8 \qquad j = 1,2,\cdots,n_4 \tag{9}$$

where $cn_{8,j}$ is the normalized cumulative counts; f_8, g_8 and h_8 are parameters obtained by performing a linear least squares regression. Repeat Step 3 and Step 4 seven times to get (f_i, g_i, h_i); herein, i = 1,2,...,7. Hence, eight groups of (f_i, g_i, h_i) can be obtained.

Step 5. There are eight groups of (f_i, g_i, h_i). Calculate $TD_{i,j}'$ based on Equation (9), cn, f_i, g_i and h_i. Each element in cncorresponds to eight groups of $TD_{i,j}'$. Calculate the mean value of eight groups of $TD_{i,j}'$. The quantity of mean values is M. Then get the values of parameters fm_9, gm_9, and hm_9 based on calculated mean values and cn by performing a linear least squares regression. Thus, parameters fm_9, gm_9, and hm_9 in $TDM_{9,j}(t)$ can be obtained:

$$TDM_{9.j} = fm_9 \cdot cn_{9.j}^2 + gm_9 \cdot cn_{9.j} + hm_9 \qquad j = 1,2,\cdots,n_3 \tag{10}$$

Similarly, model parameters of samples S1–S8 can be calculated according to step1–step5. Figure 5 shows the values of parameter am_l and bm_l, l = 1,2,…,9. Figure 6 shows the values of parameter fm_l, gm_l, and hm_l, l = 1,2,…,9.

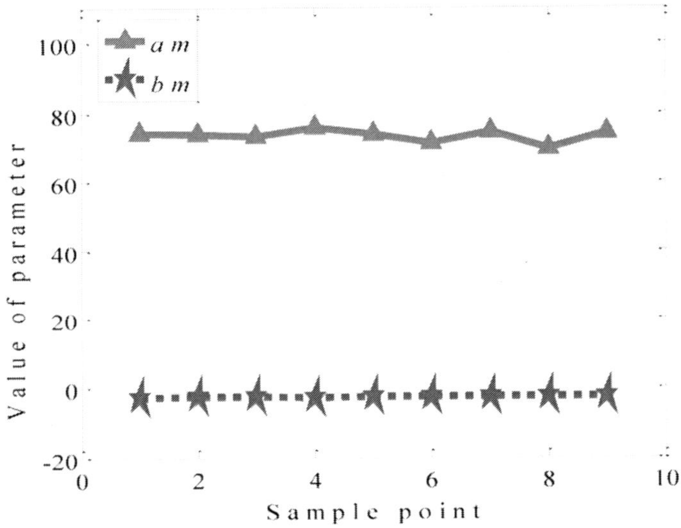

Figure 5. Values of parameter *am* and *bm*.

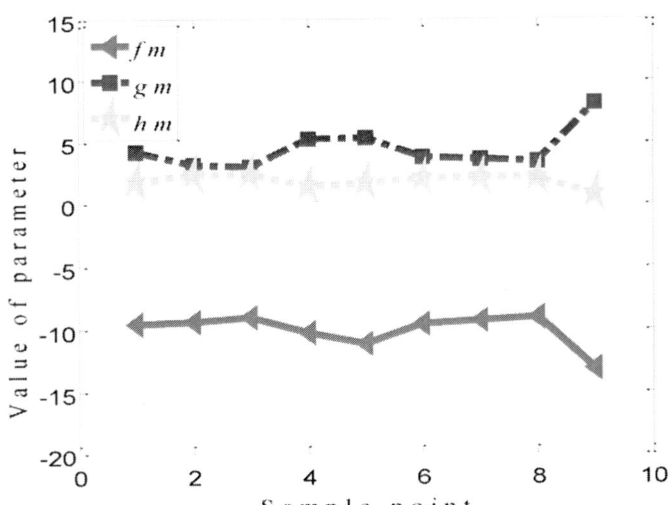

Figure 6. Values of parameter *fm*, *gm*, and *hm*.

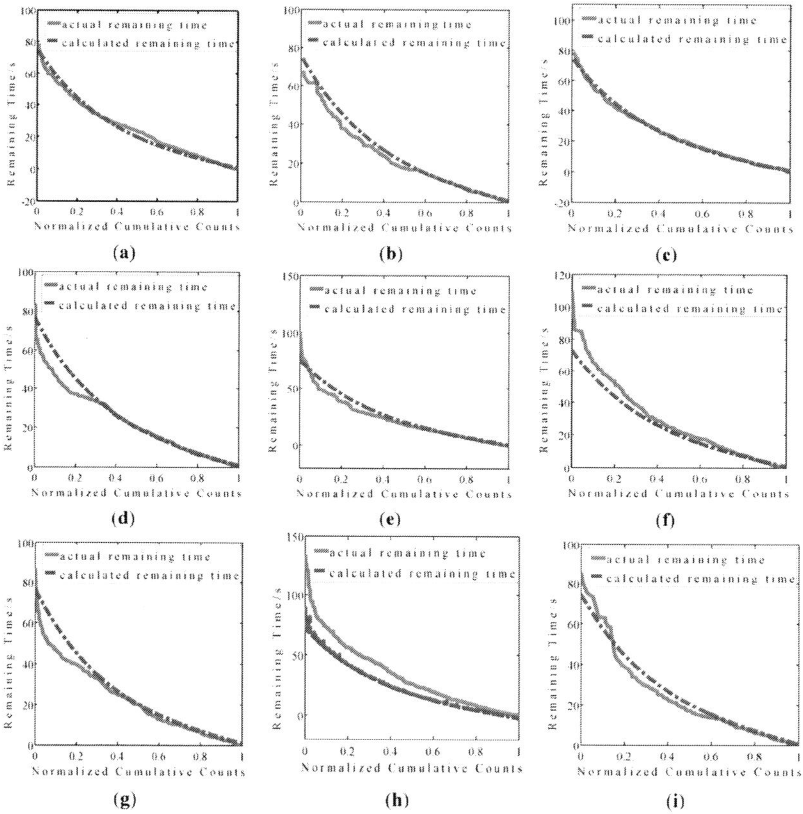

Figure 7. Comparison between the remaining time of elastic stage calculated from AE data and actual remaining time of the elastic stage in S1–S9. (**a**) S1; (**b**) S2; (**c**) S3; (**d**) S4; (**e**) S5; (**f**) S6; (**g**) S7; (**h**) S8; and (**i**) S9.

3.5. Results and Verification of the Model

Once the values of parameter am_l, bm_l, fm_l, gm_l, *and* hm_l were obtained, the remaining time of samples can be calculated based on Equation (4). Figure 7 displays the comparison between the remaining time of the elastic stage calculated from AE data and actual remaining time of the elastic stage in all samples.

It is indicated in Figure 7 that the proposed model can be used to quantify remaining time of the elastic stage of high-strength aluminum alloy A356. The error between the calculated remaining time of the elastic stage and the actual remaining time is getting smaller when the material is close to the failure point.

Table 3 shows the variance between remaining time of elastic stage calculated from AE data and actual remaining time. It is indicated in Table 3 that the maximum absolute error is 59.9 s, the maximum average absolute error is 11.0 s for all samples. The average error in Table 3 refers to the mean of variance between the remaining time calculated from AE data and the actual remaining time of all the data points in each sample. Average errors of all nine tensile samples are acceptable. In other words, the proposed tensile damage quantification model is effective to quantify material tensile damage evolution of high-strength aluminum alloy A356.

Table 3. Variance between the predicted and actual remaining life.

Sample	Maximum Error/s	Minimum Error/s	Average Error/s
S1	5.7	0.001	2.0
S2	7.9	0.003	2.6
S3	4.6	0.003	1.0
S4	14.5	0.001	3.3
S5	25.0	0.144	3.4
S6	36.5	0.001	4.1
S7	14.8	0.002	3.8
S8	59.9	2.046	11.0
S9	10.8	0.224	3.2

4. CONCLUSIONS

In this paper, a tensile damage quantification model based on the relationship between AE counts and tensile behavior was presented to quantify material tensile damage evolution of high-strength aluminum alloy A356. Specialized tensile tests with AE monitoring were developed and carried out to verify the validity of the proposed model. Based on this study, the following conclusions can be drawn as follows:

1. The correlation between tensile damage and AE signals was established by characteristic parameter $AC(t)$, which can be used to monitor material elastic deformation of tensile damage.

2. The proposed model is effective to quantify elastic deformation of tensile damage of high-strength aluminum alloy A356 of high-speed train gearbox shells.

3. Cumulative counts, as one of the most commonly-used AE parameters, can be performed in combination with the proposed model to provide warning signs for gearbox of high-speed trains when tensile damage comes to the failure point, where the final fracture will be attained quickly, and catastrophic failure may occur.

4. The method presented in this paper was a prognostic method only if data obtained from tensile tests is applied. In other words, the proposed elastic stage remaining time quantification model in this paper is offline. Hence, building an online elastic stage remaining time prediction model is work that needs to be done in the future.

ACKNOWLEDGMENTS

The authors would like to acknowledge the financial support provided by the National Natural Science Foundation of China (Grant No. 61273205), the Fundamental Research Funds for the Central Universities of China (Grant No. FRF-SD-12-028A), the 111 Project (Grant No. B12012).

AUTHOR CONTRIBUTIONS

Chang Sun and Weidong Zhang conceived and designed the study. Chang Sun and Yibo Ai performed the experiments. Hongbo Que provided the basic performance analysis of the material. Chang Sun made the data analysis. Chang Sun and Weidong Zhang wrote the paper. All authors read and approved the manuscript.

REFERENCES

1. Wang, B.M. *Overall System and Bogie of High-Speed Motor Train Units*; Southwest Jiaotong University Press: Chengdu, China, 2008.
2. Li, X.J. CRH bullet trains overview. *Railw. Tech. Superv.* **2007**, *9*, 26–28.
3. Yang, M.W. *Acoustic Emission Testing*; Machinery Industry Press: Beijing, China, 2005.
4. Haneef, T.; Lahiri, B.B.; Bagavathiappan, S.; Mukhopadhyay, C.K.; Philip, J.; Rao, B.P.C.; Jayakumar, T. Study of the tensile behavior of AISI type 316 stainless steel using acoustic emission andinfrared thermography techniques. *J. Mater. Res. Technol.* **2015**, *137*, 1–13.
5. Jalaj, K.; Sony, P.; Mukhopadhyay, C.K.; Jayakumar, T.; Vikas, K. Acoustic emission during tensile deformation of smooth and notched specimens of near alpha titanium alloy. *Res. Nondestruct. Eval.* **2012**, *23*, 17–31.
6. Luo, X.; Haya, H.; Inaba, T.; Shiotani, T.; Nakanishi, Y. Damage evaluation of railway structures by using train-induced AE. *Constr. Build Mater.* **2004**, *18*, 215–223.
7. Re, V.D. Acoustic emission applications for defect detection in steels and GFRP. *Int. J. Mater. Prod. Technol.* **1988**, *3*, 38–53.
8. Chen, H.L.; Choi, J.H. Acoustic emission study of fatigue cracks in materials used for AVLB. *J. Nondestruct. Eval.* **2004**,*23*, 133–151.

9. Johnson, M. Waveform based clustering and classification of AE transients in composite laminates using principal component analysis. *NDT E Int.* **2002**, *35*, 367–376.
10. Lugo, M.; Jordon, J.B.; Horstemeyer, M.F.; Tschopp, M.A.; Harris, J.; Gokhale, A.M. Quantification of damage evolution in a 7075 aluminum alloy using an acoustic emission technique. *Mater. Sci. Eng. A* **2011**, *528*, 6708–6714.
11. Patrik, D.; Jan, B.; Frantisek, C.; Pavel, L.; Dietmar, L.; Karl, U.K. Acoustic emission during stress relaxation of pure magnesium and AZ magnesium alloys. *Mater. Sci. Eng. A* **2007**, *462*, 307–310.
12. Godin, N.; Huguet, S.; Gaertner, R.; Salmo, L. Clustering of acoustic emission signals collected during tensile tests on unidirectional glass/polyester composite using supervised and unsupervised classifiers. *NDT E Int.* **2004**, *37*, 253–264.
13. Cousland, S.M.; Scala, C.M. Acoustic emission during the plastic deformation of aluminum alloys 2024 and 2124.*Mater. Sci. Eng.* **1983**, *57*, 23–29.
14. Wen, W.; Morris, J.G. An investigation of serrated yielding in 5000 series aluminum alloys. *Mater. Sci. Eng. A* **2003**,*354*, 279–285.
15. Bohlen, J.; Chmelík, F.; Dobrŏn, P.; Kaiser, F.; Letzig, D.; Lukáč, P.; Kainer, K.U. Orientation effects on acoustic emission during tensile deformation of hot rolled magnesium alloy AZ31. *J. Alloys Compd.* **2004**, *378*, 207–213.
16. Máthis, K.; Chmelík, F.; Janeček, M.; Hadzima, B.; Trojanová, Z.; Luká, P. Investigating deformation processes in AM60 magnesium alloy using the acoustic emission technique. *Acta Mater.* **2006**, *54*, 5361–5366.
17. Cakir, A.; Tuncell, S.; Aydin, A. AE response of 205L SS during SSR test under potentiostatic control. *Corros. Sci.* **1999**,*41*, 1175–1183.
18. Vinogradov, A.; Lazarev, A.; Linderov, M.; Weidner, A.; Biermann, H. Kinetics of deformation processes in high-alloyed cast transformation-induced plasticity/twinning-induced plasticity steels determined by acoustic emission and scanning electron microscopy: Influence of austenite stability on deformation mechanisms. *Acta Mater.* **2013**, *61*, 2434–2449.
19. Kocich, R.; Cagala, M.; Crha, J.; Kozelsky, P. Character of acoustic emission signal generated during plastic deformation. In Proceedings of the 30th European Conference on Acoustic Emission Testing & 7th International Conference on Acoustic Emission, Granada, Spain, 12–15 September 2012; pp. 1–8.

20. Zhang, W.D.; Zhang, X.W.; Yang, B.; Ai, Y.B. Damage characterization and recognition of aluminum alloys based on acoustic emission signal. *J. Univ. Sci. Technol. Beijing* **2013**, *35*, 626–633.
21. Ai, Y.B.; Sun, C.; Que, H.B.; Zhang, W.D. Investigation of material performance degradation for high-strength aluminum alloy using acoustic emission. *Metals* **2015**, *5*, 228–238.
22. Okafor, A.C.; Natarajan, S. Acoustic emission monitoring of tensile testing of corroded and un-corroded clad aluminum 2024-T3 and characterization of effects of corrosion on AE source events and material tensile properties.*AIP Conf. Proc.* **2014**, *1581*, 492–500.
23. Keshtgar, A.; Modarres, M. Acoustic emission-based fatigue crack growth prediction. In Proceedings of the Reliability and Maintainability Symposium: Product Quality & Integrity, Orlando, FL, USA, 28–31 January 2013; pp. 1–5.

Chapter 10

Acoustic Emission in Drying Materials

Stefan Jan Kowalski[1], Jacek Banaszak[1] and Kinga Rajewska[1]

[1] Poznań University of Technology, Institute of Technology and Chemical Engineering, Department of Process Engineering, Poznań, Poland

1. INTRODUCTION

Drying of wet materials is one of the oldest and most common unit operation found in diverse processes such as those used in the agricultural, ceramic, chemical, food, pharmaceutical, pulp and paper, mineral, polymer, and textile industries. It is also one of the most complex and least understood operations because of the difficulties and deficiencies in mathematical descriptions of the phenomena of simultaneous – and often coupled and multiphase – transport of heat, mass, and momentum in saturated porous materials. Drying is therefore an amalgam of science, technology, and art, or know-how based on extensive experimental observations and operating experience [Strumiłło, 1983; Mujumdar (Ed.), 2007].

Drying processes ought to be appropriately arranged and operated to obtain a high quality dried products, that is, products without excessive deformations, surface cracks, and above all crosswise fractures. A non-uniform moisture distribution in products arising during drying causes a non-uniform material shrinkage and generates stresses, which are responsible for permanent deformations and material fracture. A risk of fracture in drying samples is possible to analyze both theoretically and experimentally. Mechanistic drying models make the basis for numerical simulations of drying kinetics and analysis of the drying induced stresses [Kowalski, 2003]. In this way it is possible to determine the spots, where the drying induced stresses reach maximum and the possibly crack may occur [Kowalski and Rybicki, 2007]. The theoretical predictions are confronted with the experimental data obtained due to application of the acoustic emission method (AE), which enables monitoring *on line* the development of the drying induced fractures caused by stresses [Kowalski et al., 2000].

The acoustic emission method (AE) is a non-destructive method allowing indirect control of micro- and macro-fracture development during drying and above all the identification of the period and also the place where the fractures start to develop. In this sense the AE is a method that enables control of drying process and help to protect the material against destruction [Kowalski, 2010]. Thus, the EA provides a unique advantage of early detection of subcritical crack growth and recognize when and where the crack is growing. The cracks and deformations arising inside dried materials constitute the AE source. The intensity of AE signals, their number and energy inform about the state and magnitude of stresses [Kowalski, (2002), Kowalski et al., (2004)].

The aim of this chapter is to show the possibly using the AE method to diagnostic purposes of destruction due to monitoring materials subjected to drying. The results of the tests obtained from convective and microwave drying of ceramic and wood materials carried out in the laboratory drier equipped with the acoustic emission set-up constitute the illustrative material of this chapter.

The example under analysis concern cylindrical samples made of kaolin and wood. Based on the mechanistic drying model, the stress distribution in the samples and its evolution in time were determined. In this way the moment at which the stresses reach the critical value causing material damage was appointed [Kowalski et al. 2012]. The system of double coupled differential equations of this model, adopted to the cylindrical geometry, was solved numerically with the help of the finite element (FEM) and the finite difference (FDM) methods. Due to AE method, the number and the energy of AE hits were measured, and the crest value of acoustic waves was appointed, and these data enabled validation of the theoretical predictions. A good adherence of the theoretical and experimental results serves for identification of fractures occurring in materials during drying.

2. THE ESSENCE OF ACOUSTIC EMISSION (AE) IN DRYING

2.1. AE Descriptors

- Different regimes of compresional acoustic waves propagating through the material from the crack places to the AE detectors attached to the samples, can be identified through the proper choice of the AE descriptors. The descriptors suitable to assessment of mechanical phenomena occuring in drying materials are selected mostly to be: *the number of acoustic emission hits*, *hit rate* (showing the dynamic of the process), *the maximum energy of hits*, and *the crest value* (showing the power of AE signals).
- When applying the AE method to drying processes, a suitable selection of AE descriptors that let to obtain the most useful information about the

phenomena occurring in dried materials is an important issue. The parameters characterizing the AE signals that are recorded by the detector inform about intensity and possible size of destruction, and therefore are significant for the precise assessment of the AE occurrence. So, the appointing of the descriptors which qualitatively fit best for description of the AE occurrence is a responsible and difficult task.

- Based on the authors' experience and the performed up to now experiments, it was stated that AE descriptors best reflecting the character of mechanical phenomena occurring in drying materials, are:
- *Hits rate.* This descriptor shows the dynamics of the destruction development (e.g. a rise of temperature drying involves rapid growth of the AE hits rate). Moreover, this descriptor indicates the stages of drying, in which the reduction or increase of the AE activity takes place.
- *The hit of maximum energy.* This descriptor is more useful than that "energy of hits" as it shows the single hit with maximum energy in a given time interval. The descriptor "energy of hits" presents the energy of all hits in a given time interval.
- *Crest value.* This descriptor presents the intensity of hits in time. It is a very significant parameter illustrating the "power" of existing hits.
- *The total number of hits and the total energy of hits.* These parameters show some individual phenomena occurring during drying. Thanks to these descriptors it is possible to distinguish stages of drying in which some irregular changes of the AE energy or the AE hits rate appear. These descriptors point out the critical moments of drying, in which the fracture of drying material may occur.

2.2. Calibration of AE Energy

The registered by the equipment AE signals are characterized by two fundamental parameters: the amplitude and the time of signal duration. The relative energy of each signal, called also the acoustic energy of AE signal, is possible to determine integrating the surface under the envelope curve of this signal. It is a relevant characteristics of the magnitude and power of the AE signal source. By application of the AE method to monitoring of drying processes, which are characterized with constant reduction of moisture content (MC) in dried materials, it should be taken into account that the AE energy depends on the material MC. If assume that the sources of AE signals cause cracks of a similar size, however, occurred first in wet material and next in dry one, then, the registered by the equipment AE signals will be different for these two events. This follows from damping of the acoustic waves propagating though a not perfectly elastic material. It is obvious that a more saturated material characterizes with stronger damping properties than an unsaturated one. Therefore, it is essential to take into account the damping effects by analysis of the AE energy descriptors.

It is necessary then to carry out the calibration of the AE energy for each examined material in dependence on its MC. There is a number of calibration methods (Malecki and Ranachowski 1994,Banaszak and Kowalski 2010). Here, the mechanical method is presented, that is, the method of falling ball (Berlinsky at al. 1990, Luong Phong 1994). The calibration was carried out with the use of the equipment presented in figure 1.

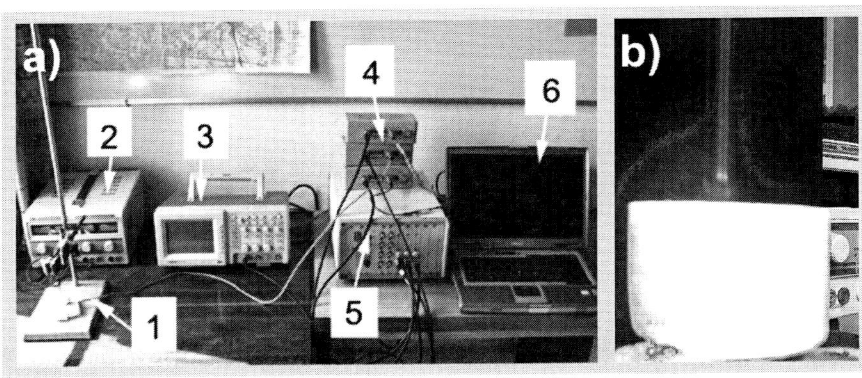

Figure 1. AE energy calibration set-up: a)1 – the ball's releasing mechanism and AE sensor, 2 – AC power supplier, 3 – oscilloscope, 4 –preamplifier, 5 – AE acquisition system AMSY5, 6 – computer, b) impact of dropping ball at the upper kaolin sample surface.

A still ball of mass *m* = 5.60 g and diameter *d* = 11 mm was used as a source of acoustic signals having known and constant energy. The ball was planted in the grip with release mechanism, being the electromagnet connected to the power supply adaptor of direct current (2). The ball was situated at height of 10 cm above the upper surface of the sample. The AE sensor (1) was attached to the bottom surface of the sample. The registered AE signals were conveyed through the preamplifier (4) to the module unit AMS-5, where they were processed by means of control unit (6). The digital oscilloscope (3) was connected to the system to analyze the course of signal appearance and to assign the level of noise.

The release from the grip ball hit the upper surface of the sample (Fig. 1b) and generated elastic wave, which experienced damping when propagating through the not perfectly elastic sample. After its arriving to the AE sensor, it became converted into the electric signal and next undergone a suitable energetic analysis. Each test was repeated five times for each sample of given MC, and the average value was taken for further considerations. On the basis of those tests the curve of attenuation of the AE energy was determined as a function of the material MC. Figure 2 presents the results of the research carried out for the kaolin and the walnut wood.

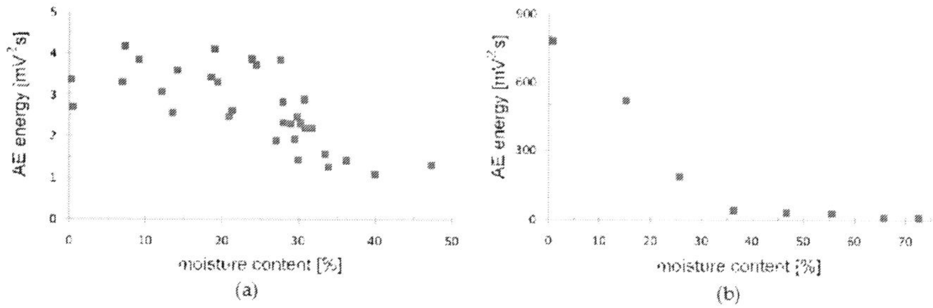

Figure 2. Damping of EA energy in dependence of material MC determined in the test of falling ball: a) kaolin, b) walnut wood.

Damping of the AE signal energy depends strictly on the material moisture content. Kaolin material becomes plastic for the moisture content over 27-29% and thus it stronger attenuate the AE signals than that unsaturated one. For the walnut wood the attenuation of acoustic waves becomes very strong for the moisture contents above the fiber saturation point (FSP) (ca. 30%), and there remain on a constant level. Dry wood is a very acoustic material and the energy of AE signal in such a material is weakly attenuated. Alongside with the increase of the MC up to the fiber saturation point the damping of EA waves increases radically. Over this critical MC wood stops to be acoustic material and utters characteristic hollow sounds. It should be noted a very intensive linear drop of energy in the range below the fiber saturation point, what means that wood is a material very sensitive to changeability of the moisture content, for example, music instruments made of wood should be always adjusted to the actual air humidity.

There is a necessity of suitable correction of the energetic values of AE signals received from the measurement set-up, dependent on the actual MC of the material. For this purpose it is necessary to construct the calibration curve of AE energy for the tested material. It can be constructed by the best adjustment of the theoretical curve expressed by the fourth order polynomial to the experimental data. (Fig. 2).

Figure 3 presents the results of measurements of the mean energy of AE events for kaolin and walnut wood during drying with and without taking into account the calibration curve.

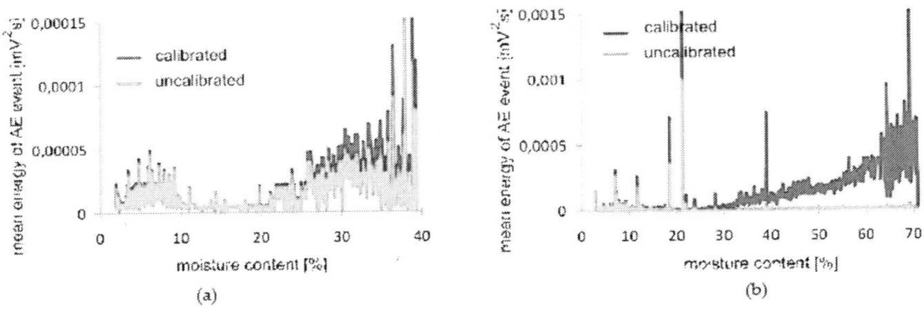

Figure 3. Comparison of mean energy of a AE hits as the function of material MC with and without taking into account of the damping effects during drying: a) kaolin, b) walnut wood.

The curve of calibrated AE energy for kaolin has not a significant impact on the final results by the analysis of the AE events. The character of plots with and without taking into account the calibration curve is similar. It follows from insignificant difference in AE signal attenuation for saturated and unsaturated materials. For walnut wood, however, at the beginning of drying (ca. 70% MC) there were registered low energetic AE signals only. High energetic signals appear else by about 20% MC. Taking into consideration the fact that wood strongly attenuates the AE signal with increase MC, then, it can be noticed (Fig. 3b) that the real annotated AE energy, measured with using calibration curve, is significantly higher. It influences then the analysis of AE signals for MC above the fiber saturation point (30%). Some signals arriving to the AE sensor are high energetic ones even for MC 70% to 60%. The initial high energetic signals for such a high MC originate from sample heating. For example, the AE signal registered for 39% MC originates from a crack in the sample. This AE event could be identified else after taking into account in the analysis the calibration curve presenting the effects of AE energy damping. The other high energetic signals, connected with the successive cracks in the sample, were registered for the sample with MC below the fiber saturation point. They are clearly visible both for the calibrated and not calibrated energy with respect to the material MC.

The presented above results show how important is taking into account the damping of AE signals in studies of drying processes in which the AE method is used for monitoring of the mechanical effects, particularly in wood. Without such an approach a part of the AE signals can be wrongly interpreted, for example, some crack occurrence could not be noticed.

3. EXPERIMENTAL SETUP EQUIPPED WITH AE MEASUREMENT INSTRUMENTS

3.1. Scheme of The Equipment

The drying tests were realized in the laboratory hybrid drier in the Department of Process Engineering, Institute of Technology and Chemical Engineering, Poznań University of Technology. Presented in this chapter experimental results has been taken for over a decade and according to technical progress our measurement instruments obviously has changed few times during those years. Figure 4 presents the photograph of the latest dryer version equipped with the acoustic emission (AE) system.

Figure 4. Photo of the laboratory hybrid dryer.

Figure 5 presents the scheme of the dryer. The cylindrical kaolin or wood samples were placed in the drier chamber on a special ceramic thimble with mandrel embedded on the balance located beyond the chamber. In this way the measurement of the sample weight was possible continuously during all kinds of drying, also during microwave drying. The AE detector was attached either directly to the sample foundation by convective drying or indirectly to the ceramic thimble with mandrel by microwave drying.

The hybrid drier enabled different combination of the three methods of drying: the convective, microwave, and infrared. The drier instrumentation enable programming and control of the velocity and temperature of the air supplied to the drier chamber, control of the microwave power, two-step control of the infrared heater, and the measurement of the sample surface temperature with the help of the optical pyrometer.

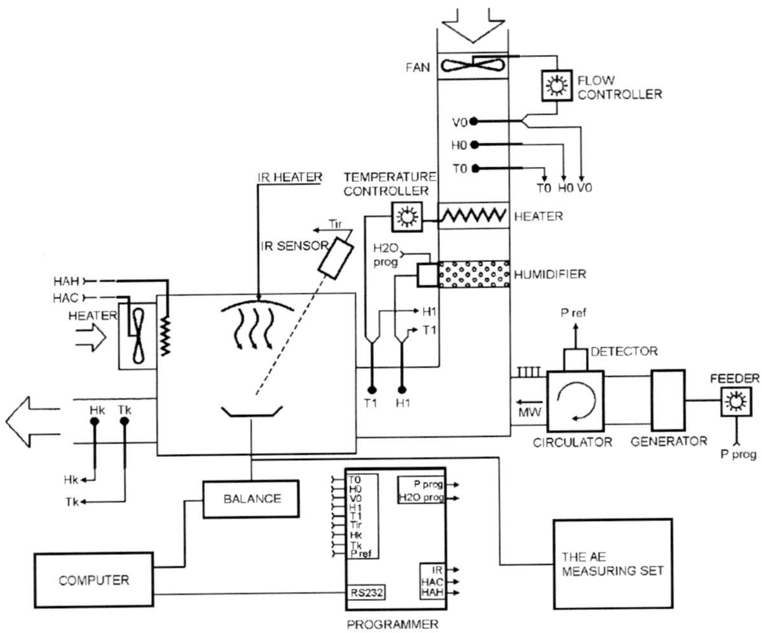

Figure 5. Scheme of the hybrid dryer.

3.2. Methodology of AE Measurement

The AMSY-5 AE system manufactured by Vallen Systeme, Gmbh is shown in figure 4. The sample subjected to convective drying was placed on the aluminum plate with fixed piezoelectric sensor. Acoustic signals generated in the drying sample are registered by broad-band transducer. Next the signals were send via insulated cable to AEP3 preamplifier unit. Preamplifiers was located close to AE sensors. The main task of preamplifier was amplifying and strength the signals enough that they could be sent to the distant main measurement unit. The AEP3 unit was equipped with 5 kHz to 1000 kHz filters. The main unit used in the tests had an M6 master unit for up to six AE channels from which three were fully equipped with ASSIP card. This high speed system could store up to 30 000 AE signals per second. The frequency filter inside the unit was used to eliminate noise sources. It can be set up for each channel separately. In our tests the low and high filters in the range from 12 to 850 kHz were applied to collect AE signals from kaolin and wood samples subjected to drying. The filtered AE signals were digitalized in A/D converter and stored in computer memory. The companion notebook PC had a software to control the whole measure system. During the tests the measured data were analyzed online and displayed on computer monitor so that it was possible to recognize the development of defects within the tested object.

4. MATERIALS AND CONDITIONS FOR AE APPEARANCE

KOC kaolin-clay from the Surmin-Kaolin SA Company, Nowogrodziec, Poland was the material investigated experimentally and theoretically in the drying tests. For this material some charakteristic data necessary for numerical calculation of drying kinetics and drying induced stresses were already given by the Surmin-Kaolin SA Company (see Table 1 in Kowalski et al. 2000). The KOC kaolin-clay is widely applied in ceramic industry for manufacturing sanitaryware and tableware. It provides a good strength and plasticity during shaping of the mentioned products and reveals a reduced amount of pyroplastic deformation in the process of their firing.

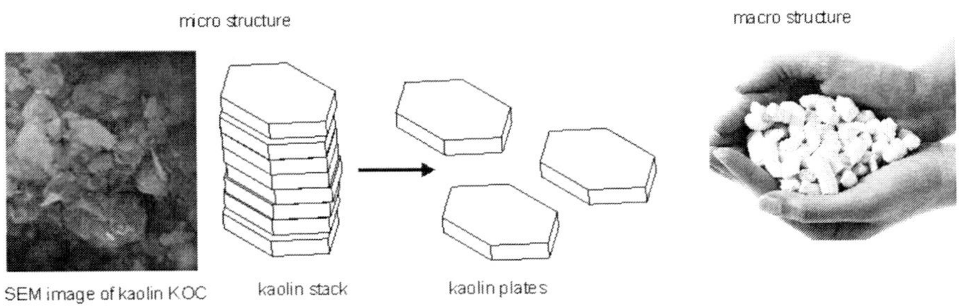

Figure 6. Image of kaolin KOC.

The kaolin-clay was delivered in a dry state, and before experiments it was grinded and wetted with a predetermined amount of water and mixed to achieve a greasy paste of initial moisture content (MC) approximately equal to 0.45 [kg water/kg dry kaolin]. The greasy paste was stored and homogenized in a closed box for 48 hours to unify moisture distribution in the whole material. The obtained in such a way soft kaolin-clay mass was used to mold cylindrical samples of 6 cm in diameter and 6 cm height. The cylindrical samples were extruded from a special instrument to preserve their regular shape (Fig. 7a), and samples of such a form were used for drying tests.

Figure 7 shows the shape of kaolin and pine wood samples applied in the studies.

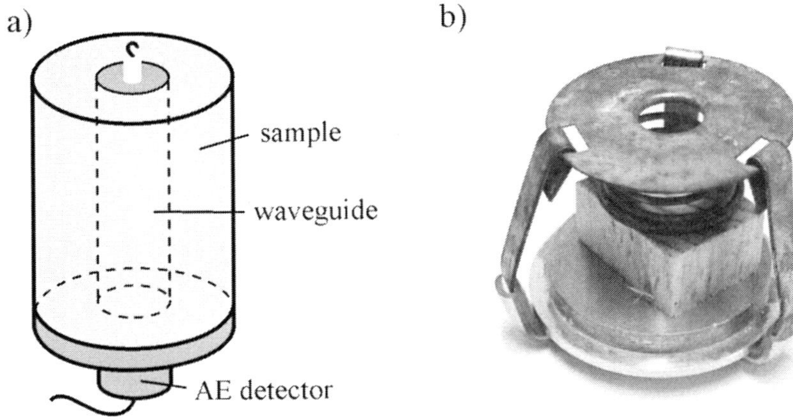

Figure 7. Image of the samples used in drying tests: a) kaolin cylindrical sample with attached AE detector, b) pine wood sample with spring pressed the sample to the AE detector.

Samples of the pine wood in the form of cuboid of dimensions about 4×4×2 cm were cut from the blade of the trunk of diameter about 30 cm and keep the symmetry with regard to the axis of the core. The walnut samples were of cylindrical form with height about 26 ÷ 27 mm and the diameter about 44 ± 2 mm. These samples were cut out from walnut branches deprived of defects in the structure. They contained 5 annual growth rings on average. The samples were placed on the aluminum support to which they were pressed with the springs to get a better contact with the AE sensor. The initial humidity of pine wood samples was about 117% and walnut samples about 85%. The samples prepared in this way were used to convective and microwave drying tests in the laboratory dryer presented above.

5. EXAMPLES OF AE IN DRYING MATERIALS

5.1. Convective Drying of Ceramic-Like Materials

One of the goals of the realized tests was to interpret the AE signals that may occur during drying of kaolin-clay (Fig. 8). The first (I) characteristic group of AE signals appears at the beginning of drying process, the second (II) one in the period when the surface layer intensively shrinks, and third (III) one is noticeable sometimes in the final stage of drying and identified as being generated by the reversed stresses.

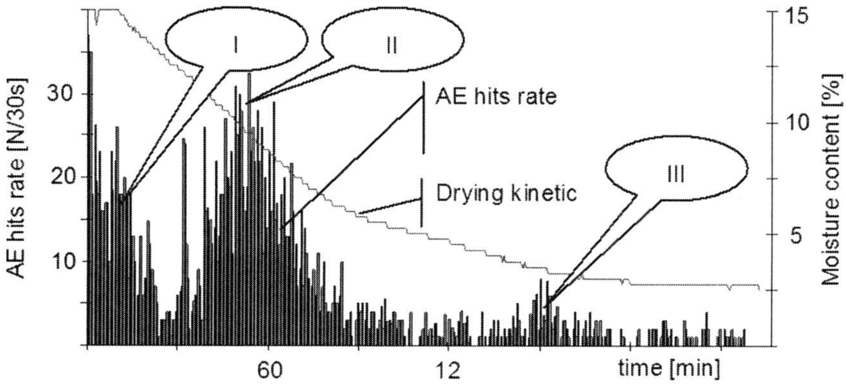

Figure 8. AE signals and the curve of drying.

As the drying body is almost fully saturated at the initial stage of drying, the number and the maximum value of AE signals in the heating period (I) is proportional to temperature of the drying process. At this stage of drying, the thermal stresses dominate in the clay sample which are rather not so much meaningful. Unfortunately, at the initial stage of drying some of AE signals come from heating aluminium probe and the AE sensor. It is hard to decide which AE signals in this stage of drying are from the investigated kaolin-clay sample and which from other sources. In tests with lower process temperatures the first maximum was also lower.

The second group of AE signals is evidenced at the end of the constant drying period (II). Their number is the highest of the whole drying process. The reason for appearance of these signals can be explained by the tensional stresses that arise at the external layers of the cylinder as a result of shrinkage. The surface of the body becomes more and more dry while its core is kept still wet. The first cracks on the external surface of the cylinder are observed when the local stress reaches the yield strength or the strain exceeds the allowable ultimate limit.

The third (III) maximum of AE signals is rather of small or moderate magnitude and depends on drying conditions. It was stated that this maximum exists usually for high temperature or low humidity of the drying medium. It holds for porous materials revealing inelastic properties (e.g. wet wood, clay, kaolin), and can be explained as follows: at the beginning of drying, the external layers are stressed in tension and the core in compression. Inelastic strains occur both in the surface layer and in the wet core. Later, under a surface layer with reduced shrinkage, the core dries and attempts to shrink causing the stress state to reverse. These new induced tensional stresses in the core cause fracture of a brittle (almost dry) structure, in what follows generate the acoustic signals.

Fig. 9 presents the rate of AE hits for the five different temperatures of drying. It is seen that for the conditions of high drying rates created by high temperatures, the rate of AE hits achieve higher values than for lower temperatures. That high active emission of AE signals has a reflection in the drying induced stresses.

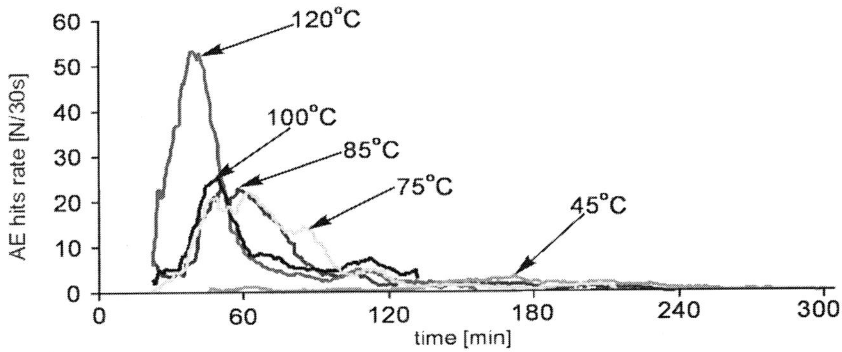

Figure 9. The rate of AE hits during drying at different temperatures.

Note that the highest peak of the AE hits, which correspond to temperature 120°C, appears earlier than the lower peaks corresponding to the lower drying temperatures. The primary peak of AE hits appeared in 50 min of drying time, that is, when the tensional stresses at the cylinder surface reached maximum. The secondary peak is visible about 110 min drying time for temperature 120°C. In this time the core of the body starts to dry. The wet core wants to shrink but the surface layer is not able to deform itself because it is almost dry. So, in these circumstances the tensional stresses arise in a core. The secondary maximum is, of course, much lower than the first one.

Another AE descriptor termed *the total energy* is useful in analysis of fracture phenomenon by drying. It measures successively the total energy released during the whole process of drying. Figure 10presents the rate of AE hits and the curve of total energy versus time.

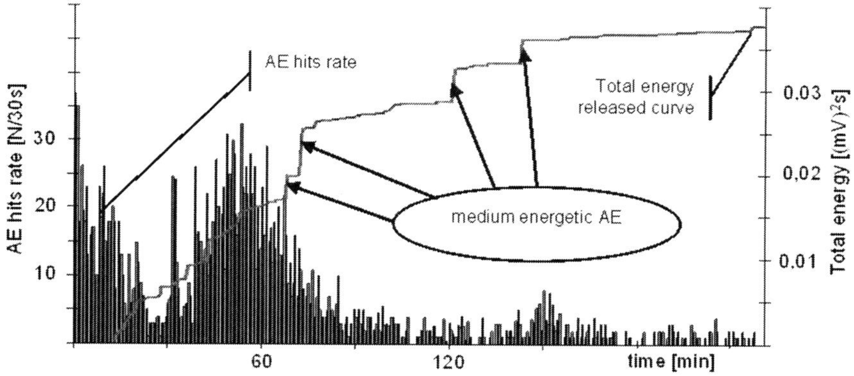

Figure 10. Total energy of AE signals released during drying.

The curve of summarized in time acoustic energy released during drying can deliver essential information on fracture dimension occurring in the investigated body. A highly fractured material is qualified as a bad quality one. The descriptor of total energy released may serve as an indicator, whether the dried product is of good or bad quality at the end-state. Strong cracks of body structure release high energetic EA signals. The high or medium energetic signals are evidenced in figure 10 as the strait upright lines. In some cases, the high energetic signals denote macro-cracks or splits that are visible on the sample surface. Taking into considerations the above presented curve of total energy one can state that it represents a dried product of bad quality.

The strength of material with damaged structure is impaired. Often a number of internal small micro-cracks arising during drying may nucleate and create macro-cracks during utilization of dry products, so that an unpredictable total damage of the body may take place in any time after drying.

The reason for fracture of materials under drying results mostly from not proper drying conditions (e.g. too high temperature or too low drying medium humidity). By optimal drying process the high energetic AE signals ought to be eliminated or minimized. The majority of registered AE signals ought to be low energetic (horizontal or almost horizontal lines in figure 10). Low energetic AE signals mean lack of destruction in dried products.

Figure 11 shows several curves of total EA energy released from kaolin cylinders during drying at different temperatures. Each EA signal carries a certain portion of energy. The flat horizontal lines represent the low energetic signals. Hits of high energy create sudden vertical lines as, for example, those visible on the energy curves obtained for drying at temperatures 100 and 120°C. These very energetic signals are generated by strong material cracks.

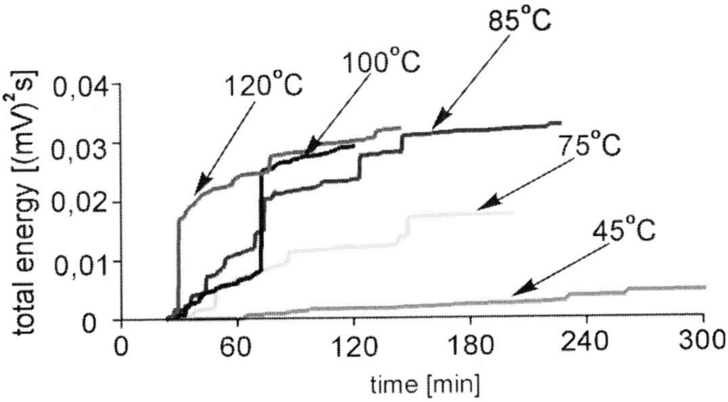

Figure 11. Total energy of hits during drying process for various conditions.

Analysing figure 11 one can see the differences in released energy for different drying conditions. The curve of 45 °C, being almost horizontal, represents low energetic AE signals. It means that drying at this temperature is unpropitious for creation of material fracture, so the manufactured product is of good quality and without residual stresses. Unfortunately, drying at such a low temperature takes a long time and is unsatisfactory from the economic point of view.

The total energy curve for drying at 75°C looks a little different from that of 45°C. It contains a greater number of signals in the time period from 40 to 100 min but these signals are not so much energetic either, some of them might be generated by invisible micro-cracks. This curve is slightly inclined upwards.

The next curve (85°C) has similar character as that of 75°C one, however, represents more energetic signals. The energy released here is much higher than that represented by the two previous curves. The dried product at this temperature may have a number of micro-cracks that can expand into visible cracks under the action of residual stresses.

The most energetic AE signals are represented by the curves characterizing severe drying conditions (100°C and 120°C). So, high temperatures together with low humidity of the drying medium (air) are propitious to high drying rate. This produces quickly dry and brittle surface layers while the wet core becomes still wet and deformable. The drying induced stresses cause damage of the fragile surface.

5.2. Microwave Drying of Ceramic-Like Materials

Figure 12 presents the photographs of the cylindrical samples after microwave drying with 300 W of the microwave power (MWP). The damage of the cylinder is occurred in its center and looks like an explosion caused by a high vapor pressure inside the cylinder due to intensive phase transitions of water into vapor. This proofs that by microwave heating the highest temperature arises inside the material. It is confirmed by the picture presented in figure 12c made with infrared (IR) camera (Flir therma-cam B2).

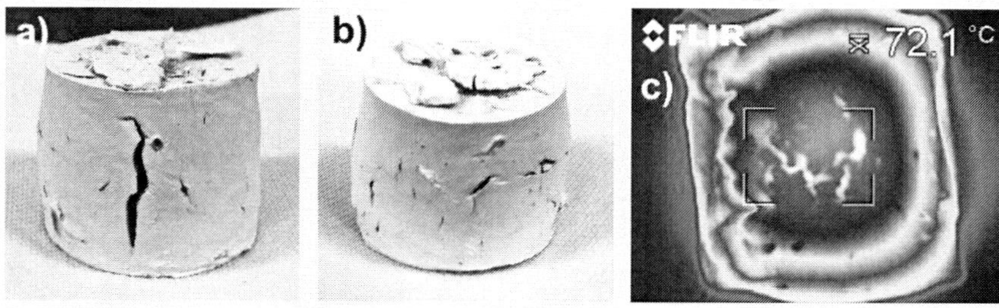

Figure 12. Kaolin sample subjected to 300 W MWP a) front view, b) rear view, c) IR camera picture of temperature distribution in the sample longitudinal cross-section.

The sample subjected to 300 W of MWP has a huge vertical slit, almost 3 cm long and 2.5 cm deep (Fig. 12a). It was caused by the explosion after 20 min of drying time. The barrel shape of the samples is very clearly visible in both figures 12a and 12b. The picture of temperature distribution in the sample longitudinal cross-section (Fig. 12c) shows that the temperature reached about 90°C in some hot spots, although the mean temperature in the central part was about 72°C. This indicates the existence of places where the water was rapidly changed into vapor. The rapidly increased vapor pressure created the big slit, so that the water vapor had found the way out.

Figure 13 presents AE signals acquired during microwave drying of kaolin cylinders.

For 300W of MWP the huge acoustic energy wave was generated during the explosion. It is visible infigure 13 as a high single signal at 20 minutes of drying. Drying with lower MWP (180 W and 240 W) induces signals of lower AE energy.

The number of AE signals induced during microwave drying depends on the MWP (Fig. 14). By lower MWP the number of AE signals is greater, but they are of lower energy than those of higher MWP.

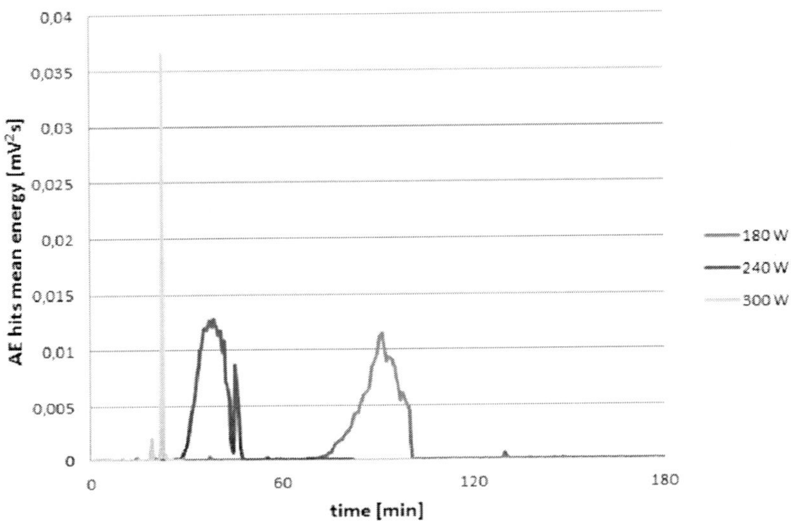

Figure 13. AE hits mean energy acquired during microwave drying of kaolin cylinders.

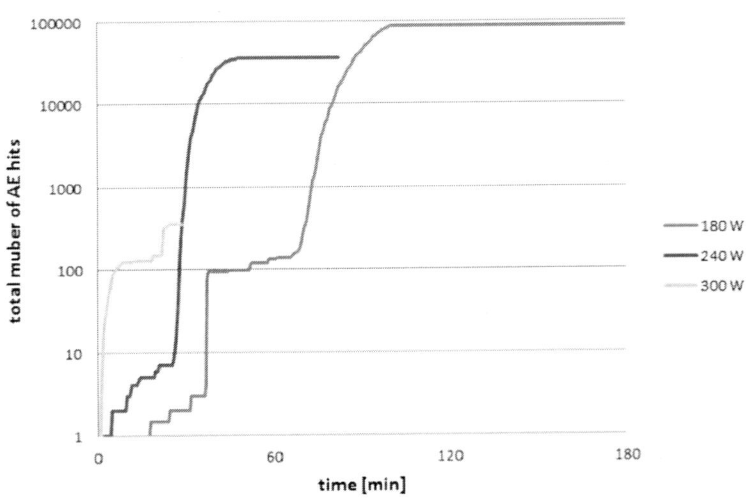

Figure 14. AE results for microwave drying of kaolin: the total number of AE hits

The destruction of samples raised by microwave drying proceeds mainly in the second drying rate stage, except the one dried in 300W MWP. For samples dried in 180 W and 240W MWP-s the destruction was observed as step by step splitting small parts (Fig. 15).

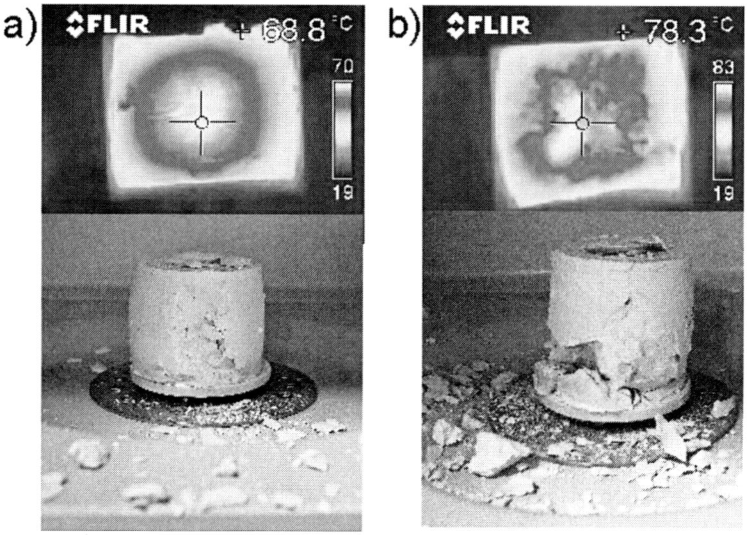

Figure 15. The kaolin sample dried in microwave oven by 240 W MWP.

The convective and microwave drying methods differ from each other in the way of heat supply. By convective drying the heat is delivered from the surroundings through the material surface from the hot air of temperature higher than the temperature of drying material. By microwave drying on the other hand the heat is generated volumetrically as a result of the dispersion of the high-frequency monochrome waves of the order 2.45 GHz. The microwave power is absorbed mainly by the water present in the material pores.

The different ways of heat supply affects the directions of heat and mass fluxes. During the convective drying the heat flux is in opposite direction to the mass flux, and this causes a decrease of moisture removal, what is favorable to non-uniform distribution of the moisture inside the material. Such a negative thermodiffusive effect not appears or is minimal in microwave drying by which the heat flux coincides with the mass flux. The interior of the material usually has the temperature higher than the surroundings. The moisture distribution in the material in this case is more uniform than during the convective drying. So the drying induced stresses should be smaller. Diagrams get from measurements of total numbers of AE signals and the AE energy confirm this prediction.

Figure 16 presents the AE hits in kaolin sample dried convectively. It appears from the presented diagrams that the maximum number of AE signals was occurred in the middle of the constant drying rate period (CDRP).

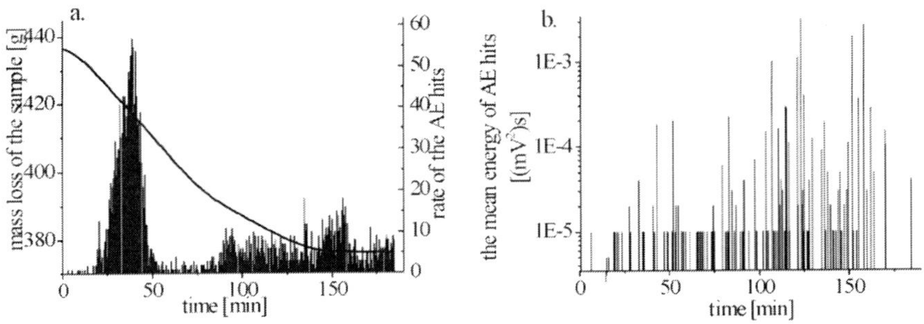

Figure 16. The AE hits for convective drying of kaolin sample: a) drying curve and AE hits rate, b) mean energy of AE hits.

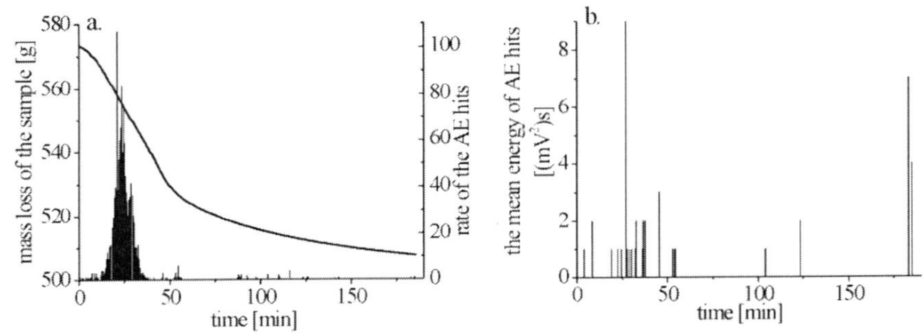

Figure 17. The AE hits by microwave drying of kaolin sample: a) drying curve and AE hits rate, b) mean energy of AE hits.

Kaolin clay sustains the greatest shrinkage in the initial period of drying (15 – 90 min). External layers of the cylindrical sample are dried first and shrink generating stresses and cracks, which is manifested by an increased number of AE signals. The density of these signals decreases with the course of drying, however, their energy becomes more and more greater. Figures 16a presents the rate of AE hits and 16b the mean energy of AE hits. One can see that the initial great number of hits is of relatively low energy but the subsequent ones revealed much bigger energy. In the latter cases a number of distinct scratches was visible on the surface of dried samples and in some cases even micro and macro cracks were formed, particularly by drying in more severe drying conditions.

Figure 17 presents the AE hits in a kaolin sample that occurred by microwave drying. In the case of microwave drying one can see that both the number of AE hits (Fig. 17a) and the emitted AE energy (Fig. 17b) are much lower in comparison to the convective drying. The lower number of AE hits in microwave drying can be justified by the coincidence of the heat and mass fluxes and thus more uniform distribution of the moisture through the material in this kind of drying, which consequently resulted in reduction of stresses. It is important that a significant increase of drying rate was noticed by microwave drying. The CDRP amounted only from 10 to 70 min (Fig. 17a), dependent on MWP.

5.3. CONVECTIVE DRYING OF WOOD

Drying of pine wood samples was conducted at three different temperatures: 60, 80 and 100°C. Figure 18 presents the rate of AE hits, i.e. the density of AE signals per a time interval (e.g. 30 s) (Fig. 18a), and the drying curves (Fig. 18b), for different drying temperatures.

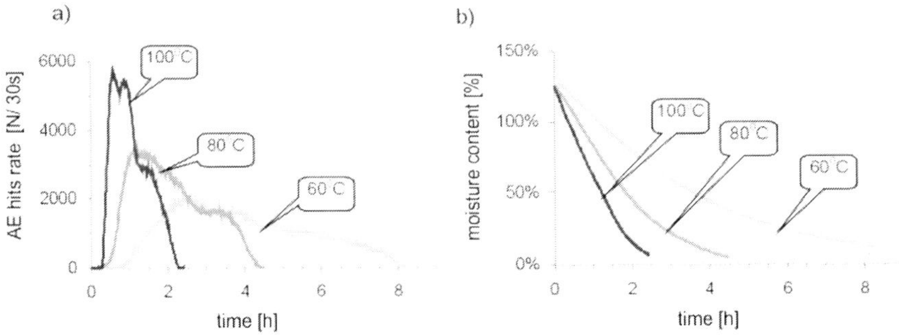

Figure 18. Results of pine wood drying: a) the rate of AE hits, b) drying curves.

One can state based on the experimental results that the rate of AE hits in drying wood depends on the number and size of fractures occurred in wood samples. The shrinkage of wood begins at the moment, when the MC reaches the fiber saturation point (FSP) (c.a. 30%). In the initial drying period, when the MC in wood is higher than FSP, the AE activity is insignificant. Only when MC in the surface layers drops below FSP, and the MC inside the samples exceeds this value, the AE start to reveal greater activity, in particular for high drying temperatures. The considerable increase of the rate of AE hits corresponds then with drying temperatures and is attested by drying stresses and cracks of different sizes, which generate acoustic wave.

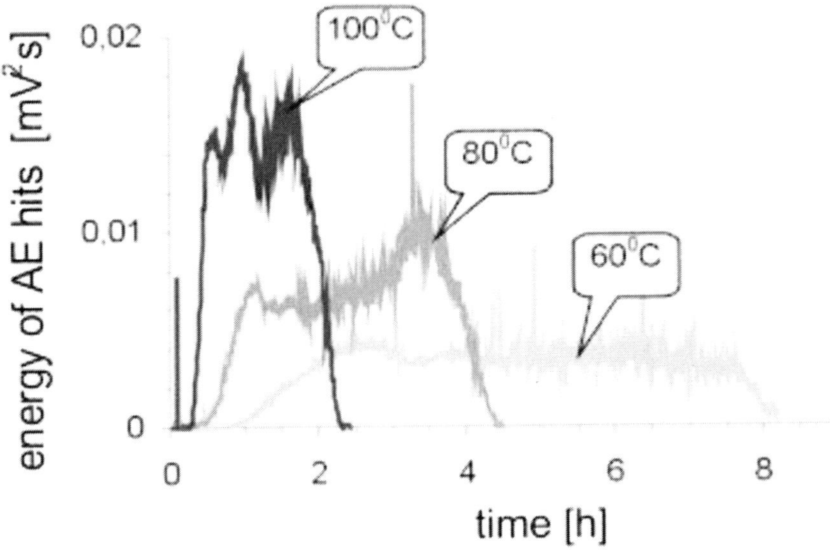

Figure 19. The energy of AE hits for pine wood sample dried at various temperatures

Figure 19 shows the energy of AE hits for pine wood for different drying temperatures. The graphs presented in this figure, and in particular this referring to drying at temperature 100°C, point out a danger of wood destruction at the moment when the AE energy reaches maximum.

As the plots in figure 19 show, the rate of AE hits grow rapidly at the beginning but after some time they decrease also quite rapidly. The rate of AE hits decreases in the period that refers to drying of the sample core, however, the emitted AE energy stays on the same level or even grows, as it is seen on the graph for 80°C. Incurred earlier micro-cracks start expanding and linking. A danger of wood destruction could be then even greater than in the initial period, because of increase of the summarized energy of AE signals. Dried wood becomes more and more rigid and the risk of the brittle cracking becomes more and more probable. Observation of the mean energy of AE hits presented in figure 20confirms it.

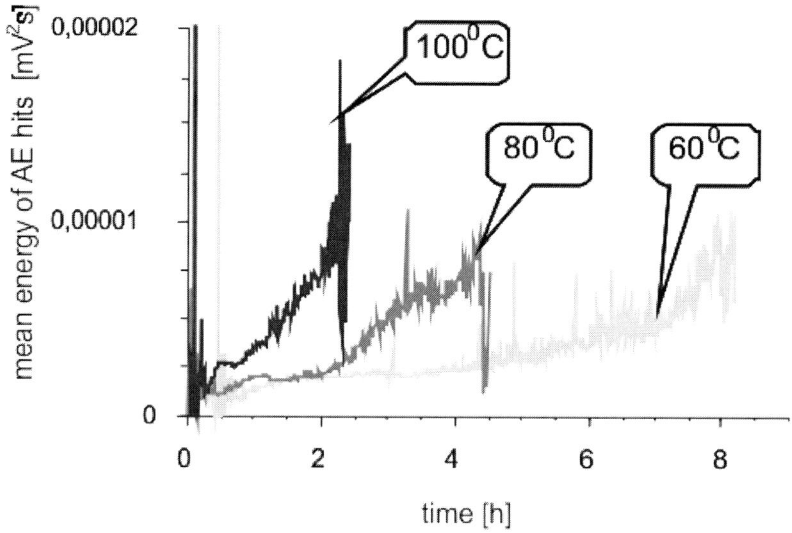

Figure 20. The mean energy of AE hits.

The energy is released up to the end of drying process. Incurred earlier micro-cracks grow further. The mean energy of the AE hits is a measure of the progressing decomposition of wood. It is bigger for higher temperatures of the drying medium. However, in the conducted series of drying tests in no case occurred a visible crack of dried sample. This fact is evidenced on the curves of total energy presented in figure 21, that is, the energy summed up during the whole course of drying.

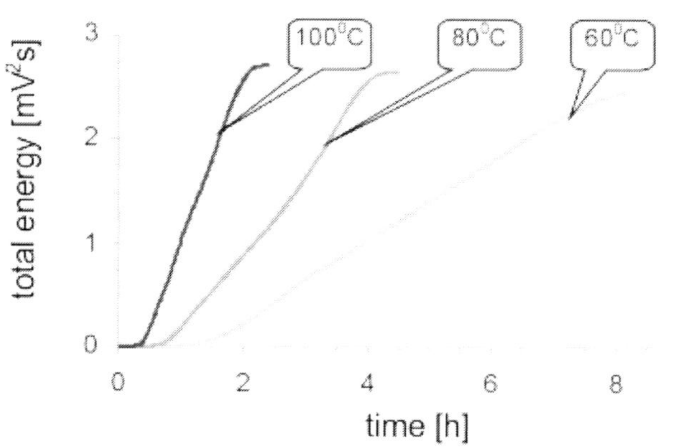

Figure 21. The total energy of AE hits for the whole drying process.

One can see that the curves of total energy are growing smoothly, that is, without violent jumps. Violent jumps on graphs of total energy indicate just fractures visible even with a naked eye.

5.4. The Phenomenon of Stress Reverse

The phenomenon of stress reverse results from a constrained shrinkage in dried material. It can happen when the surface of material becomes deformed permanently due to intensive shrinkage. The stresses arose on the surface are tensional and in the material interior compressive ones. When the drying is progressing deeper into the material and its core starts to shrink but is hindered by the surface shell previously stretched in a permanent way, then the stress on the surface becomes compressive and that in the material interior tensional, so thus the stress reverse take place.

Modeling of the stress reverse phenomenon requires taking into account inelastic properties of the materials. This issue was already considered in the work by [Kowalski, 2001, 2002; Banaszak and Kowalski, 2002; Kowalski and Rajewska, 2002; Kowalski et al., 2002].

Figure 22 shows the time evolution of the circumferential stresses distributed in the cylindrical kaolin sample dried convectively by the assumption that kaolin is viscoelastic and obey Maxwell model [Kowalski, 2001, 2002; Kowalski and Rajewska, 2002].

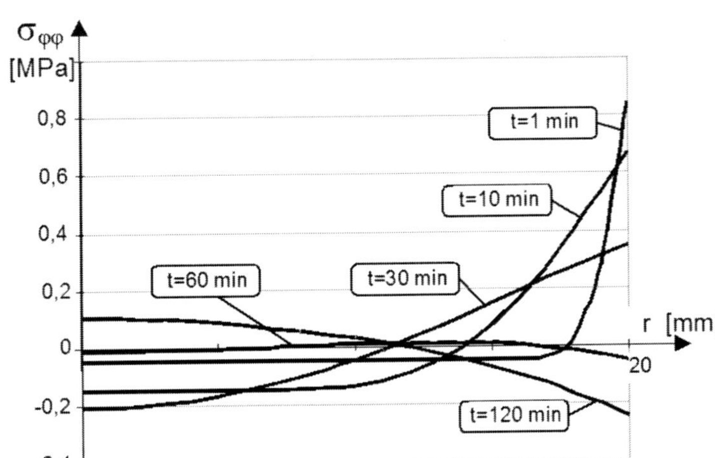

Figure 22. The evolution of circumferential stresses in kaolin cylinder dried convectively.

We can see that the circumferential stresses are compressive in the core and tensional in the boundary layer at the beginning and reversibly signed at the end of the drying process. We suppose that the tensional stresses in the core of the cylinder may cause structural fracture and thus the emission of the third group of acoustic signals at the final stage of drying.

Figure 23 shows the evolution of circumferential stresses in wooden cylinder (birch) dried convectively [Kowalski et al., 2002]. Here the Maxwell model was used for wood.

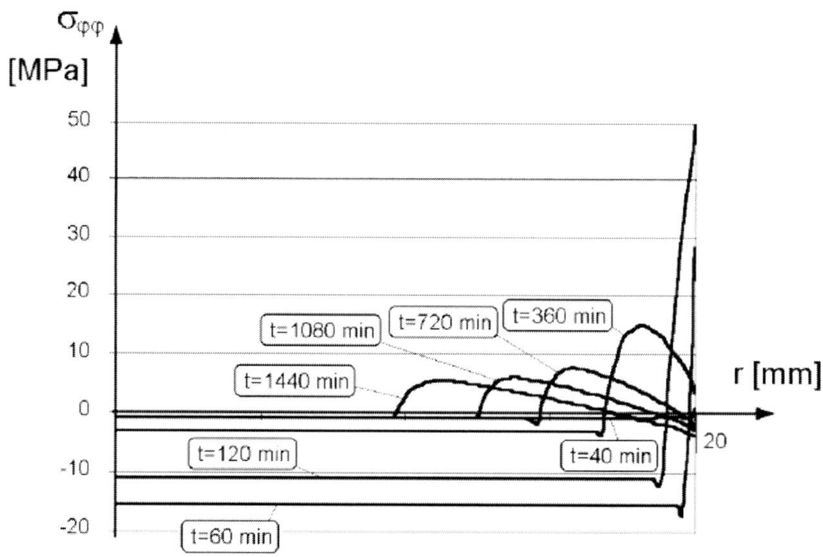

Figure 23. The evolution of circumferential stresses in wood cylinder dried convectively

At the first stage of drying the stresses for the viscoelstic model run in a similar way as for the elastic model. After some time, however, when the dry zone extends deeper towards the wet core, the circumferential stresses start to change their sign at the boundary from tensional to compressive. Note that the maximum value of the tensional circumferential stresses is moving during drying from the boundary surface towards the interior of the cylinder.

Figure 24 shows the evolution of the number of AE hits in time during convective drying of kaolin sample. The plot of the number of AE hits is confronted with the curves presented the circumferential stresses determined on the basis of elastic Hooke model (dashed line) and viscoelastic Maxwell (solid line).

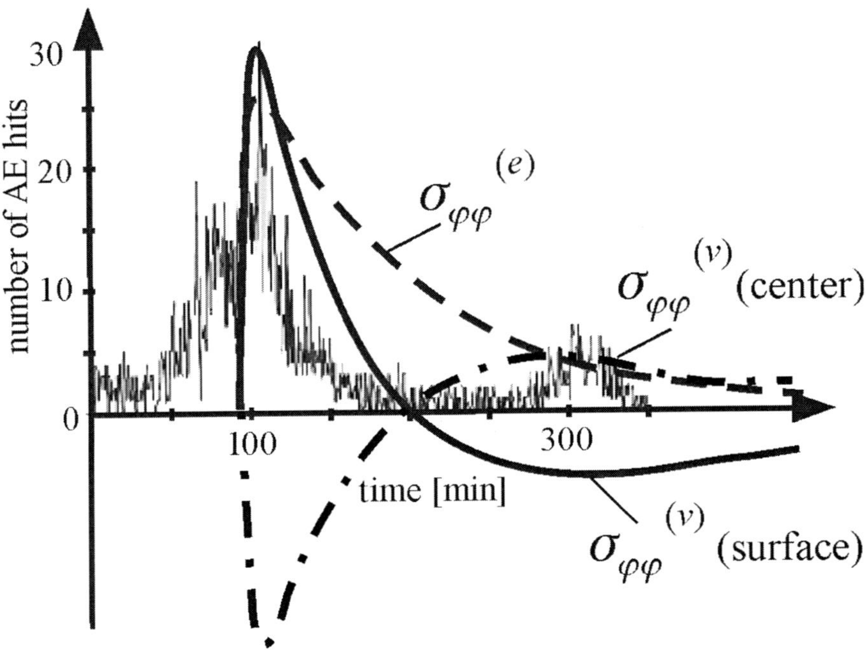

Figure 24. Hits rate and the theoretical curve of circumferential stresses as a function of time.

As it is seen from figure 24, Hooke model does not reflect the occurrence of stress reverse. It demonstrates only the compliance of the first maximum of stresses with the increased AE activity. The consideration allow us to appreciate the meaning of the mathematical model adequacy for the description of the mechanical phenomena by drying of wet materials. The experimentally observed enhanced emission of acoustic signals, betoken the enhanced destruction of the material at the final stage of drying. Using the elastic model one obtains the stress development, which rises from the beginning, then reaches a maximum at point in time, and next, as the process proceeds further, disappears totally. Otherwise, the model that takes into account the permanent deformations of the dried material, allows us as in the case of viscoelastic model to observe the stress reverse, and in particular the appearance of tensional stresses inside the dried material at the final stage of drying. The tensional stresses in the core at the final stage of drying cause an increase of AE signals, which is visible in figure 24.

6. CONTROL OF MATERIAL DAMAGE WITH THE HELP OF AE

6.1. Avoiding Material Fractures Through Changes Drying Conditions

The non-stationary (intermittent) convective drying denotes drying with different drying rates in several periods. The results of the drying studies presented in this chapter allows to state that the intermittent drying can be recommended above all to drying of materials, which have a tendency to cracking during drying as, for example, ceramics and wood. Through changes of drying conditions in the right moments one can avoid material fracture and thus preserve a good quality of dried products. Thus, one can state that intermittent drying positively influences the quality of the dried materials without significant extension of the drying time.

In these considerations the intermittent drying was realized through periodically changing both temperature and humidity of drying air. The results of intermittent drying are compared with adequate processes of stationary drying to show the profits resulting from the former.

Apart from the visual assessment of the quality of dried products, the acoustic emission (AE) method was applied for monitoring of the micro- and macro-cracks developed during drying [Kowalski and Pawłowski, 2010]. In those studies it was measured the total number of AE hits and the total amount of AE energy emitted. The descriptor of total AE energy is the sum of energy of all acoustic signals emitted by the dried sample from the beginning to the end of drying. It denotes the energy released due to material cracking. These descriptors show the moment at which the AE becomes intensive and how big is the AE intensity. Knowing these descriptors, one can assess the intensity of micro- and macro- cracks that arise in dried materials as well as their magnitude. The "intensity" quantifies the number of cracks per 30 s intervals or the total number of cracks in the whole process. The crack "magnitude" is evidenced as the vertical straight line on the descriptor of total AE energy curve. In this way we can estimate the degree of destruction and how fast the destruction advances in dried materials.

Figure 25 presents the total number of AE hits and the total AE energy emitted during stationary drying of the kaolin cylindrical sample at temperature 100 ^oC.

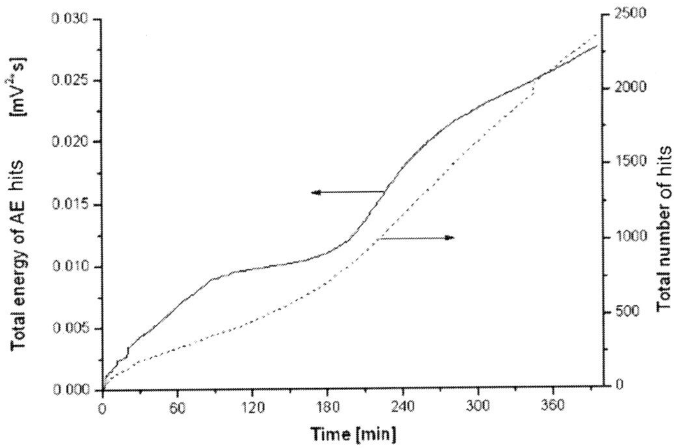

Figure 25. Total number of AE hits and total AE energy in stationary drying of cylindrical kaolin sample at 100°C.

The plots in figure 25 show a continuous increase of the number of AE hits and the AE energy. The flatness on the energy plot that begins at about 100 min of drying follows from the release of the elastic energy accumulated in the stressed material due to material fracture, and in this way a reduction of the stress state occurs.

Figure 26 presents the total number of AE hits and the total AE energy emitted from the cylindrical sample during drying with periodically changing temperature between 50 and 100 °C in the falling drying rate period (FDRP).

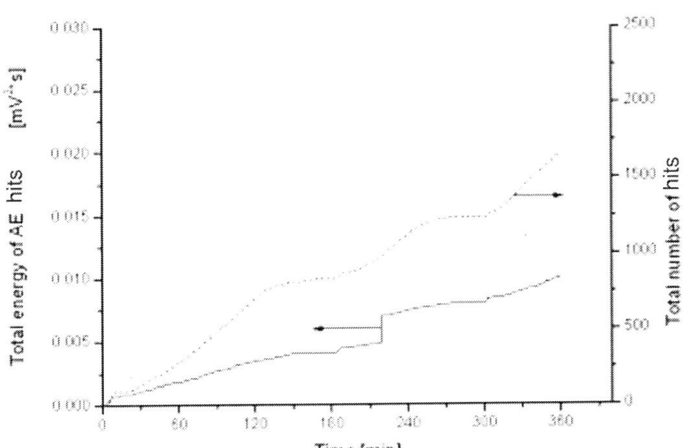

Figure 26. Total number of AE hits and total AE energy by convective drying of kaolin cylindrical sample at periodically changing temperature between 50 °C and 100 °C.

Note that the total number of AE hits and the total AE energy emitted during drying with periodically changing temperature is less than those in stationary drying. The plots in figure 26 are not so smooth as those in figure 25, which follows from the variable air temperature, and strictly, by switching off and on the air heater and cooler.

Figure 27 presents the total number of AE hits and the total AE energy emitted during drying with variable air humidity. The number of emitted AE hits and the total AE energy in this kind of drying is lesser than those during stationary drying and also lesser than during drying with variable temperature.

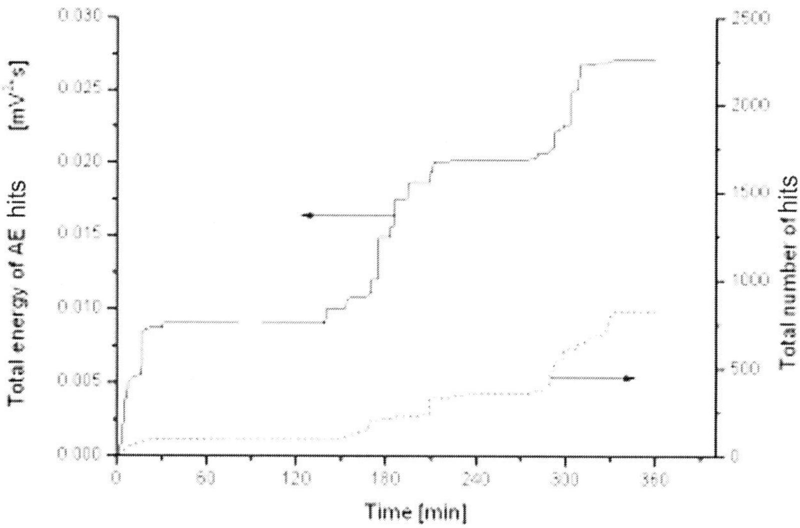

Figure 27. Total number of AE hits and total AE energy by convective drying of kaolin cylindrical sample at periodically changing air humidity between 4 % and 60 ÷ 80 %.

The humidification of air in the chamber dryer caused the plots of acoustic emission to be very rugged. Nevertheless, we can state that drying in intermittent conditions accomplished through periodically changing temperature and air humidity is accompanied by a smaller number of AE signals and smaller value of AE energy. This denotes less micro- and macro-cracks in dried material and simultaneously better quality of dried products.

6.2. Reduction of Material Fractures Through Surfactant Application

In order to improve moisture transport inside the dried body, and thus to assure more uniform distribution of moisture in the material and thus avoid its cracking, the authors proposed wetting the raw kaolin-clay with water containing surface active agents (surfactants). These agents have the ability to stimulate the surface tension between water and the pore walls and thus to improve moisture transport inside the material [Cottrell, 1970; Wert & Thomson, 1974].

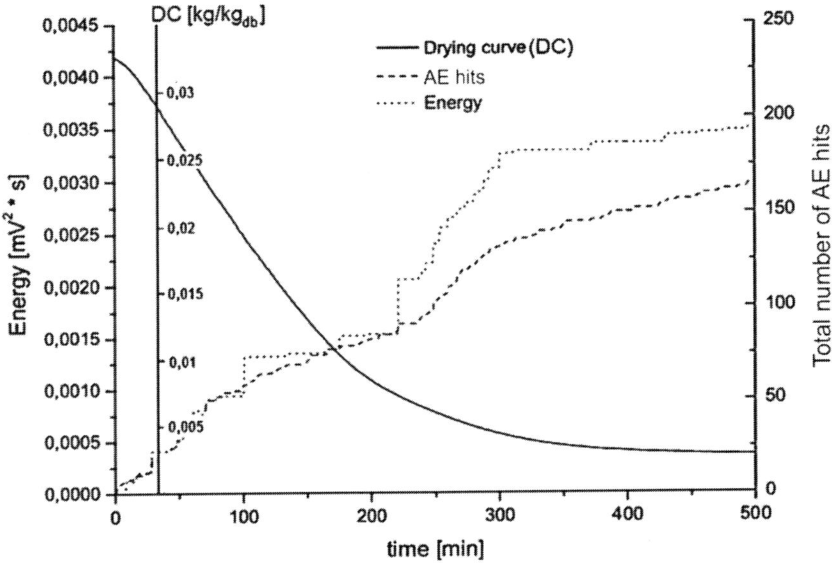

Figure 28. Drying curve, total AE energy and total number of AE hits in clay samples saturated with pure water of clay samples at air temperature 120 °C.

Figure 28 presents the typical drying curve of clay samples with the CDRP (0 – 180 min) and the FDRP (180 – 400 min), and the descriptors of total AE energy and total number of AE signals emitted during drying of clay saturated with pure water.

Each rapid increase of the total AE energy visible on the AE curve denotes a crack occurring of in the clay sample at a given moment. As seen in this figure, the biggest cracks were formed in the second stage of the CDRP and in the first stage of FDRP. At this stage the sample surface became dry while the core of the sample was still wet, and material cracks occurred at this stage.

Figure 29 presents comparison of total AE energy for dried clay saturated with water containing different concentration of surfactant SDS (0, 0.001, 0.01, 0.1 and 1%).

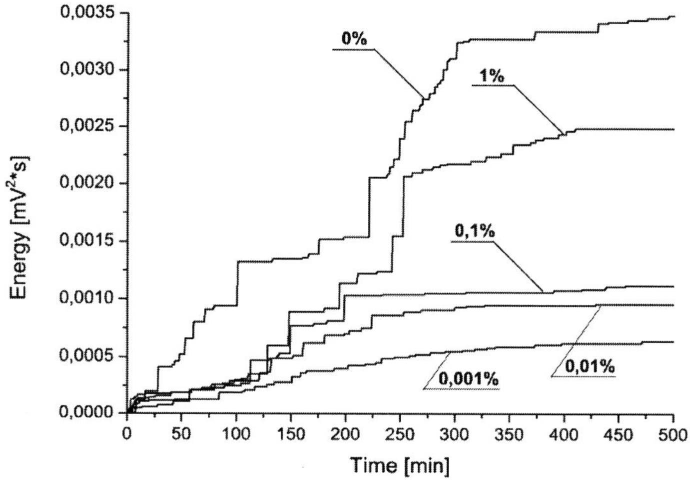

Figure 29. Comparison of total AE energy emitted by clay with different concentration of SDS.

It is seen that different amounts of SDS added to water solutions used for clay saturation differentiate the total AE energy emitted by kaolin-clay during drying. As it follows from this figure, the greatest energy is emitted for clay saturated with pure water (0% SDS) and for the greatest surfactant concentration (1% SDS). It means that there is a SDS concentration at which the AE energy reaches minimum. The limit value of the surfactant concentration, at which the drying results are efficient is called *the critical micelle concentration* (CMC).

Figure 30 shows that the quality of sample with 0.01% surfactant concentration is much better than that with 0% (pure water) concentration.

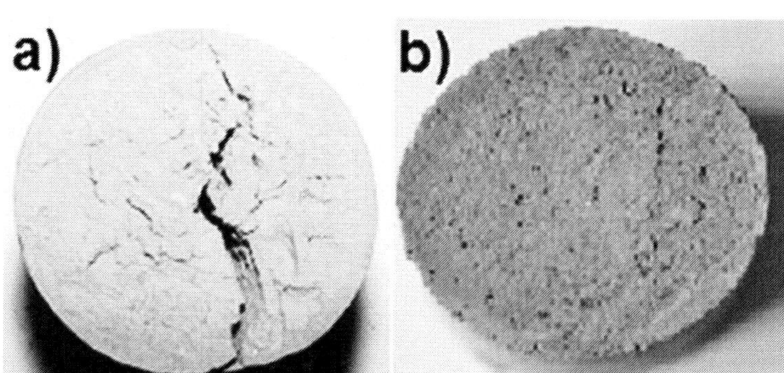

Figure 30. Bottom surface of kaolin-clay samples after drying: a) without surfactant, b) with 0.01% of SDS.

Figure 30 proves that surfactant concentration of value close the CMC have a meaningful influence on moisture transport inside capillary-porous materials. These conclusion is confirmed by the good quality of dried product visualized on the photo of samples bottom surface presented in figure 30b.

7. CONCLUSIONS

The presented in this chapter results of research concerning analysis of the AE activity in dried materials allows to state that the AE method can indeed support the control of the drying process and facilitate the guidance for the purpose of avoiding destruction of materials during drying. Comparison of the drying induce stresses simulated numerically on the basis of mechanistic drying model with the experimentally measured descriptors of AE activity reveal an excellent adherence of the theoretical and experimental results. Although with the AE method we are not able to estimate strictly the magnitude of generated stresses, however, the assessment of material destruction intensity caused by the stresses and the time and place of their occurrence is very helpful for control of drying processes. Besides, monitoring of the AE events can be helpful also for validation of the failure criterion formulated on the basis of the mechanistic theory of drying, which is used for estimation of the magnitude and location of maximal stresses as well as their time evolution during drying.

It is worth to point out here the importance of the acoustic emission method that allows us observations*on line* the development of the acoustic signals connected with the destruction of the materials. The possibility of the registration of various descriptors such as: the intensity of acoustic signals, the energy of emitted signals and the total number of signals or total amount of energy allows the current control of drying processes.

8. ACKNOWLEDGEMENTS

This work was carried out as a part of research project No. N N209 031638 and N N209 104337 sponsored by the Polish Ministry of Education and Science.

REFERENCES

1. Y. M Berlinsky, J Rosen, J Simmons, Wadley HNG, (1990A Calibration Approach to Acoustic Emission Energy Measurement, Journal of Nondestructive Evaluation, 10 (1).
2. J Banaszak, S. J Kowalski, 2002Drying induced stresses estimated on the base of elastic and viscoelastic models. Chem Eng J 86139143

3. J Banaszak, S. J Kowalski, 2010Acoustic methods in Engineering Applications, Publisher: Poznań University of Technology, Poznań (in Polish).
4. J Banaszak, S. J Kowalski, 2005Theoretical and experimental analysis of stresses and fractures in clay like materials during drying, Chem. Engineering and Processing, 44497503
5. A. H Cottrell, 1970The mechanical properties of matter. Warsaw: PWN (in Polish).
6. S. J Kowalski, 2002Theoretical study of stress reversal phenomena in drying of porous media. Dev. Chem. Eng. Mineral Process., 10(3/4): 261-280.
7. S. J Kowalski, 2003Thermomechanics of Drying Processes, Springer, Berlin, Heidelberg.
8. S. J Kowalski, 2010Control of mechanical processes in drying. Theory and Experiment, Chemical Engineering Science, 65890899
9. S. J Kowalski, J Banaszak, A Rybicki, 2012Damage Analysis of Microwave-Dried Materials, AIChE Journal.
10. S. J Kowalski, G Musielak, 1999Deformations and stresses in dried wood. Transport in Porous Media 34239248
11. S. J Kowalski, W Molinski, G Musielak, 2004Identification of fracture in dried wood based on theoretical modeling and acoustic emission. Wood Sci Techn 383552
12. S. J Kowalski, A Pawlowski, 2010Drying of wet materials in intermittent conditions,, Drying Technol. 28 (5). 636643
13. S. J Kowalski, K Rajewska, 2002Dried induced stresses in elastic and viscoelastic saturated materials. Chem Eng Sci 5738833892
14. S. J Kowalski, K Rajewska, A Rybicki, 2000Destruction of wet materials by drying, Chem. Eng. Sci. 5567556762
15. S. J Kowalski, A Rybicki, 1996Drying stress formation induced by inhomogeneous moisture and temperature distribution. Transport in Porous Media 24139156
16. S. J Kowalski, A Rybicki, 2007The vapour-liquid interface and stresses in dried bodies, Transport in Porous Media 664358
17. Luong Phong M1994Centrifuge simulation of Rayleigh waves in soils using a drop-ball arrangement, ASTM Special Technical Publication, (1213), 385-399.
18. I Malecki, and J Ranachowski, 1994Acoustic emission: Sources, Methods Applications, Pascal Publications, Warszawa (in Polish).
19. A. S Mujumdar, Ed.) (2007Handbook of Industrial Drying (Third Edition), Taylor & Francis Group, New York.

20. Strumiłło Cz1983Fundamentals of the Theory and the Technology of Drying, WNT Warszawa, (in Polish).
21. C. A Wert, R. M Thomson, 1974Physics of solids, National Scientific Publishers, London, UK

Chapter 11

Application of Acoustic Emission for Quality Evaluation of Fruits and Vegetables

Artur Zdunek[1]

[1] *Institute of Agrophysics, Polish Academy of Sciences, Lublin, Poland*

1. INTRODUCTION

Food crushing sound is one of the main factors used for food quality evaluation. Crispness and crunchiness are attributes of high quality product and are usually pointed on the top of a list of consumer preferences. However, the meanings of crispness and crunchiness are still imprecise. Its perception varies from country to country and from individual to individual. Despite of this there is a general consensus that crispy and crunchy sensation is related to fracture properties. Crispy product is mechanically brittle, firm and acoustically noisy as a result of large number of small fractures. Crunchiness is probably related to events (fractures) occurring on subsequent layers in a cell structure what gives the sense of extension of sound duration in time.

In spite of sensory and subjective nature of food quality evaluation by human senses, a big effort is put for objective sound properties analysis during biting and chewing and for developing instrumental methods for human independent food evaluation. The first instrumental analysis of sound was published by Drake in 1963, who found that crisper products emit louder sound and an average amplitude of successive bursts during mastication decreases [1]. Then, several authors used different sound descriptors for judging a chewing sound, as the number of sound burst in a bite *n*, the mean amplitude of the burst *A* or the products of these values *nA* or *nA/sound duration* [1, 2, 3]. The first hypothesis was that the sense of crispness is an auditory phenomena, i.e. is the air-conducted sound. However, work done by Christensen and Vickers in 1981 showed that crispness may be a vibratory phenomena, i.e. is the bone-conducted sound [4].

Most of studies on crispness and crunchiness concern dry food products, like cakes, chips, etc.. However, this problem has been found as important also for fruits and vegetables called as wet food products. In 2002, presumably for the first time, Fillion and Kilcast stated that crispness and crunchiness are very complex concept containing sound, fracture characteristic, density and geometry of fruit and vegetables [5]. They found that crispy wet food would refer to a light and thin texture producing a sharp clean break with a high-pitch sound mainly during the first bite with the front teeth. Crunchy wet food would be hard and has a dense texture producing loud, low-pitch sound that occurs over successive chews.

Many instrumental methods have been applied for crispness and crunchiness evaluation of fruits and vegetables. Due to the fact that mastication is a highly destructive process, mechanical tests are the most popular to simulate the biting. Results of such tests, like texture profile analysis, compression, tension, twist or three-point bending, show correlation with properties of a material thus can be also used for its texture evaluation. One of the simplest is a puncture test where probe is pushed into tissue and a maximum force is used as a firmness value.

As it was obtained by Christensen and Vickers [4] crispness may by the bone-conducted phenomena. Therefore, the acoustic emission (AE) method where a sensor is in a contact with material investigated is promising tool for food properties evaluation. In general, the acoustic emission is monitored during deformation of a material to provide information on cracking and internal friction of material pieces by analysis of AE descriptors like: amplitude, frequency, energy, counts, events, etc..

2. TISSUE MODEL AND SOURCE OF AE IN FRUIT AND VEGETABLES

The largest volume of fruit and vegetable tissue is taken by parenchymatous cells therefore this part of fruit is particularly often studied, also from the reason that it is relatively easy to cut a sample for testing. A simplified model of parenchyma tissue, which considers the most important mechanical actors, is shown in Fig. 1. The mechanical model is built of cells which are fluid filled and walls which are elastic-like. Cell walls are made of polysaccharides: cellulose fibrils network embedded in a matrix of pectins and hemicelluloses. Cells adhere to each other through middle lamellas which are made of amorphous pectins. Tissue structure also contains intercellular spaces, which for some fruits, like apple, can even take 25%. It is generally agreed that cell walls have elastic properties whereas pectins in the middle lamella are plastic-like.

External forces which cause deformation of such structure increase a pressure inside the cells and tension in the cell walls. Simultaneously, shearing forces between cells increase. Thus, two failure modes are possible: cell wall rupturing and cell-cell debonding when strength of cell walls or/and intercellular adhesion is overcome, respectively. These two processes can be the sources of acoustic emission in the case of cellular plants. However, acoustic emission from cell-cell debonding due to plastic character is less likely compared to sudden rupturing of elastic cell walls.

Studying of cracking process of plant tissue is thus indeed important for understanding of quality of fruits and vegetables. When tissue cracks between cells, for example for fruit and veggie which have been stored too long, the material may show a mealy character whereas the fresh one, just after picking up, cracking through cell walls reveals juicy and crispy properties which are usually desired by consumers. Mechanical strength of plant tissue at microstructure level is also important not only from the point of view the sensory properties. For example, microcracking is important for blackspot bruising of potato which starts enzymatic browning of the tissue under the skin.

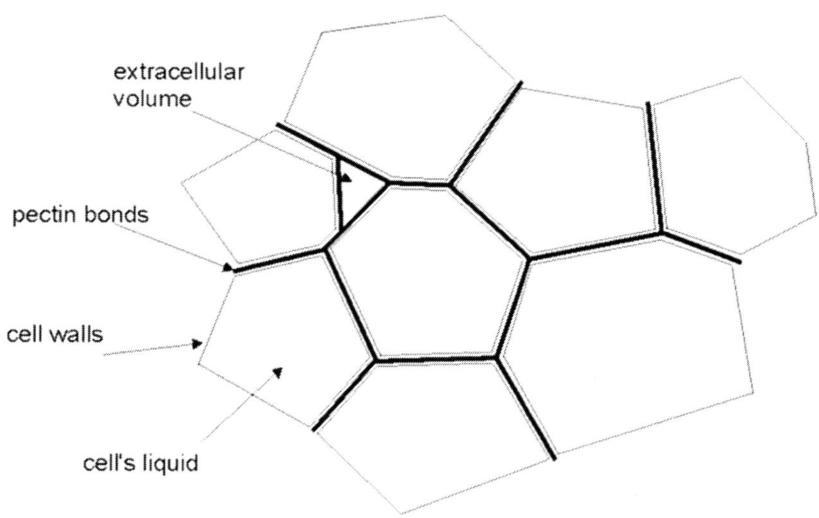

Figure 1. A simplified mechanical model of parenchyma tissue of fruits and vegetables

3. ACOUSTIC EMISSION (AE) FOR EVALUATION OF THE ONSET OF PLANT TISSUE CRACKING

As it is for other solid materials, acoustic emission has been found as a very useful method for monitoring of cracking of plant tissue under external loading. AE has been applied for this group of materials for the first time in the late 90's [6]. The first attempt was aimed on observation of the AE signal from damaged sample of potato tissue. Figure 2 presents scheme of the first apparatus used. A one-column and low noise testing machine was used for studying the mechanical properties and forces that participate in the process of plant tissue cracking. In that work a wide-band piezoelectric sensor was used for the recording of the acoustic emission signal with high-sensitivity in the frequency range from 25 kHz to 1 MHz. Due to a small size of the samples which are usually used in experiments with plants and relatively large deformations which can cause friction between sensor and sample, fixing the AE sensor directly on the sample was impossible. This problem was solved by fixing the sensor to the jaw of the testing machine, like it is shown in the Fig. 2. Since the AE signal passes from the material with lower density and enters into the material with higher density at the border between the sample and the jaw (sample of tissue – steel), sensitivity of measurement is sufficient to record even small AE events. In order to eliminate any friction and improvement of sound conductivity, silicon grease was applied on each boundary on the way of elastic waves from sample to sensor. A set consisted of a pre-amplifier (40dB) with a high-pass filter (25 kHz) and a low-noise amplifier with adjustable gain was used for signal conditioning. The set was supplemented with a high speed transducer card A/D that allowed for the recording of counts, events and energy in time intervals from 0.001 to 1 second or fast sampling with 2.5MHz short samples 0.25 s long.

The most useful method of analysis of AE signal bases on the transformation of the time signals to the form of descriptors recorded in time intervals. If a certain threshold is established for the amplitude of the received signal, called a discrimination threshold, then every time the amplitude goes above this signal is recorded as one count. Groups of the AE signal with the characteristics shape of a damped sinusoid curve are called events (Fig. 3). Instead of analysis the shape of event, it is possible to define an AE event as a group of counts recorded in consecutive samples. The number of counts and the number of events recorded in time gates are called count rate and event rate. Registered AE signal, presented in Fig. 3 in amplitude – time coordinates can be also characterized by AE energy E. Assuming that a signal event of duration t and of peak amplitude V of an event, then energy of each event can be evaluated as:

$$E = 0.5V^2t \tag{1}$$

Besides the parameters mentioned above, other parameters based on the transformations of time or spectrum of the AE signal could be also used. Detailed definitions and descriptions of the AE signal descriptors can be found in literature [7]. The system presented in Fig. 2 also allowed for simultaneous measuring mechanical and acoustical properties. For example, stress and AE counts as a function of strain could be recorded together as it is shown in Fig. 4.

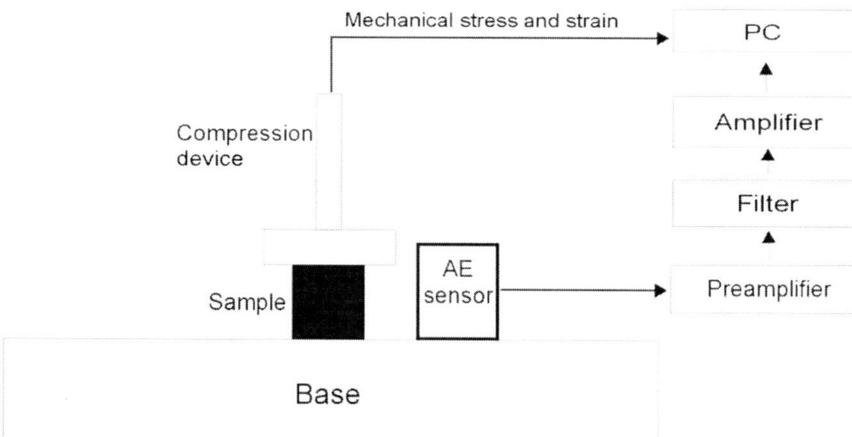

Figure 2. Scheme of the first apparatus used for acoustic emission recording from deformed plant tissue (based on the work of Zdunek and Konstankiewicz [6]). The EA 100 was the analyser which allowed both recording samples with 2.5MHz and conversion of the signal to AE descriptors.

In Fig. 4 it is easy to notice that for a decrease in the slope of the stress-strain curve (bioyeld and rupture points in this case), a high value of AE counts was observed. The highest values, however, appear at the moment of sudden rupture. Macro-cracks of the samples are clearly visible on the cross-sections of the sample after rupture, and sometimes are even audible to human ears. Before rupture, presumably some cracks also appear, particularly in the region close to bioyield, however it is difficult to detect them visually. The acoustic emission signal shown in Fig. 4 unambiguously proves that bioyield point and rupture could be assigned to tissue cracking, however maybe at different scales. Moreover, a long before bioyeld, AE signal has been detected too, although with significantly lower number of counts. In this region of deformation no noticeable decrease in the slope of the stress-strain curve was observed. Generally, the AE counts before bioyield is more irregular and have lower values in comparison to acoustic signal after bioyield point. It is believed that microcracking is developing gradually due to heterogeneous structure of the plant tissue even before it could be noted from force-deformation curve. This is particularly important because even a small crack of tissue, for example damage of intercellular plasmalemma, could start irreversible biochemical processes which decay quality of the material or affect its function.

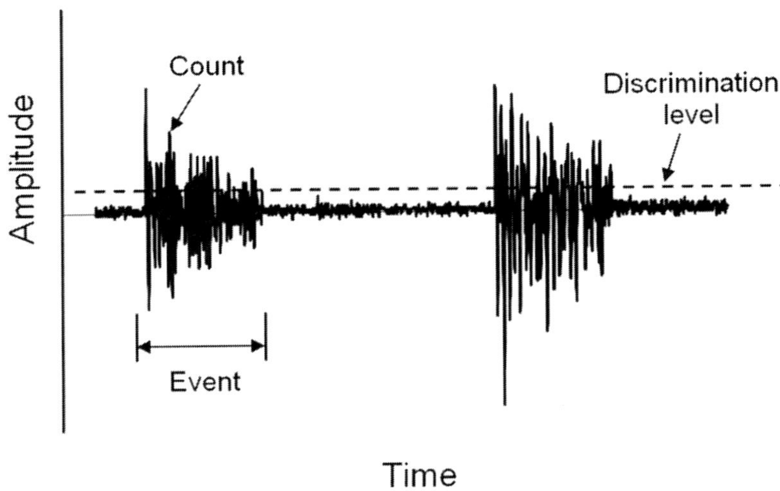

Figure 3. The method of determination of AE descriptors: counts and events from time-amplitude signal. The discrimination level is found experimentally to avoid external noises, for example from the loading device.

In the first paper on application of AE for potato tissue, new mechanical parameters have been proposed. Critical stress (R_c) and critical strain (ε_c) have been defined as the mechanical conditions at the onset of acoustic emission. In further studies critical stress and critical strain was analysed under different conditions of mechanical test (Fig. 5) and samples itself (Fig. 6).

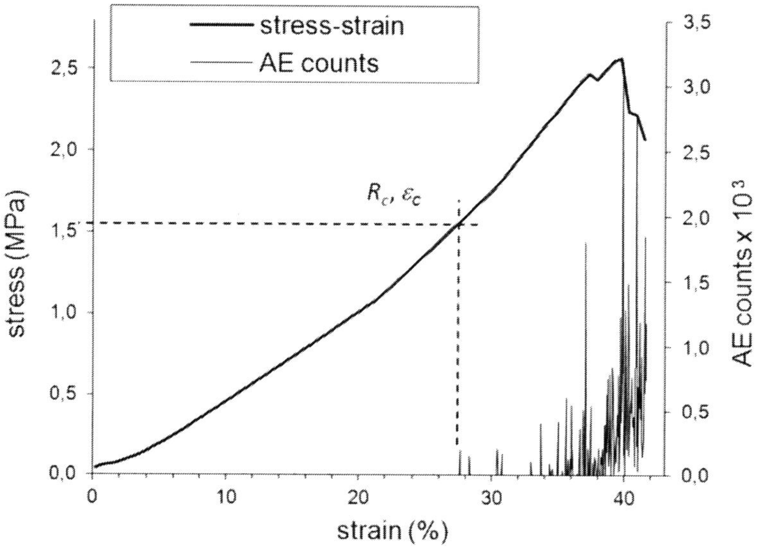

Figure 4. Examples of simultaneous recording of stress, strain and AE counts for potato tuber tissue. Bioyield point is visible as the short drop down of the stress-strain curve, rupture is visible as major final fall down of the stress, and the critical point as the onset of the acoustic signal (AE counts). The critical stress (R_c) and critical strain (ε_c) define the critical point (result obtained by the author)

Fig. 5 presents how the critical stress and critical strain change with strain rate of two potato cultivars (*Solanum tuberosum* cv. Danusia and Kuba). An increase in the strain rate decreases exponentially both parameters. From microscopic point of view, deformation of plant tissue causes changes in the cell shape. Since initially cells of parenchyma tend to have rounded shape due to incompressibility of the intracellular fluid, the ratio of cells surface to cells volume increases. This means that cell walls are generally stretched during deformation. The tension force in the walls is a function of strain rate and wall permeability. In a simplified model, low strain rate has an effect similar to that of high permeability in the model. At a relatively high strain rate, a seepage of the intracellular fluid through the walls is limited and leads to a higher tensions at the same cell deformation. When the strain rate is low, the intracellular fluid has relatively more time to flow out of the cells, and this produces smaller increases of the tensile forces in the walls. Thus, in this case the strength limit of the cell wall can be reached at higher cell deformation and higher external forces. For relatively slow cell deformation, the cell can even be completely compressed without wall rupture. This explains the pronounced increase of the critical values at very slow rates (Fig. 5).

The relations between critical stress, critical strain, and osmolality of mannitol solutions in which the samples were hydrated or dehydrated are shown in Fig. 6. Higher osmolality of the mannitol solutions corresponds with lower turgor of the tissue reached after 24h treatment. A strong influence of tissue turgor on critical stress and failure stress was observed. Both critical stress and failure stress increase in a linear manner when turgor decrease (direction of turgor change is shown in Fig. 6). The turgor effect can be interpreted in terms of a model of a single cell. Before deformation, higher turgor causes larger preliminary tension in the cell wall. Thus, the additional cell deformation or the additional external force necessary for wall rupture are lower.

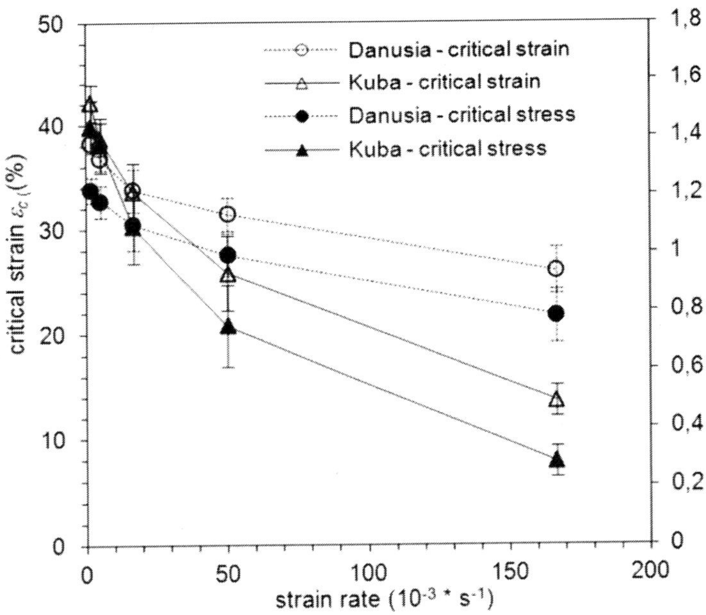

Figure 5. Critical stress and critical strain obtained as the onset of AE during compression of two potato cultivars 'Danusia' and 'Kuba' with different strain rates (result obtained by the author).

Application of the acoustic emission method has proven that micro-cracking of tissue starts significantly earlier than it can be observed on the stress-strain curve. However, no correlation between the critical values and the failure values of samples tested under the same conditions (the same strain rate or turgor) has been observed [8]. This means that observation of the critical values does not allow prediction of the bioyielding conditions for example. This is presumably result of the fact that critical point and the bioyield point are different stages of cracking. Between them cracking is developing, from the first local micro-cracks to large macro-cracks. This propagation mat be very chaotic and accidental due to heterogeneous microstructure of a tissue.

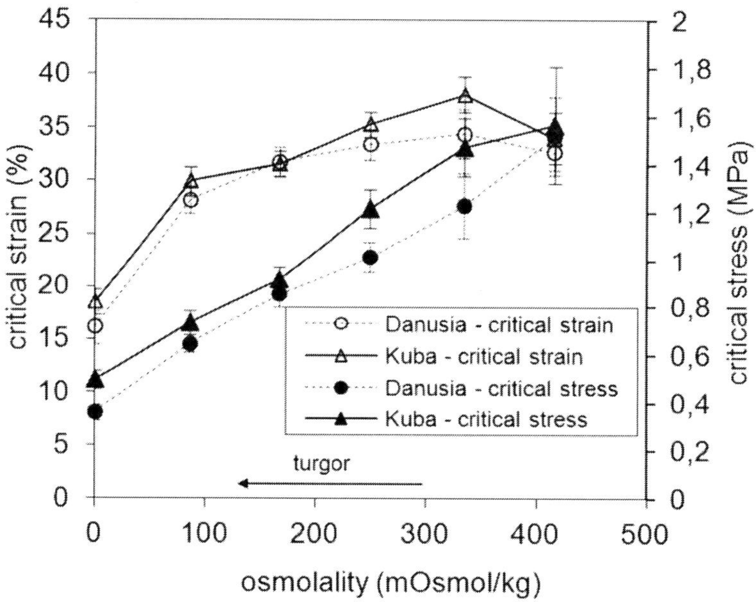

Figure 6. Critical stress and critical strain obtained as the onset of AE during compression of two potato cultivars 'Danusia' and 'Kuba' dehydrated and hydrated in different mannitol solutions (result obtained by the author).

4. ACOUSTIC EMISSION AND DIFFERENT MECHANICAL TESTS

4.1. Puncture Test

Application of acoustic emission is possible in any mechanical test which are used commonly for plants. The key issue is to apply AE sensor to the sample which in the case of plants is usually largely deformable. Solution presented previously uses indirect attachment through solid body. This solution was proven to be very effective in various mechanical tests. Sensors of AE could be placed inside device for mechanical tests, exactly in the probe for material deformation. Such device for puncture test is presented in Fig. 7. Here, acoustic emission during the puncture test is caused and recorded by head with two sensors placed inside. The head consists two parts. Top part is made of ertacetal and the bottom part is made of duraluminium which effectively conducts elastic waves. The application of two different material was intended to limit eventual disturbances from mechanical system. They are screwed to each other. Acoustic emission sensors are glued (or it can be screwed also) to the top surface of the metal part. In the system presented in Fig. 7, one sensor works in audible range 1-16kHz (SA), whereas the second sensor has maximum sensitivity in ultrasound range 25-100kHz (SU) to cover as wide as possible frequency region. The sensors are connected to individual amplifiers with adjustable amplifying. The signal is filtered in the range 1 kHz -20 kHz for lower band and 10kHz-900 kHz for higher band. Next, the analogue signals are converted into digital one by A/D boards. Sampling rates per channel: 44 000 and 150 000 samples per second are more than double of the frequency range of the sensors used. The second channel of each card is used for recording an analogue signal of force delivered from universal testing machine to synchronize both acoustic and mechanical signals.

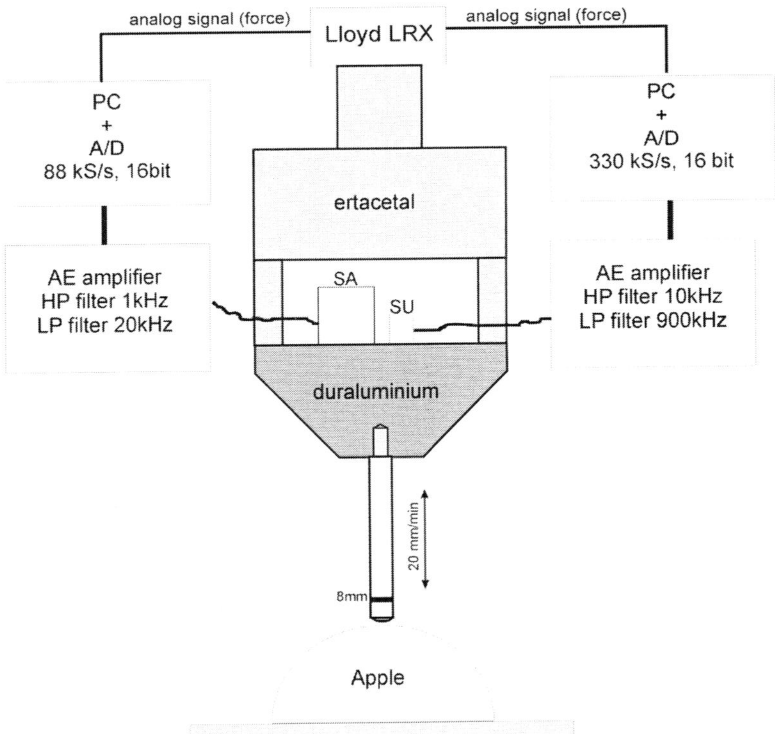

Figure 7. Scheme of the system for recording of acoustic emission from punctured fruit. Two channels were used: 1-16kHz and 20-75 kHz. Puncturing was performed with Lloyd LRX device. SA is the sensor for audible frequencies, SU is the sensor for ultrasound frequencies. Dimensions of the puncture probe are exemplary (scheme by the author).

In Fig. 8, typical profiles of acoustic emission counts recorded during puncturing of apple flesh are shown. In the case of apple puncturing, the AE signal starts just from the moments of touching puncture probe to tissue. The number of counts increases progressively up to a moment when force-deformation curve yields. At this moment the whole curved part of the probe is in a contact with the tissue. When the probe goes deeper into apple tissue the acoustic activity decreases. This could be result of damping of acoustic waves by surrounding tissue and already damaged tissue layers under the probe. In many studies, softening of apples during ripening and storage has been reflected in a lower penetration force (lowering firmness). The integrative use of contact AE and the puncture test showed that major acoustic signals are observed together with drops of force. The coincidence was interpreted as an energy release in the form of sound as a result of material fracturing. In the case of plant tissues, AE signal comes mainly from rupture of the cell wall because of its somewhat elastic properties, whereas the middle lamella due to plastic properties rather do not generate sound. This hypothesis is supported by analysis of AE during ripening of apples, which will be discussed later on in this chapter.

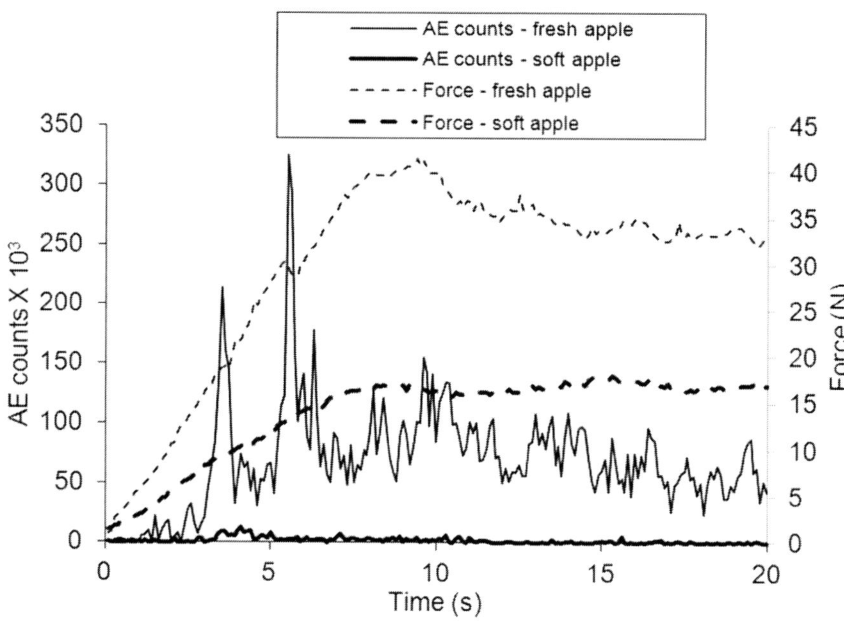

Figure 8. AE counts and force recorded during puncturing of fresh (1 day of shelf life) and stored (10 days of shelf life) apples (result obtained by the author).

In Fig. 9, spectrum of the signal is presented from two frequency ranges in a form of "acustograms". Colors in the acustograms represent a power of the signal in time-frequency coordinates. A few dominant frequencies can be found: 5.5 kHz, 9.5 kHz,15 kHz, 32 kHz, 44 kHz and 56 kHz. They constantly appear during puncturing. Precise analysis showed that they are also characteristic for the system used because no significant changes were found with properties material used. The only one change observed with change of properties of the material was an overall change in amplitude that occurred uniformly for all bands. Additionally, Fig. 10 presents relation between AE energy in 1-16 kHz and 20-75 kHz obtained for a large set of apples in puncture test. It is shown that the relation is very linear and in the case of this material, investigation in higher band does not provide any additional information.

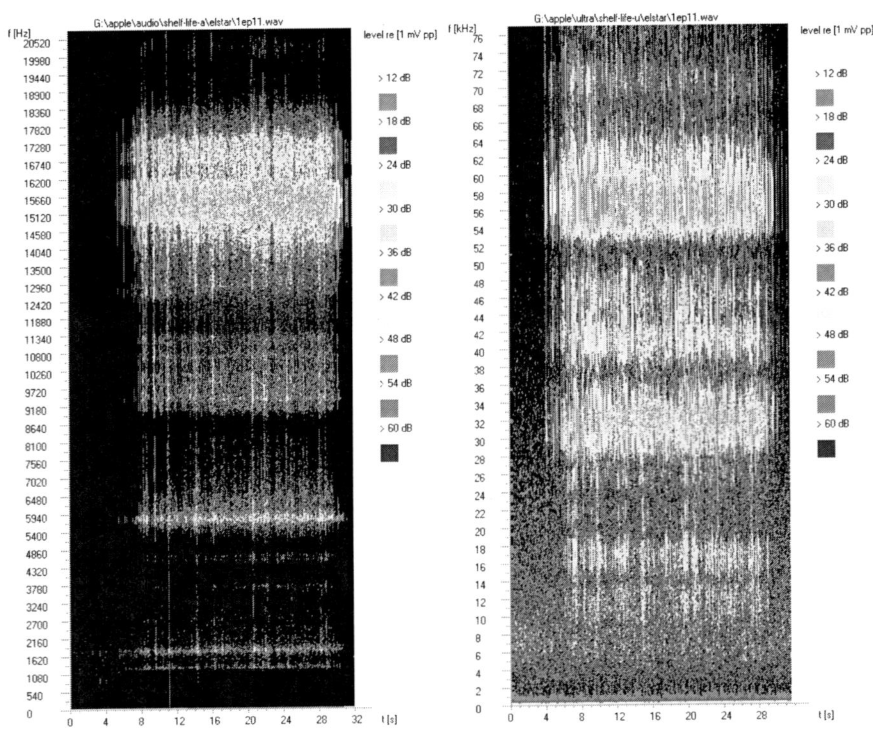

Figure 9. Acustograms of apple tissue in puncture test within frequency range 1-16 kHz (left) and 20-75 kHz (right), f – frequency, t – time (result obtained by the author).

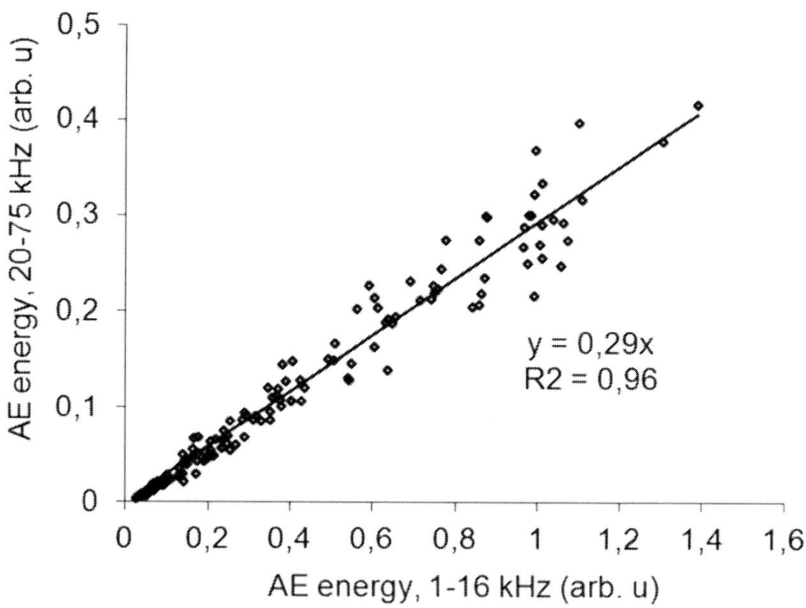

Figure 10. Relationship between the total energy of acoustic signal recorded within frequency range 1-16 kHz and 20-75kHz for apples in puncture test (result obtained by the author).

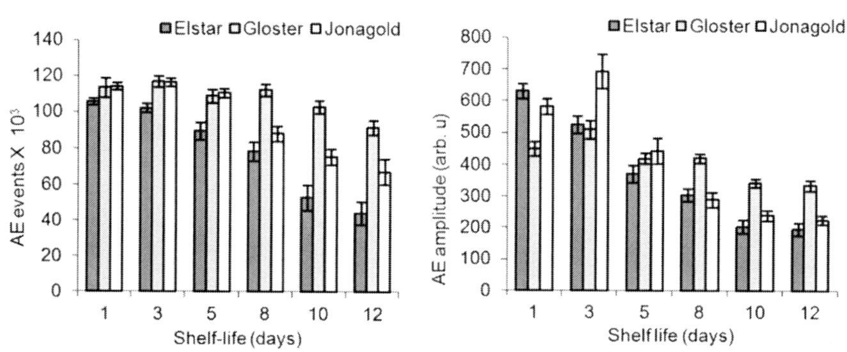

Figure 11. Total number of AE events and mean AE amplitude registered in puncture test of apples (three cultivars) during shelf life storage (result obtained by the author).

Fig. 11 presents changes of total AE events and mean AE amplitude during shelf life storage of three apple cultivars. Data was obtained in puncture test. AE descriptors decrease almost linearly during shelf-life storage revealing large Pearson's correlation coefficients *R* with time of storage (Table 1). InFig. 11 is visible that the acoustic emission method is very sensitive for registering changes that occur during postharvest storage of apples. Total number of AE events and mean AE amplitude usually shows higher *R* value when compared to firmness from puncture test (Table 1).

Table 1. Correlation coefficients *R* of changes of AE events, AE mean amplitude and firmness, in puncture test of apples with shelf-life days for three cultivars 'Elstar', 'Gloster' and 'Jonagold' (results obtained by the author).

Variable	R		
	Elstar	Gloster	Jonagold
AE events	-0.80	-0.54	-0.81
AE amplitude	-0.90	-0.70	-0.88
Firmness	-0.80	-0.40	-0.82

The postharvest softening of apples is caused by biochemical processes. During apple ripening two major processes occur that affect the mechanical properties of the tissue including its fracturing mechanism. Pectin degradation during ripening causes a decrease of adhesion between cells leading to tissue softening and changes the fracturing mode toward cell–cell debonding. As a result of respiration and metabolism during storage, turgor pressure can decrease, which has consequences for the fracturing process. The lower tension of the cell wall at low turgor causes a greater deformation that leads to wall fracture. Another consequence of low turgor is a decrease of cell–cell adhesion. Thus, in general, degradation of pectins and lower turgor lead to changes of fracturing mode toward cell–cell debonding. Due to the microstructure, the thin cell wall in plants is considered as an elasto-plastic material. The elasto-plastic character of the cell wall is responsible for the brittle fracturing necessary for sound generation. The intercellular lamella between cells consists of amorphous pectin and are considered as plastic. It is probable that the pectin plasticity causes slow dissipation of strain energy and no brittle fracturing without sound generation. Therefore, it is most likely that the sound made during puncturing is generated mostly when cell walls fracture. In material science terms, a crack propagates if there is a stress concentration into a small tip zone. Thus, propagation is ineffective if there is any plastic zone (in the case of apples it would be pectin). Other features that halt crack propagation are the cell interiors or intercellular spaces. There is ~ 25 % space within apple tissue, and this amount increases with ripening. Thus, ripening attenuates conditions of cracking propagation. Again, in terms of material science, the cell wall (as the material where the stress can concentrate) has a key role in crack propagation.

The maximum number of acoustic events recorded after pushing the probe into the apple flesh was about 10^5 (Fig. 11), which was obtained for fresh apples immediately after harvest. This number agrees roughly with the number of fractured cell walls during the test estimated on the basis of two assumptions; that the mean cell diameter is ~0.25 mm, which is true for apples, and all cells in the path of the probe are damaged (i.e., all cell walls are broken). This result shows that the breakage of each cell wall would be the source of single acoustic event. The mean AE amplitude depends on a stress value in the source of cracking (i.e. the strength of the cell wall) and on the attenuation of the elastic waves on the way from the source to the sensor. As mentioned above, pectin degradation occurs during apple storage, which can decrease the strength of the cell walls due to an increased mobility of cellulose fibrils in the pectin matrix. On the other hand, softening of the bulk tissue caused by pectin degradation in general and the decrease of turgor, increases the attenuation of the elastic waves. Both causes that amplitude decrease with softening of apples.

4.2. Texture Profile Analysis (TPA)

Texture profile analysis (TPA) is used for simulation of eating process. Compression test is performed in two cycles to the same deformation level of a sample. Scheme of TPA test and graphical representation of TPA descriptors used for sample characterization are shown in Fig. 12. Texture profile analysis is performed on cylindrical samples in two cycles. Maximum deformation applied could be about 20-40% of initial sample height for both cycles, depending on sample strength (20% for apple, 40% for potato). Maximum deformation I should be close to failure points of investigated material. TPA requires cracking of the test material to simulate the destructive process during eating. On the other hand, deformation should not be too far to prevent the compression of the small pieces of the initial sample in the second cycle which causes springiness and cohesiveness to become physically meaningless. The probe always returns to the trigger point after the first cycle. No rest periods is programmed between the TPA compression cycles to avoid material relaxation. The textural parameters are calculated in the following way (Fig. 12). Hardness 1 is the force peak of the first cycle. Hardness 2 is the force peak of the second cycle. Cohesiveness is calculated as the ratio of the area under the curve of the second cycle to the area under the curve of the first cycle. Springiness is the ratio L2/L1, where L2 is the time or distance from the beginning of the second cycle to hardness 2 point and L1 is the time or distance from the beginning of the test to the hardness 1 point. Acoustic emission during the TPA could be recorded using the same head as for the puncture test described above.

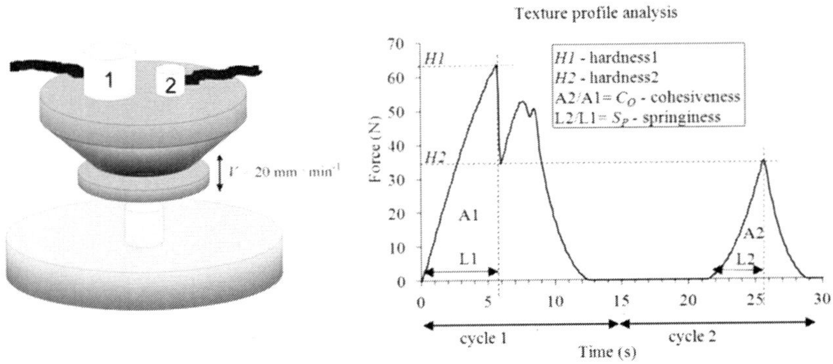

Figure 12. Scheme of TPA test performed on cylindrical sample of plant tissue. Proposed positioning of the AE sensors (1 and 2 with different frequency range) is shown. The upper plate for compression is removable to change a probe for other mechanical tests. Graph to the right presents typical TPA curve with the method of calculation texture descriptors (by the author).

Typical TPA curves together with acoustic emission counts for apple and potato are shown in Fig.13. For comparison, results for two apples, fresh and soft one are plotted, and for hydrated potato sample. The fresh tissue has higher hardness 1 which was reached earlier than in the case of the soft tissue. The range of macro-cracking, visible as the gradual force decrease in the first cycle, is longer and more jagged. The second cycle of TPA also shows larger forces in comparison to soft sample. It is typical that for fresher samples and with higher turgor, the failure occurs at a lower deformation or earlier on the time axis. Acoustic signal appears earlier and it has higher values in the case of fresh apple than in the case of soft one. In apple, acoustic counts are recorded almost from the beginning of the compression. This would be a result of both weaker cell walls and intercellular bonds than for potato which is actually very dense and strong tissue. Failure is accompanied by high acoustic emission counts for both materials as a result of macro-cracking (Fig. 13). This moment is also usually air-conducted and audible. The acoustic emission in TPA is recorded mainly during the downward movement of the machine probe. During the upward movement, a small signal is only observed just after the probe starts returning. It disappears at the end of the returning stage. The second cycle of TPA may also cause acoustic emission. However, the signal is usually weak especially in the case of apple. The second cycle in TPA starts from the trigger point of the first cycle. Thus, the time of deformation during the second cycle is related to recovery of the material after the first cycle. If a crack occurs during the first cycle, it can propagate during the second one. If the material failed during the first cycle (macro-cracking occurred) relaxation of the material is less and deformation in the second cycle is also smaller. Therefore cracking propagation is less and, as consequence, acoustic emission is low. In other words, the weakening of the material in the first cycle causes only small acoustic emission due to the propagation of already existing cracks within the material during the second cycle.

Fig. 14 and Table 2 show decrease of acoustic descriptors during shelf life and that the correlation with days of storage is in general similar as for mechanical descriptors from this test.

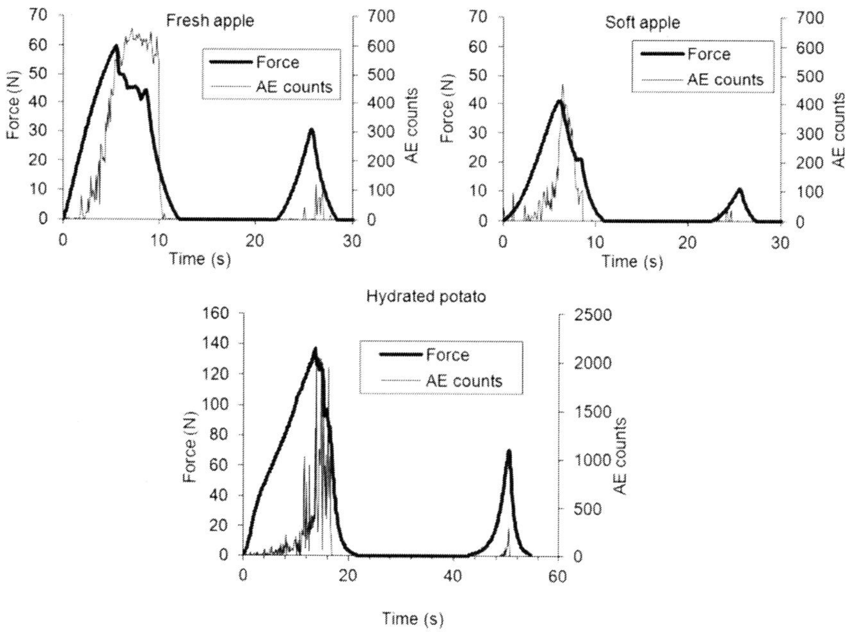

Figure 13. Typical TPA curves with AE counts for fresh and soft apple, and for hydrated potato sample (result obtained by the author).

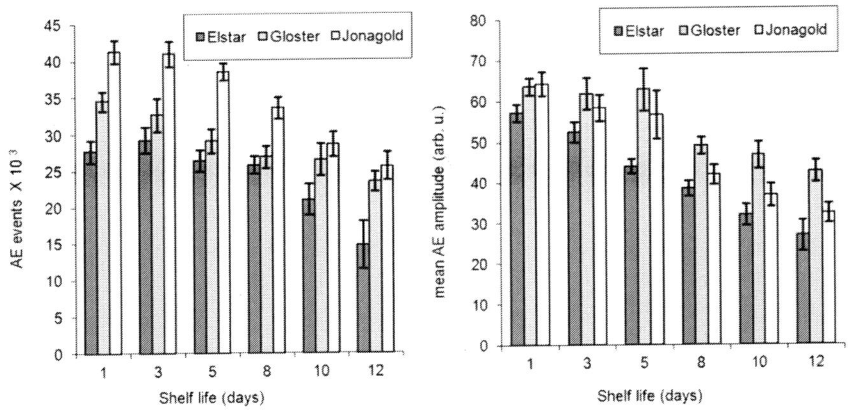

Figure 14. Change of acoustic descriptors obtained in TPA test during shelf life storage for three apple cultivars (result obtained by the author).

Table 2. Correlation coefficients between AE events, mean AE amplitude, hardness H1 and H2, cohesiveness C_o, springiness S_P in TPA test and shelf life days for three apple cultivars 'Elstar', Gloster' and 'Jonagold' (result obtained by the author).

Variable	R		
	Elstar	Gloster	Jonagold
AE events	-0.58	-0.64	-0.78
Mean AE amplitude	-0.83	-0.70	-0.82
H1	-0.78	-0.19	-0.85
H2	-0.78	-0.33	-0.83
C_O	-0.45	-0.19	-0.44
S_P	-0.23	0.27	0.38

4.3. Single Edge Notched Bending (SENB)

Recently, new engineering mechanical tests has been introduced for analysing the fracture properties of plant tissue, so called single edge notched bending (SENB) [9]. In the test rectangular sample with a notch is bended to breaking up. From sample geometry and from failure force obtained from force-bending curve, a critical stress intensity factor *Kc* can be calculated. This material parameter is tried to correlate with textural properties of a tissue, like crispness or crunchiness. However, from mechanical point of view, the critical stress intensity factor is a force criterion for starting cracking propagation up within material.

Single edge notched bending is performed on rectangular beams according to the ASTM Specification E-399 standard. It is very often that fruit or veggies size or geometry does not allow cutting desired by the standard sample dimensions, which should be also sufficient to produce detectable acoustic emission. According to the standard, S/W=4 (span/height) is suggested [10]. Sample of potato tissue of height W=16 mm and width B=8mm emits strong enough signal in the system showed in Fig. 7 andFig. 15. Although, to keep the ratio the span should be 64mm, it is usually difficult to cut samples longer than L=40mm from typical potato or apple.. Therefore, the span most often must be shortened to S=32mm for example, which is reasonable and gives S/W=2 ratio. According to standard, in the middle of the sample a notch with depth of a=8mm is cut.

Scheme of SENB test configuration is presented in Fig. 15. Acoustic emission during the SENB could be recorded using the same head as for the puncture test described above where one or more AE sensors could be placed. Sample is placed on support with the notch to the bottom. Bending is performed up to fracture of the sample.

SENB allows determination of a critical stress intensity factor *Kc*. The *Kc* can be obtained using formula:

$$K_c = \frac{P_C S}{BW^{\frac{3}{2}}} f\left(\frac{a}{W}\right). \tag{2}$$

where: *S*- span, *Pc* is a failure force. Function *f(a/W)* is given as:

$$f\left(\frac{a}{W}\right) = \frac{3A\left(\frac{a}{W}\right)^{1/2}}{2\left(1+\frac{2a}{W}\right)\left(1-\frac{a}{W}\right)^{3/2}}. \quad (3)$$

where:

$$A = 1.99 - a/W\left(1 - \frac{a}{W}\right)\left(2.15 - \frac{3.93a}{W} + \frac{2.7a^2}{W^2}\right). \quad (4)$$

Kc has a physical meaning if following formula is true:

$$B_C \geq 2.5\left(\frac{K_C}{\sigma y}\right)^2. \quad (5)$$

where: *Bc* is minimal width of a sample used, *σy* is a failure stress in uniaxial compression of the same material.

Figure 16 presents typical SENB curve and acoustic emission events for fresh and soft apple. AE signal starts just from the beginning of bending which suggest that cracking propagation also starts from the tip of the notch. For fresh apples, which is also harder and has higher *Kc* value, AE is significantly larger than for the soft material however in both cases acoustic emission lasts up to sample fracture.

SENB test, similar to puncture and TPA, is able to distinguish sample according to its softness. Fig. 17presents example for three apple cultivars which were stored at shelf life conditions. It is visible that acoustic descriptors diminishes during storage. Table 3 presents correlation coefficients of parameters obtained from SENB test with time of shelf life storage. The coefficients for acoustic descriptors are higher than these for mechanical descriptors which shows again that AE method is very suitable for monitoring properties of fruits.

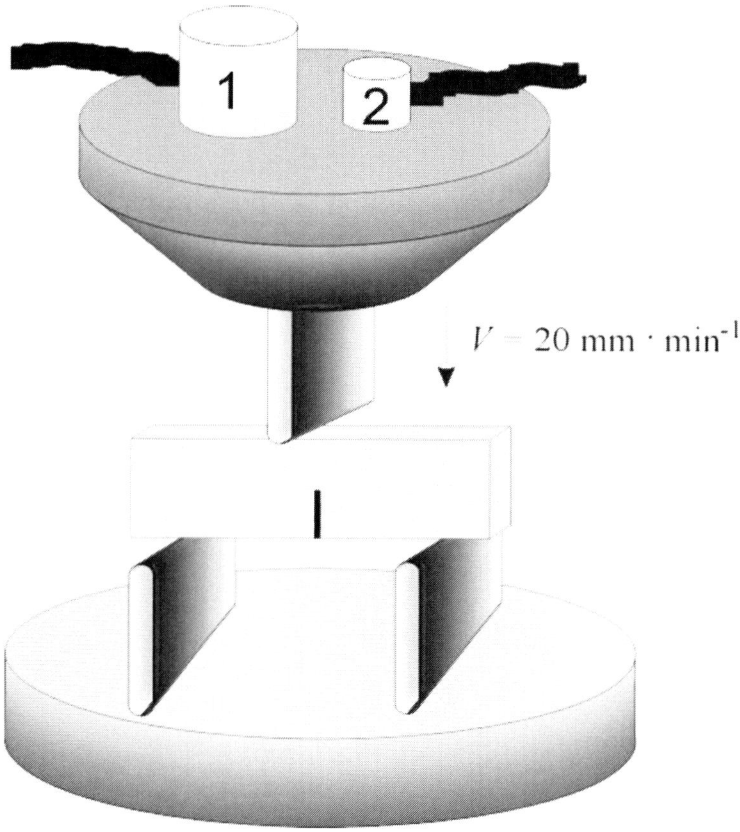

Figure 15. Scheme of the single edge notched bending SENB test for plant tissue with locations of AE sensors (1 and 2, for audible and ultrasound range for example), (by the author).

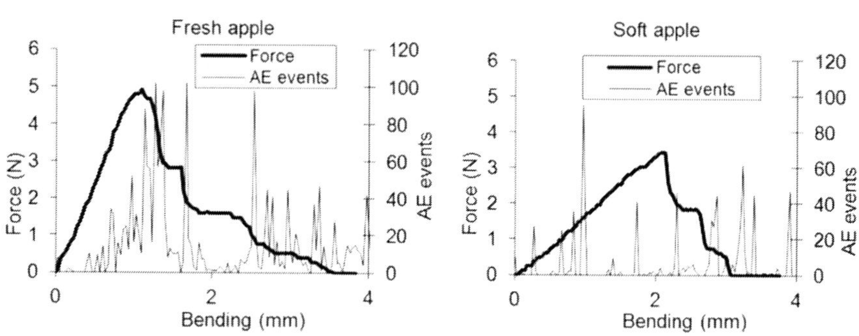

Figure 16. Typical SENB curves with acoustic emission for fresh and soft apples (result obtained by the author).

Table 3. Correlation coefficients *R* for changes work to maximum force, *Kc*, AE events, mean AE amplitude in SENB test and days of shelf life for three apple cultivars 'Elstar', 'Gloster' and 'Jonagold' (result obtained by the author).

Variable	*R*		
	Elstar	Gloster	Jonagold
AE events	-0.77	-0.43	-0.83
Mean AE amplitude	-0.78	-0.47	-0.84
Work to maximum force	-0.52	-0.14	-0.64
K_c	-0.70	-0.18	-0.73

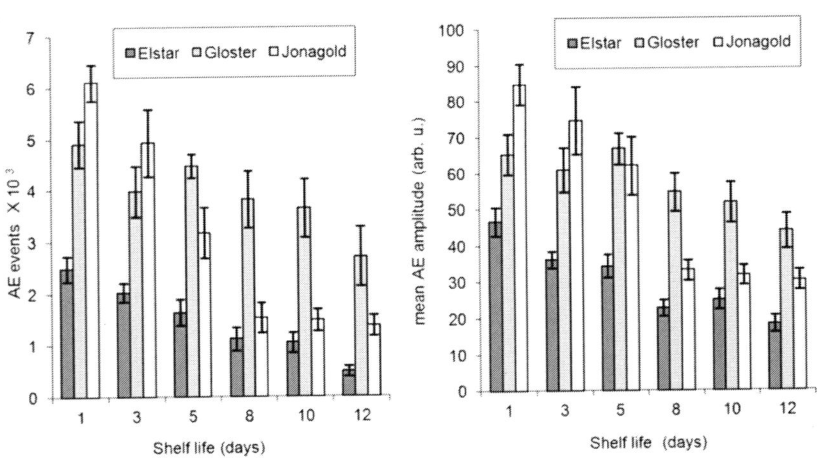

Figure 17. Change of acoustic descriptors obtained in SENB test during shelf life storage for three apple cultivars (result obtained by the author).

5. EVALUATION OF SENSORY PROPERTIES WITH ACOUSTIC EMISSION

There are four key factors of food quality: appearance, smell, texture, and nutritional value [11]. The first three are known as sensory acceptability factors, since they are perceived by the human senses and can be evaluated directly by the consumers. Sensory acceptability of food products is incredibly important, since people want to enjoy eating their favorite products. It can also be difficult to convince consumers to eat healthy products that are unappealing in terms of appearance and texture. Food gives us pleasure not just through its flavour or fragrance; we also want to be aware that what we are eating is fresh. In case of fruit, we associate the latter with mechanical qualities; fruits are desirable when their texture is crunchy, crisp and juicy, and less so when they are mealy.

From a mechanical perspective, crispness, juiciness, and mealiness are all associated with how the cellular structure is broken down. If biting into an apple causes the cell walls to rupture releasing intracellular juices, it makes the apple feel juicy and crispy. This is because of the acoustic signal generated as part of the process, which is perceived positively by our auditory system. It is believed also that crispness can be perceived as a combination of acoustic impressions and the strength required to break down the product, while the acoustic signal is largely perceived as vibrations by the jaw bone (bone-conducted sound). Once the cellular walls rupture, the fruit takes on a mealy quality and the fruit is generally perceived to be overripe.

Texture is a sensory characteristic; assessing it objectively is extremely difficult since consumers' personal and cultural predispositions vary greatly, and perceptions can even depend on the person's mood or frame of mind at the time. Texture of fruits and veggies is also not a constant feature, and is affected by many factors, such as natural biological variability, treatment prior to picking, time of picking, and method and duration of storage. This is why it should be monitored on an ongoing basis, while at the same time the measurements should be simple, repeatable, and low-cost. Unfortunately sensory assessment conducted by a professional panel or representative group of consumers does not meet these criteria.

Since crispness may be the bone-conducted phenomena the approach of utilizing acoustic emission with use of sensor in contact with sample is appropriate way of instrumental analysis of the sensory texture sound-related properties. An advantage of this approach is relatively low sensitivity to external noises comparing to air conducted methods, like these ones which use microphones placed close to sample. The use of the "contact" acoustic emission while mechanical test has also advantage of recording both important for consumers attributes: acoustic and mechanical ones. Typically, system used must be calibrated with reference to standard sensory analysis. Descriptors from instrumental method, independently or as a combination, should be compared with sensory texture attributes to provide the most robust calibration model as possible.

Despite of various mechanical methods used for quality testing of fruits, described previously, the puncture method is still the most popular. This simple puncture test has been used for a long time in laboratories, orchards and industry. The output of the test is firmness value expressed in Newton (N) defined as the maximum force needed to push probe into fruit flesh. In the most common configuration of the test, probe of 11.1 mm with dome-shaped ending with a radius of curvature of 8.73 mm is pushed 8 mm into the fruit. These settings are valid especially for apples. For other fruit they can be adjusted according to their hardness and dimension.

For acoustic emission, the system presented in Fig. 7 may be used. It could be a laboratory system with commercial universal testing machine (machine noise at desired speed should be considered) completed with a low noise set up for AE recording and the most important: correctly chosen sensor. Since the goal is to relate sensory perception with the instrumental method, the frequency range of sensor used can be limited to the audible range: 1-16 kHz. This range can be covered easily by one sensor only. The use of commercial devices provides possibility of easy adjusting of settings to different materials and application of different mechanical loadings programmes, however it is relatively expensive solution. Recently the first simplified system has been developed for apples only (Fig. 18). The CAED (contact acoustic emission detector developed by the author) has a fixed puncture probe and the parameter of the puncture test adjusted exactly for apple. The device uses an accelerometer with sensitivity within the audible frequency range. To avoid large data sets, electronic converts time-amplitude signal into counts in 0.1s time intervals. Counts and actual force can be exported to ASCII whereas sum of all counts (called total AE counts) in the test and firmness are displayed after each test.

A different instrument for texture evaluation was proposed by N. Sakurai's team from Japan (Fig. 19). The device uses a piezoelectric element, attached between wedge type probe for inserting into investigated material and piston driven by hydraulic mechanism [12]. Vibrations, caused by destruction by the wedge type probe of investigated fruit or veggies, are detected by piezoelement. The absolute amplitude in Volts (V) and time of duration of the signal (T) has been used for definition so called Texture Index (TI) according to the formula:

$$TI = \frac{\sum |V|}{T}. \quad (6)$$

TI could be determined within several frequency bands to check witch of them could discriminate a sample. TI has been used for many fruits and vegetables as well for dray food products which showed that TI has frequency related pattern characteristic for different objects. TI was also compared with sensory texture of persimmon which showed that correlation of TI with several texture attributes (sweetness, juiciness, thickness, hardness, fragrance, appearance, and overall acceptability) can reach 0.8, particularly in the frequency range lower than 3 kHz [13].

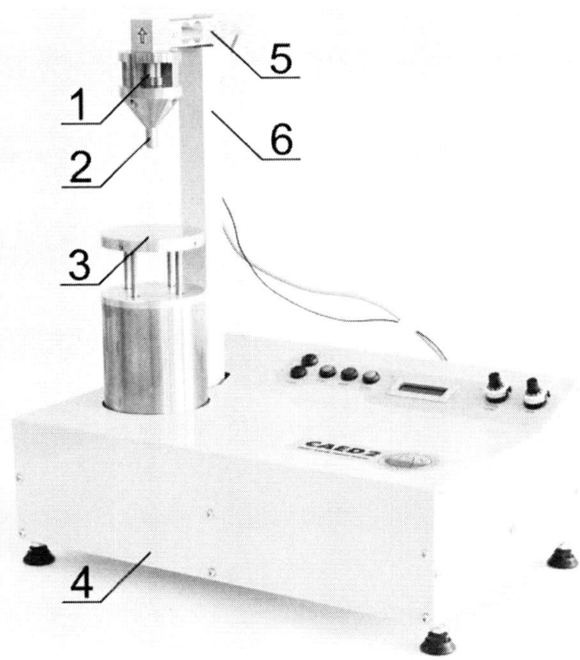

Figure 18. Contact acoustic emission detector (CAED) for apple testing developed by the author. Device uses AE sensor (1) which is placed in the AE head ended by the puncture probe (2). Apple is lift up by a motorized stage (3) to puncture probe. Force is recorded by the force sensor (5) with capacity of 200 N. Electronic (4) calculates on line AE counts and records the actual force within 0.1s time intervals. Sum of counts and firmness (N) are displayed on the screen after the test (photo by the author).

To calibrate the instrumental method with the use of acoustic emission, a generic descriptive analysis is a suitable method for obtaining sensory texture attributes. Sensory testing laboratory should fulfils the general requirements of a standard, as an example ISO 8589:1988 standard for sensory testing conditions. Each test booth should be equipped with a system for data acquisition from panellists. The expert panel usually consists more than 6 trained persons selected on the basis of the ability of individuals to discriminate tastes and texture attributes. Before the experiment, the panellists usually take part in a training session, where definitions of attributes are discussed and clarified (as in Table 4). For the experiment pieces of fruits are assigned a code and the samples are presented to panellists in random order. During the experiment the panellists determine the perceived intensity of texture attributes using linear, unstructured scale with a range of 0 – 100 points. After the test, the results are often converted to the most frequently used: 10-point scale.

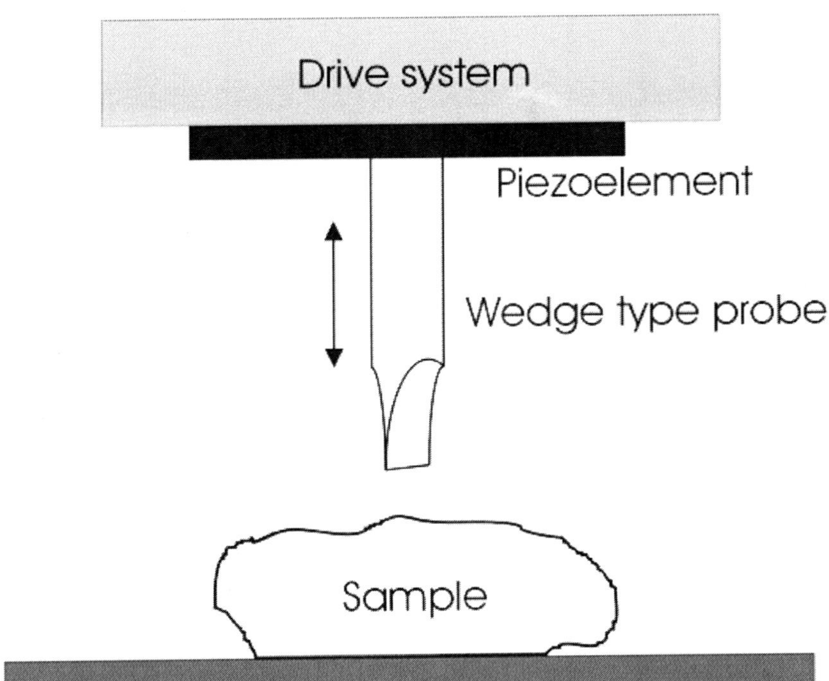

Figure 19. Scheme of device for texture index (TI) evaluation (scheme based on Taniwaki et al. [13]).

Table 4. Definitions and scale of some sensory texture attributes.

Sensory texture attribute	Definition	Scale
Crispness	The sound intensity during the first bite with the front teeth	0 = no sound.100 = very noisy
Hardness	The resistance during the first bite with the front teeth	0 = very soft. 100 = very hard
Juiciness	The sense of juice release during biting	0 = no juice. dry. 100 = very juicy
Mealiness	The mealy sense, especially on the tongue and the palate	0 = not mealy. 100 = very mealy
Overall texture	The overall sensory harmonization of textural attributes	0 = bad. 100 = very good

In the case of CAED which provides mechanical and acoustic indexes, for construction calibration models, several methods can be used: simple linear regression, multiple linear regression or multivariative regressions. For construction the models, averaged values from 10 apples (totally 244 samples from 19 apple cultivars) were taken, as it is usually assumed for sensory analysis to minimalize individual preferences. Examples of statistics for different calibration models are presented in Table 5(after Zdunek et al. [14]). These data were obtained for different 19 apple cultivars, which were stored in various ways. This example shows that the performances of the linear regression models are satisfactory for crispness and hardness prediction by both firmness or by total AE counts however quantitative prediction is impossible in any case using this modelling approach. Crispness is slightly better predicted by total AE counts than by firmness when these individuals are taken for simple linear model whereas hardness is apparently better predicted by firmness than by acoustic variable. It is presumably due to different origins of the variables: sensory crispness is governed mostly from auditory phenomena whereas sensory hardness from mechanical one. Table 5 presents also performance statistics of multiple regression models (MLR) where both firmness F and total AE counts were considered in the linear model. General improvement of models is observed in the case of each sensory attribute. Furthermore, multivariative principal components regression (PCR) models, where total AE counts and firmness are used as the predictors of a group of sensory variables, show remarkable improvement of calibration performance comparing to linear regression and multiple regression models. Full cross validation (CV) in the PCR for showed that satisfactory prediction is possible in the case of hardness. The models allow for prediction also crispness and overall texture with slightly less accuracy. In the case of juiciness, successful prediction seems to be doubtful whereas mealiness prediction is impossible. Test set validation (TSV) method showed apparently better model performance in the case of crispness and slightly better in the case of juiciness whereas for the rest of sensory attributes performance from TSV method is worse that from CV method. In general both validation methods show satisfactory prediction of crispness and hardness from multivariative PCR calibration models.

The model improvement, when both acoustic and firmness are considered in calibration models, agrees with the hypothesis that crispness perception should be interpreted as counteraction of acoustic and mechanical phenomena. It is usually observed that firmer apples are also more crispy. In Fig. 8 it is visible that firmer apple has more jagged force-deformation (FD) profile during puncturing whereas soft apple has more smoother one. It was accompanied with higher AE counts at the each force dropping down. One can say that firm apple is also more brittle. The jaggedness of the FD is important from the point of view crispness because humans can detect loads of less than 0.1 N. Such interpretation is especially true for dry food stuff however there is no reason to refuse it for plant tissue where sound is produced mainly from cell wall breakdowns and it could cause the momentary force dripping down. It has been shown for dry food that acoustic and mechanical parameters related with saw like force profile could be used for sensory crispness measurement [15], thus presumably in a future it will be the case also for fruits and vegetables.

The above calibration models for CAED were obtained with use of averaged values from 10 apples for the each calibration point. Taking into account that RMSEP value of the calibration models is slightly less than 1, an error of prediction is not larger than ±1. Since descriptive sensory analysis uses the 10 grade scale, the PCR calibration models allow for classification of sensory attribute to one of the 5 grades. This is very satisfactory results taking into account that the results obtained is less expensive and testing of the 10 apples lasts less than 10 minutes only. This means that instrumental evaluation of fruit texture with use of combination of sound-related descriptors and mechanical descriptors could replace soon sensory panels as it is faster, and – as is usually the case with technology – it is objective and does not suffer from fatigue.

Table 5. Performance statistics of linear regression models, multiple regression and principal component regression models for prediction sensory texture attributes of apples by CAED (after Zdunek et al [14]). Ncal – Number of samples used for calibration. Ntest – Number of samples used for validation, F-firmness, C_{AE} – total AE counts, CV-cross validation, R^2 – determination coefficient, RMSECV - root mean squared errors of cross validation or RMSEP - root mean squared error of prediction, RPD - ratio of prediction to deviation calculated as the ratio of standard deviation of validation data set to RMSECV or RMSEP. If the RPD was below 1.5 the model is not useful, and when the value was higher than 2, the model can predict quantitatively sensory attributes [16].

Variables used for calibration	Validation method	Performance statistic of validation	Calibrated sensory texture attribute				
			crispness	hardness	juiciness	mealiness	overall texture
Linear regression F	CV	R^2	0.57	0.68	0.40	0.38	0.52
		RMSECV	1.10	0.86	0.95	0.83	1.05
		RPD	1.53	1.75	1.29	1.27	1.45
Linear regression C_{AE}	CV	R^2	0.62	0.60	0.48	0.33	0.52
		RMSECV	0.98	0.95	0.89	0.87	1.06
		RPD	1.72	1.63	1.45	1.27	1.50
Multi-linear regression F and C_{AE}	CV	R^2	0.71	0.77	0.53	0.43	0.62
		RMSECV	0.90	0.73	0.84	0.80	0.96
		RPD	1.87	2.07	1.46	1.32	1.59
Principal component regression F and C_{AE}	CV Ncal=244	R^2	0.72	0.77	0.53	0.43	0.63
		RMSECV	0.90	0.73	0.84	0.80	0.93
		RPD	1.87	2.12	1.53	1.38	1.71
Principal component regression F and C_{AE}	TSV Ncal=187 Ntest=57	R^2	0.90	0.77	0.67	0.25	0.51
		RMSEP	0.53	0.66	0.68	0.96	1.01
		RPD	2.91	2.04	1.61	1.15	1.44

REFERENCES

1. Drake BK. Food Crushing Sounds: An Introductory Study, Journal of Food Science 1963; 28, 233-241.
2. Chen J, Karlsson C, Povey M. Acoustic envelope detector for crispness assessment of biscuits. Journal of Texture Studies 2005; 36, 139-156.

3. Vickers Z. Crackliness: Relationships of Auditory Judgments to Tactile Judgments and Instrumental Acoustical Measurements. Journal of Texture Studies; 1983, 15, 49–58.
4. Christensen CM, Vickers ZM. Relationships of Chewing Sounds to Judgments of Food Crispness. Journal of Food Science 1981; 46, 574.
5. Fillion L, Kilcast D. Consumer Perception of Crispness and Crunchiness an Fruits and Vegetables. Food Quality and Preference 2002; 13, 23–29.
6. Zdunek A, Konstankiewicz K. Acoustic emission as a method for the detection of fractures in the plant tissue caused by the external forces International Agrophysics 1997; 11 (3) , 223-227.
7. Malecki I, Ranachowski J. Acoustic emission, sources, methods, applications. Biuro Pascal, Warsaw; 1984. (in Polish).
8. Zdunek A, Konstankiewicz K. Acoustic Emission in Investigation of Plant Tissue Micro-Cracking. Transaction of the. ASAE 2004; 47(4), 1171-1177.
9. Alvarez MD, Saunders DEJ, Vincent JFV, Jeronimidis G. An engineering method to evaluate the crisp texture of fruit and vegetables. Journal of Texture Studies 2000; 31, 457-473.
10. Williams JG, Cawood MJ. European Group on Fracture: Kc and Gc Methods for Polymer. Polymer Testing 1999; 9, 15-26.
11. Bourne MC. Food Texture and Viscosity: Concept and Measurement. Second Edition. Academic Press, London; 2002.
12. Taniwaki M, Hanada T, Sakurai N. Device for Acoustic Measurement of Food Texture Using a Piezoelectric Sensor. Food Research International 2006; 39, 1099–1105.
13. Taniwaki M, Hanada T, Sakurai N. Postharvest Quality Evaluation of "Fuyu" and "Taishuu" Persimmons Using a Nondestructive Vibrational Method and an Acoustic Vibration Technique, Postharvest Biology and Technology 2009; 51 80–85.
14. Zdunek A, Cybulska J, Konopacka D, Rutkowski K. Evaluation of Apple Texture with Contact Acoustic Emission Detector: a Study on Performance of Calibration Models. Journal of Food Engineering 2011; 106, 80-87.
15. Luyten H, Plijter JJ, Van Vliet T. Crispy/Crunchy Crusts of Cellular Solid Foods: a Literature Review with Discussion. Journal of Texture Studies 2004; 35, 445–492.
16. Saeys W, Mouazen AM, Ramon H. Potential for Onsite and Online Analysis of Pig Manure Using Visible and Near Infrared Reflectance Spectroscopy. Biosystems Engineering 2005; 91 (4), 393–402.

Chapter 12

Otoacoustic Emissions

Giovanna Zimatore[1], Domenico Stanzial[2] and Maria Patrizia Orlando[1]

[1] CNR-IDASC – Institute of Acoustics and Sensor "Orso Mario Corbino", Rome, Italy
[2] CNR-IDASC - Institute of Acoustics and Sensor "Orso Mario Corbino", c/o Physics Department University of Ferrara, Italy

1. INTRODUCTION

In this chapter, we present a very special kind of acoustic emissions, coming from inside the cochlea and generated along the basilar membrane by the electro-motile (active) vibrations of outer hair cells of the organ of Corti. They are called OtoAcoustic Emissions (OAE) and are detected in the ear canal by means of microphones which are usually assembled as part of earphone-like probes. Since their discovery by Kemp [1], the study of otoacoustic emissions has become an hot topic both in basic and clinical research, due to OAE unique feature to inform directly about the normal and pathological functions of the cochlear receptors mechanisms, thus like the efficiency of the middle ear transmission chain.

From the signal point of view, the most interesting characteristics of OAE is their broad band frequency spectrum so rousing also a new interest for broad band ear immittance measurements and interpretation [2]. In this respect, this chapter will focus the reader's attention on two very innovative topics to improve objective and non-invasive audiological tests: the potentiality of Transient-Evoked otoacoustic emissions (TEOAE) to detect hearing impairment and the availability of a new microprobe able to capture directly both the pressure and velocity acoustic signals in the ear canal so allowing the direct measurement of ear immittance.

2. INSIDE COCHLEA

The cochlea is located in the inner ear, consisting of the front labyrinth and rear labyrinth, the latter having peripheral vestibular formations. The cochlea has quite a complex structure, just as complex as the Organ of Corti, contained inside the cochlea that with its neuro-epitherial hair cells makes up the first mechanical-electrical transformation stage of the sound impulse; it permits the stimulation of the afferent neural structures and the transmission of the information contained in the sound input through the acoustic canals right up to the cerebral cortex.

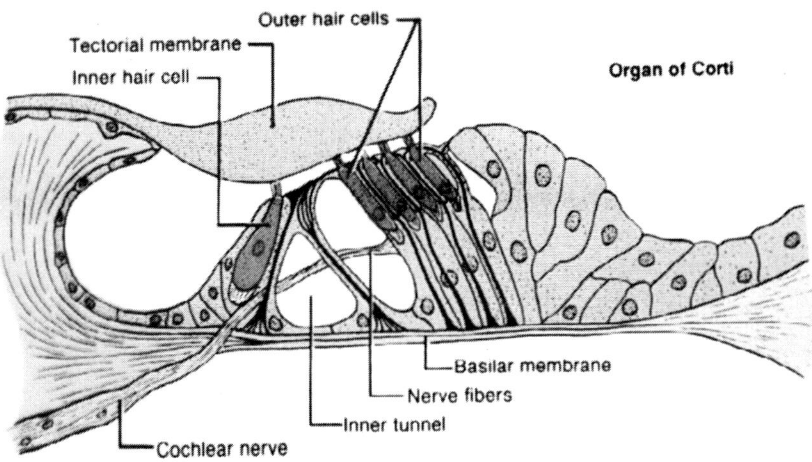

Figure 1. The Organ of Corti The organ of Corti is attached to the basilar membrane on the side of the aqueous fluid of the scala media. It is comprised of the supporting cells for the hair cells, the hair cells themselves, and the tectorial membrane (TM).

The Organ of Corti is made up of the Basilar Membrane, hair cells, support cells, Deiter, Hensen and Claudius cells, and the Tectorial Membrane. The hair cells can be divided according to their position in respect of the cochlea canal, whether outer or inner. The outer cells are more numerous and are placed along three lines; their hairs contact directly with the Tectorial Membrane and are very sensitive, are mainly stimulated by the efferent medial olivocochlea system of control. Acetylcholine (Ach) is their principal chemical mediator. The internal hair cells are arranged in a single line, don't have direct contact with the Tectorial Membrane, are less vulnerable and are supplied by afferent medial olivocochlea nerves, whose first nerve cell is within the Organ of Corti, itself enclosed within the bony labyrinth inside the cochlea. The glutamate is the main neuro–transmitter of the Internal Hair Cells. Given their afferent innervations they make up the actual sensorial cells.

The mechanical-electrical transduction of the cochlea takes place through a series of bio-chemical and bio-mechanical mechanisms. The sound impulse is transmitted from the movement of the stirrup bone on the oval window to the endolymph fluid creating a deformation of the Basilar Membrane on which the Organ of Corti rests with its hair cells that also create a deformation of the auditory cells who in turn are partially in direct contact with the Tectorial Membrane and so generating a deformation wave in the Basilar Membrane (travelling wave) as a result of the sound wave. The amount of deformation that the travelling wave produces along the Basilar Membrane will be evident at different points according to the frequency of the topical tone sound. The result of such mechanical modifications by the Basilar Membrane and hairs is the releasing of neuro-receptor neurohumours located in their synaptic vesicles inside the Hair Cells and so generating the bio-electrical impulse. The Basilar Membrane has different physical and elastic properties along the cochlea spiral from its base, through the intermediate part, to the apex. Even resonance properties vary along the cochlea. This makes one part of the Basilar Membrane resonate and deform according to the frequency of the sound rather than another part of the Basilar Membrane and consequently the Organ of Corti, the activating groups of Hair Cells and their nerve fibres based on the different frequencies contained in the sound. The part of the Basilar Membrane that is most sensitive to low frequency sounds is the apex of the cochlea whilst the part most sensitive to high frequencies is the widest part of the coil that is the base of the cochlea.

The first neuron of the auditory system is contained in the Corti Gland inside the cochlea where we find T cells whose peripheral extensions come from the Internal Hair Cells whilst the central extensions together make up the eighth cranial nerve and connect to the pontini bulb centres. It is important to keep in mind, according to the most recent theories, that inside the Organ of Corti at the External Hair Cell level there is an important active magnifying process of the signal that produces significant amplification, definition and resolution in the frequency of the sound inputs and a notable refinement of the auditory threshold. The fine longitudinal and transversal motility of the Outer Hair Cells, both spontaneous types and those stimulated externally, motility modulated by the efferent olivocochlea system, are the basis of such important functions. From this it can be deduced that a loss of Outer Hair Cells would produce a series of auditory problems more critical and complex in respect of damage to the Inner Hair Cells. Hearing loss (reduction of auditory function) connected to changes in analysis, peripheral translation and conduction of apparatus is defined as neurosensory and gives way to distortions in frequency, intensity such as recruitment, a phenomenon that distorts the subjective sound intensity (loudness), in phase, exertion and auditory conformation.

Outer Hair Cells are cells that belong to and are controlled by the efferent system more than sensor cells. They are more sensitive to auditory stimulation in respect of Internal Hair Cells which are anatomically connected to the afferent or sensorial system as previously stated. The particular sensitivity is mainly mechanical in nature and Is connected to 1) the presence of direct tectorial hair connections between stereohairs and the Tectorial Membrane and 2) their "active" vibratory motility, electrically and chemically mediated, that translates into acoustic phenomenon that can be picked up and recorded by a microphone positioned in the external auditory canal: the Otoacoustic Emissions.

3. WHAT OAE ARE

The discovery of otoemissions is attributed to the English physics professor David Kemp at the end of the '70s. He is merited with first putting forward the idea and then introducing clinical diagnosis using investigative methodologies capable of non-invasive exploration, in humans, the Organ of Corti functions and in particular the Outer Hair Cells. The basis of this methodology has produced a series of new and surprising evidence regarding the cochlea physiology that integrates, contradicts and supersedes the consolidated theories of von Békésy, Nobel Prize winner in 1960.

The direct contact between the stereo cilias of the Outer Hair Cells and the Tectorial Membrane create mechanical-electrical type reactions that transfer to the entire cell connected by ATP (Adenosin-TriPhosfate). The typical cytoskeleton-like network of muscle (actina–miosina) of which the cell is made, makes use of the electric charge originated at the level of the stereo cilia and moves either slowly or rapidly. These movements are modulated and regulated by the medial olivocochlea system, a true servo-system of control through various synaptic neurohumours and in particular Acetylcholine. The function of Outer Hair Cells is fundamental in conferring on our hearing the elevated threshold characteristics, the increased dynamics between minimum audible threshold and the perceptible maximum and frequency selectiveness.

A cochlea system with dysfunctional Outer Hair Cells rapidly loses these properties even if in theory the Inner Hair Cells are healthy. The information received mechanically from the Outer Hair Cells is transmitted in electric form as well as in mechanical form to the Inner Hair Cells and so to our proper sensory auditory system. To stress again, the Outer Hair Cells are particularly vulnerable, their high characteristic sensitivity to which are connected elevated bioenergetic and metabolic requests such that any cochlea noxae that is infected, toxic, traumatised or suffering from a metabolic disorder can bring about a lesion and become apparent prematurely. The study of Otoacoustic emissions appears significant and effective in the majority of auditory problems of peripheral receptors.

The otoacoustic emissions (OAE) are recorded by a particular probe positioned in the external auditory canal. If it is necessary to create responses the probe, other than being a receiver that records the emissions from the cochlea, contains a transducer capable of sending stimuli to the cochlea. These days it is possible to study the OAE mainly in one of three ways:

1. Recording the spontaneous emissions produced by the cochlea in the absence of any acoustic stimulus. Such emissions are called 'Spontaneous Otoacoustic Emissions' (SOAE).
2. Recording the emissions produced inside the cochlea through the sending of temporary acoustic stimuli, such as clicks, that are able to involve synchronously and globally a large number of the acoustic cells from the base to the apex. These emissions are known as 'Transient Otoacoustic Emissions' (TEOAE).
3. Cochlea emissions created by pairs of tonal stimuli of differing frequency for intermodulation phenomena, so-called 'Distortion Product Otoacoustic Emissions' (DPOAE).

Apart from the SOAE method of recording whose clinical value is unfortunately less, we shall focus on the TEOAE and DPOAE recording methods. The first method involves sending a series of clicks from a probe and recording the acoustic response from the hair cells. The acoustic response is normally represented graphically by oscillations based on a time period (milliseconds), as well as by a spectrogram that traces the size and frequency of the response. The DPOAE instead operates by way of sending a pair of pure tones (F1 and F2), with very small value frequency differences between them for example F1=1000 Hz, F2=1220 Hz, a ratio of F2/F1 = 1.22. The two tones, so-called primary tones, give rise to distortions in the cochlea deriving from their combination. The phenomenon of the combination of tones is mostly connected to the peripheral processing mechanisms of the signal which is still not wholly understood but that resides in the internal ear and in particular is connected to the active processes of the cochlea. So if two tones of differing frequency are sent simultaneously, the ear might perceive one or more tones superimposed that are the sum of the two tones or else are the difference (simple, cubic, quadratic, etc.) of the two primary tones. The response traces the form of a DP-gram, showing the extent of the response derived from the frequency of the primary tones.

Nowadays the major diagnostic clinical function is mostly engaged in the TEOAE and DPOAE being the spontaneous emissions less subject to interpretation despite having a notable scientific interest. Dedicated software systems permit the execution of a rapid measurement statistically adapted to the cochlea response. As regards the DPOAE it is interesting to note that it establishes a modern method to survey one of the more characteristic psychoacoustic phenomena: combination tones. The study of DPOAE in particular allows the design of cochlea responses in an audiometric-like way, frequency by frequency, on a graph that shows on the vertical axis the frequencies of stimulation and on the horizontal axis the intensity levels of the received Otoacoustic emissions showing immediately if the audiological threshold is within normal limits or not.

The operating range is important in identifying the dysfunction of the cochlea in Ménière's disease, in evaluating damage from noise, in ototoxic type changes, in the study of some genetic and immunological cochlea alterations, in the differential cochlea diagnosis against retro cochlea diagnosis and the identification of new pathologies such as Auditory Neuropathy. Finally, the range of neonatal auditory screening establishes the most sensitive and specific means of recognising premature infantile deafness. Auditory screening is carried out at birth before the new-born baby is discharged from hospital normally the second day after birth and, given the simplicity and speed of testing, is the best method for definitive diagnosis or alerting and preparing for further diagnosis and rehabilitative therapy within a few months and before the child's first birthday, a period of great neuroplastic and linguistic activity. It's therefore possible to control and limit the damage from auditory sensorial deprivation, language disorders, communication and behaviour disorders.

OAEs provide objectivity and greater accuracy, representing a non invasive tool for the assessment of OHC and the functionality of the cochlear amplifier, as demonstrated by experimental and clinical studies [3-5]; furthermore, the cochlear effects of exogenous factors, such as ototoxic drugs, solvents and high-level sound exposure [6-8], can be monitored by OAE. It has been suggested that OAEs may provide early indication of cochlear damage before evidence for NIHL appears in pure-tone audiometry [9-10]. Recently, TEOAE have been used to study in tinnitus subjects with normal hearing to assess whether a minor cochlear or efferent dysfunction might play a role in tinnitus [11].

One of the few limitations of OAE is related to the extent of hearing loss that we can explore: infact, already for cochlear hearing loss above 50 dB the OAE, just because otoacoustic emissions are produced by the activation of CCE, are no longer evoked.

4. TEOAE RECORDING

To record the TEOAE signals the Otodynamic Analyzer (ILO92, Otodynamics Ltd, Hatfield, United Kingdom), was widely used, by inserting a SGS-type general purpose TEOAE probe into the external ear canal. The TEOAE recordings were carried out in a standard hospital room, corresponding to the usual clinical setting for these measurements. The automated differential non-linear test paradigm was used: the stimulus was characterized by a train of four clicks, three with the same amplitude and polarity, followed by a fourth one with a 3-

fold amplitude and opposite polarity with respect to the preceding ones. The 80 µs clicks presented at 50/s were 75–85 dB SPL. The responses were obtained evaluating an average among 260 stimuli trains (1040 clicks) stored into two different buffers (A and B) for a total of 2080 clicks. The value of the automatically computed correlation or reproducibility between the two obtained waveforms (A and B) of an OAE signal is named Repro or whole waveform reproducibility (REPRO) (Pearson correlation coefficient *100) (see in Figure 2, on the right, Repro=99%).

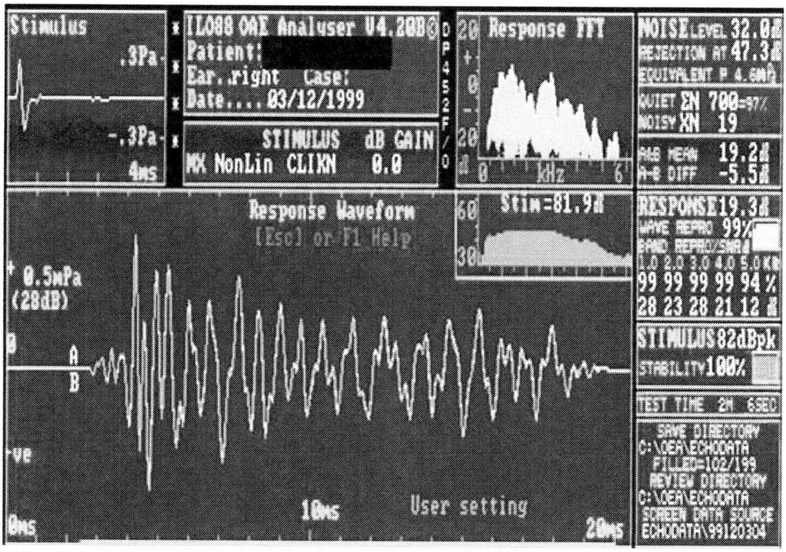

Figure 2. TEOAE signals (ILO92, Otodynamics Ltd)

5. BROAD BAND MEASUREMENT OF EAR IMMITTANCE AND PERSPECTIVE FOR IMPROVING TEOAE DETECTION

The most innovative application of micro-electro-mechanical systems (MEMS) technology to acoustic sensors is the manufacturing of thermo-acoustic velocimeters based on the two-wire anemometric transduction principle. These new sensors allow to capture directly the acoustic particle velocity signal v, and thus, by coupling and assembling them with standard microphones which are instead sensitive to the pressure signal p, a new generation of pressure-velocity (p-v) micro-probes is nowadays made available (see Figure 3).

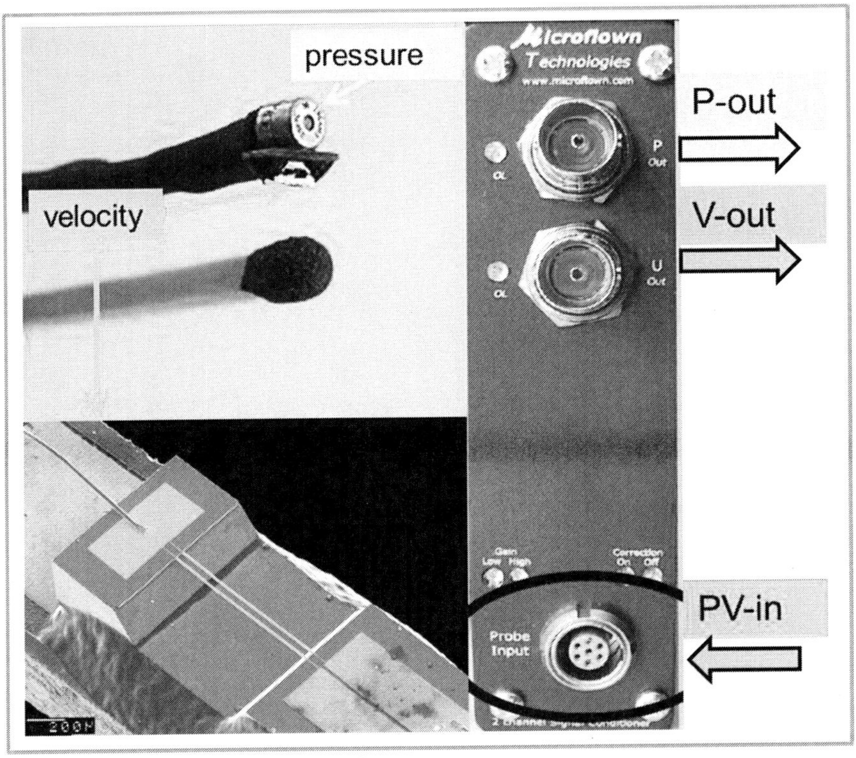

Figure 3. A p-v sound intensity micro-probe consists in the assembly of a miniaturized pressure microphone and a MEMS technology based velocimeter in a single measurement system. While the pressure sensor is a standard electret one, the velocity signal is transduced thanks to the differential anemometric principle applied to two closely spaced heated wires 10 μm apart, 1mm long and 5μm large suspended in parallel in order to form a bridge. The wire composition is 200 nm platinum (Pt) on a silicon nitride (Si3N4) substratum 150 nm thick. The captured pressure and velocity analog signals are conditioned through a common probe input and handled in output as two separate voltage signals. (The commercial system shown in the figure is by courtesy of Microflown®: www.microflown.com).

These micro-probes are clearly the ideal device for carrying out advanced direct measurements of the sound field energetic properties like sound intensity j=pv or acoustic impedance Z=p/v. To this aim, an accurate calibration procedure [12] is needed (see Figure 4).

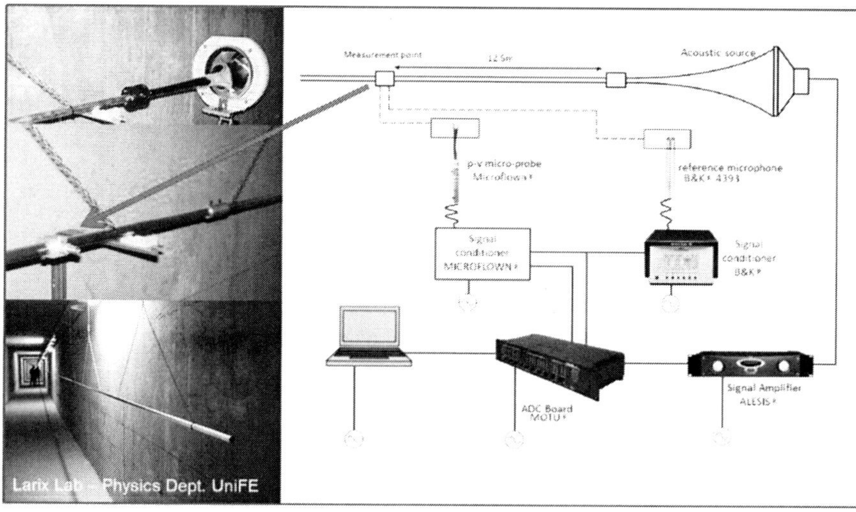

Figure 4. The facility for sound intensity micro-probes calibration installed at the Larix Lab of the Physics Department of University of Ferrara consists in a 48 m long wave guide where a progressive plane wave is generated through a bi-conical loudspeaker in the [50, 10000] Hz frequency range. The p-v micro-probe under calibration is inserted at a distance of 12.5 m from the source and is calibrated by comparison with a reference pressure microphone using the correction function Γ(ω) defined inEquation 13 of Ref. [12].

Of course, the calibration filtering process can be implemented at post-processing level but, with few engineering effort, the calibration filters can also be programmed at hardware level so making, in particular, the measurement of acoustic impedance, a completely automatic task. The technological innovation driven by MEMS application to acoustic sensors can be easily transferred to audiometric devices so transforming for instance a traditional tympanometric probe in a new setup for p-v tympanometry (see Figure 5). The main advantages of a p-v tympanometric test with respect to a traditional one are: a) the direct measure of ear immitance for more precise results; b) the test is completely non-invasive for static pressure external pumping is no longer necessary (p-v test measurements are performed in standard pressure conditions); c) the test produces wideband results in the typical frequency range of multi-tonal tympanometry [100, 1200] Hz; d) the p-v audiometer provides sophisticated sound energy analysis capability for hearing models validation (see Ref. [13]).

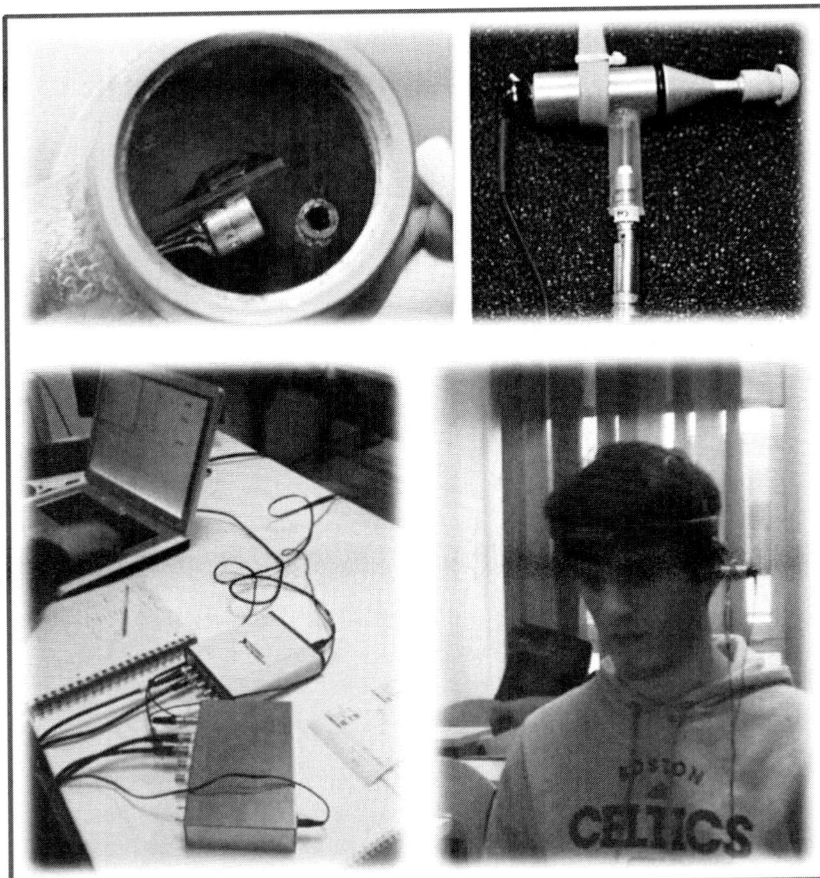

Figure 5. A p-v tympanometer is designed as a laptop based dual channel analyzer (lower left) able to record both the Impulse Responses (IRs) of pressure and velocity signals captured with the p-v tympanometric probe shown in the upper part of the figure. Once the p-v IRs of the ear canal have been measured in atmospheric pressure condition (lower right), the system calculates the external/middle ear specific immittance and displays its magnitude in dB relative to the frequency dependent baseline Y_0 obtained by plugging up the probe.

As an example of results obtained with p-v tympanometry, wideband p-v tympanograms measured in dB for 26 left and right normal ears belonging to 13 voluntary students are clustered and reported in Figure 6.

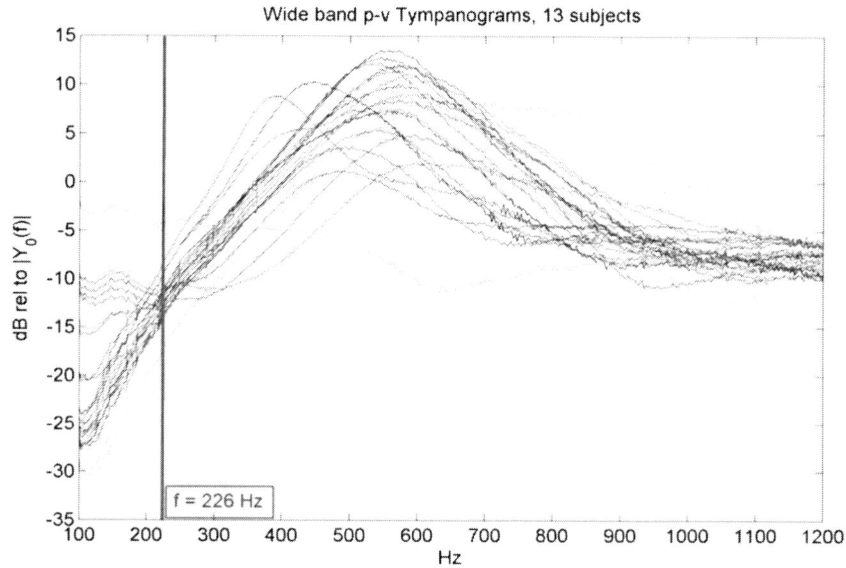

Figure 6. Wideband p-v tympanograms measured in dB for 26 left and right normal ears belonging to 13 voluntary students. One clearly see that all tympanograms converges between -10 and -15 dB for the standard frequency of 226 Hz used in traditional tympanometry. The mean value found at -12.7 dB can thus be considered the "normal" value of the immitance magnitude measured by p-v tympanometry at 226 Hz.

As the primary data collected by the p-v tympanometry are basically the measurement of the pressure and velocity ear canal IRs, a completely new perspective also for OAE studies is also opened. Specifically for the TEOAEs which could be simply detected as the non-linear byproducts of DSP algorithms used in the ear-canal immitance function calculations.

6. TEOAE POST-PROCESSING ANALYSIS

The Recurrence Quantification Analysis (RQA) and Principal Component Analysis (PCA) have been carried on TEOAE waveforms [14-17] (Zimatore et al. 2000, 2001 2002, 2003) to extract new descriptors that could enlighten an early diagnosis of hearing loss.

In the last few years, a new parameter has been introduced to analyse TEOAE, to improve the specificity of diagnostic tests and to reduce inter-subject variability. The work was concentrated on the analysis of the TEOAE focusing on their dynamics by the Recurrence Quantification Analysis (RQA). RQA is a post-processing analysis that is extremely fit to non-stationary signals and represents a valid alternative to Wavelet analysis used by other researchers. In fact, the embedding procedure allows to expand a mono-dimensional signal into multidimensional space, thus permitting the identification of fine peculiarities of the sampled series that in turn are described by few global parameters allowing for a synthetic patient description.

RQA in summary:

- RQA introduces few parameters descriptive of the global complexity of a signal, starting from what is called "recurrence plot"
- RQA descriptors are calculated on the basis of the number and location of dots in the recurrence plot
- RQA dynamic features are independent from signal amplitude

The results obtained demonstrate how proposed new global index can recognize even mild hearing loss and that an assessment of the severity of cochlear damage can be realized.

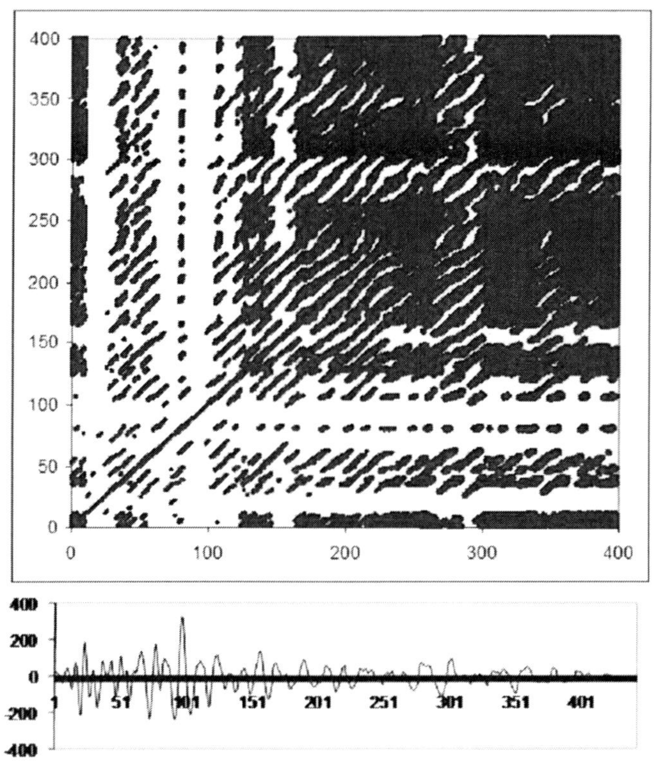

Figure 7. Recurrence plot (top) of a typical TEOAE recorded in a Normal ear (%Det=88.89) (bottom).

To build the recurrence plot, the time behavior of the original signal was represented by a series of 512 points equally spaced in time (e.g. {a1 a2 .... a 512} where ai represents the value of the signal corresponding to the i-th time position). Then, the series was arranged in successive columns (the columns number is defined by the "embedding dimension" parameter, N), each-one obtained by applying a delay in time (lag parameter) to the original sequence, in this way an "embedding matrix" was created.

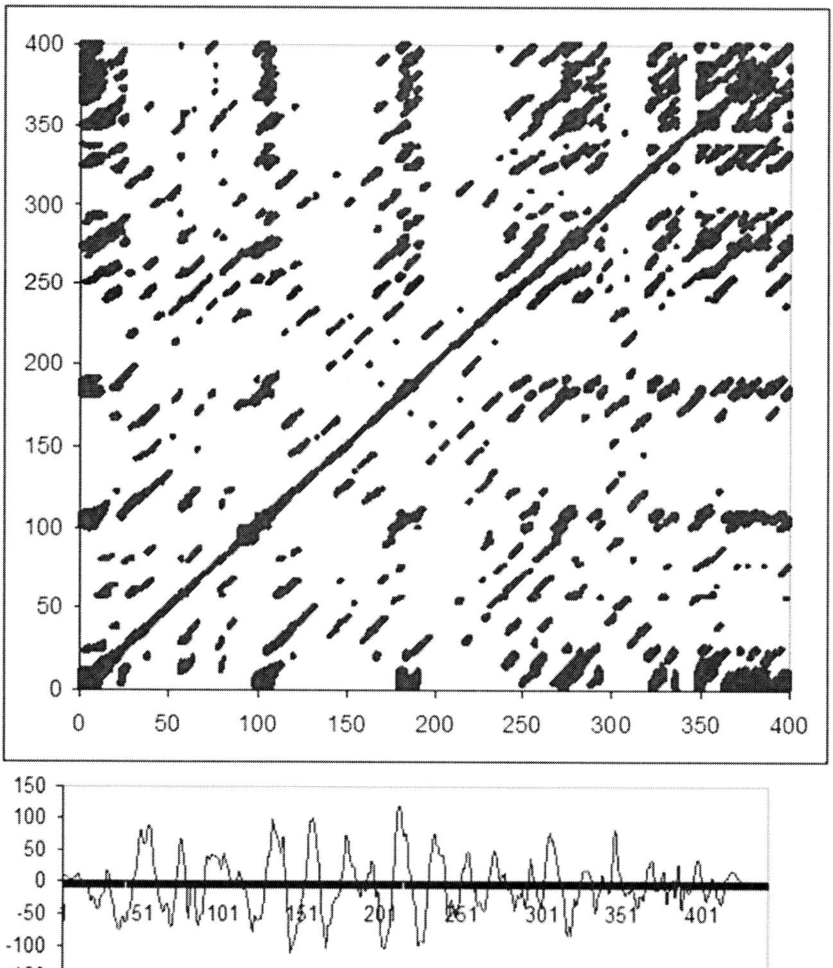

Figure 8. Recurrence Plot (top) of a representative Impaired Hearing (IH) TEOAE waveform (bottom) (% Det = 62.89).

Finally, the recurrence plot was built, drawing a black dot (named "recurrent point") in the represented space if the distance between the corresponding rows (the distance between the j-th and the (j+1)th row is of the embedding matrix was lower than a fixed value (radius). In the obtained plot, the horizontal and vertical axes represented the relative position of the 512 points into the TEOAE waveform. RQA descriptors were then calculated on the basis of the number and the location of dots in the recurrence plot. In particular, percent of recurrence (Rec) is the percentage of recurrence points in a recurrent plot; percent of determinism (Det) is the percentage of recurrence points which form diagonal lines and it indicates the degree of deterministic structure of the signal; entropy (Ent) is the Shannon entropy of the probability distribution of the diagonal line lengths and is linked to the richness of deterministic structure [16-17] (Zimatore et al. 2002 and 2003). The presence of horizontal and vertical lines in the recurrence plot shows that part of the considered signal matches closely with a sequence farther along the time (for more details see http://www.recurrence-plot.tk).

In TEOAE analysis the delay in the embedding procedure (lag) is set to 1; the number of the embedding matrix columns (embedding dimension) is set to 10; and the cut-off distance (radius) is set to 15; to eliminate the initial linear ringing, the first 2.8 ms of the recorded TEOAE signals are excluded.

Comparing Figure 7 and 8, it is clear that recurrence plots distinguish between normal hearing and impaired hearing TEOAEs especially in terms of a reduction in the deterministic structure.

As a further step of the post-processing analysis, the well known Principal Component Analysis (PCA), was applied on the obtained RQA descriptors. Briefly, PCA is a common statistical technique which provides the possibility to reduce the starting data set dimension without consistent loss of information and with a separation of the different and independent features characterizing the data set. PCA describes the original data set with a lower number of new parameters named main components (PC1, PC2) which explain more than 90% of the total variability in the data set. Having, by construction, PC1 and PC2 zero mean and standard deviation equal to 1, if a set of TEOAE signals from normal ears are studied, 96% of them will fall within a circle centered in the origin of the PC1/PC2 plane, and with a radius equal to 2 (reference circle in figure 12). The PC1/PC2 plane is defined starting from a representative data set made by 118 signals measured from normal hearing subjects [18]. The representative data set was used to define the circle in the PC1/PC2 plane in which the majority of TEOAE signals recorded in normal hearing subjects will fall. Mathematically, the parameter RAD2D is defined in the PC1/PC2 plane as the Euclidean distance of one point representing a TEOAE signal from the plane origin.

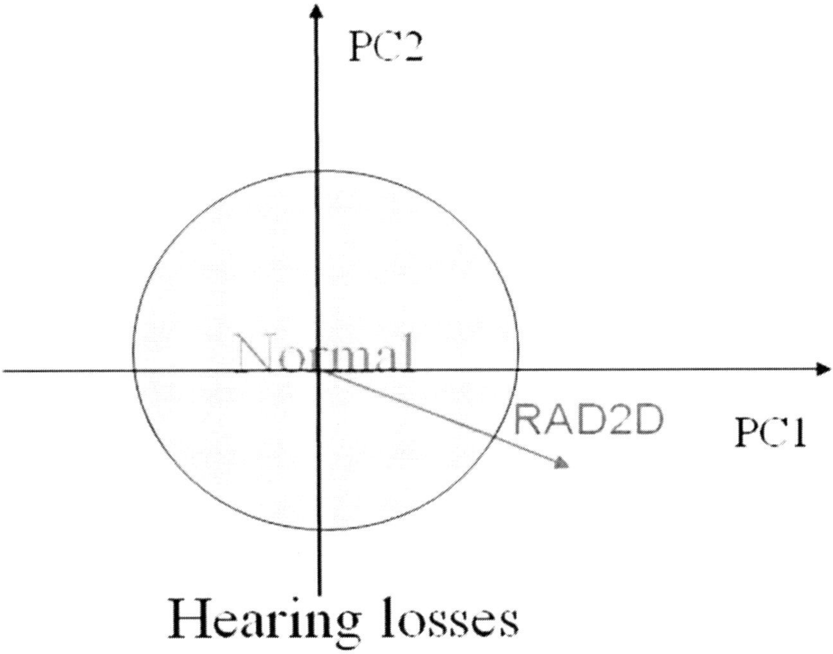

Figure 9. RAD2D is defined in the Principal Components plane as the Euclidean distance from the plane origin; the points representing the normal TEOAE signals fall in the yellow reference circle and TEOAE signals recorded form subjects with hearing losses fall outside.

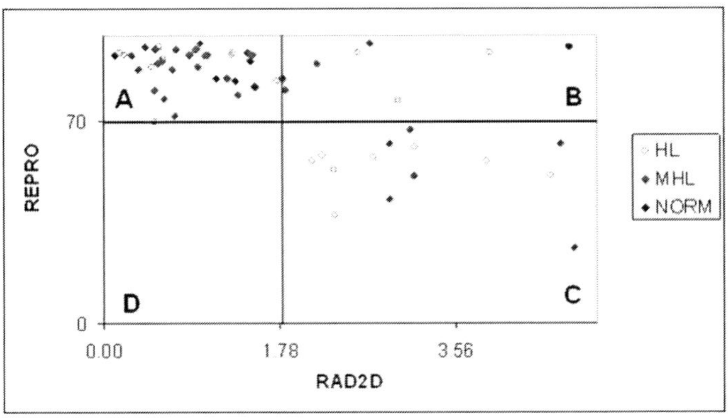

Figure 10. REPRO *vs* RAD2D from RQA parameters of TEOAE signals.

The relation correlating the RAD2D obtained for all the measured signals with the entity of cochlear damage is tested. Specifically, RAD2D was evaluated for real TEOAEs by applying the same procedure as for simulated signals combining RQA and PCA techniques.

Furthermore, the post-processing analysis proposed is useful in screening of adults, in longitudinal studies, in test to evaluate the efficacy of new pharmacological treatments, in conservation program in presbycusis and in protection program in noise induced hearing losses.

Figure 10 illustrates REPRO plotted *vs* RAD2D considering 30 subjects from Florence area (Italy). The examined ears will be classified as normal hearing (NORM) or mild hearing losses (MHL) or (severe) hearing losses (HL) ears according to their pure tone thresholds at 0.250, 0.500, 1, 2, 3, 4, 6 and 8 kHz. The three groups according to the maximum hearing threshold level are: NORM, with threshold <10 dB at all audiometric frequencies, MHL, with threshold <20 dB at all audiometric frequencies and >10 dB at least at one frequency, and HL, with threshold >20 dB at least at one frequency. In Figure 10 the HL patients (white circles), in the MHL patients (blue diamonds) and in NORM subjects (black diamonds): each point corresponds to the recorded TEOAE waveform. A very simple and immediate description is available by observing the areas identified by threshold of REPRO (the horizontal line at 70%) and of RAD2D (the vertical line at 1.78). The points above the horizontal line indicate pass signals. To the left side of the vertical line, the points indicate signals that fall inside the normality circle, that is pass signals;. the main result is illustrated in the right upward rectangle ofFigure 9 where the ears that have both high REPRO and high RAD2D are shown: these points-signals indicate 8 ears (3 HL, 4 MHL and 1 NORM) screened as pass by REPRO but identified as "fail" by our TEOAE parameter (possible false-negative of ILO test).

The 8 points-signals that fall in the B area correspond to 8 different subjects: 6 are hunters or they shoot for hobby and 2 work often with tractors or lawn mowers. The combined use of the two global parameters REPRO and RAD2D can enlighten points corresponding to the subjects with high risk of environmental noise exposure.

In this chapter the application of technique such as RQA is proposed because, it allows the quantification of the *fine-structure* of TEOAE signals without any *a priori* hypothesis and any data manipulation; moreover, the dynamical structure of signals can be investigated without taking into account the signal-amplitude differences.

7. TEAOA SIMULATION BY MECHANISTIC MODEL

An electronic model of human hearing system is used to test and improve new hypothesis of cochlear mechanisms and to anatomically distinguish different contributions to ear pathologies [18-19].

An electronic model of human hearing system can be used to test and improve new hypothesis of cochlear mechanisms and to anatomically distinguish different contributions to ear pathologies.

The considered ear model is directly inspired to the so called "travelling wave" representation of the cochlear function mechanism and is able to simulate the TEOAE responses; the electric model of the whole ear, originally introduced by Guiguère and Woodland [20-21] and used in TEOAEs analysis [15, 18, 22], has been implemented into PSpice®. PSpice® is a standard electrical simulation tool for dc, transient and ac analyses [23] (see Figure 11). The input circuit can be defined by using a graphical interface or by compiling a list representing the circuit topology. The outputs of the system are current and voltage values within the circuit which can be displayed in both tabular and graphical formats. PSpice® has been already used to study an electric model of the cochlea [24] due to the possibility to relate the model parameters to physical and physiological issues. In [24], the used lumped parameter model is entirely passive, made of a resistive network combined with two capacitances in order to model the Reissner's membrane and the OHC in the Corti Organ.

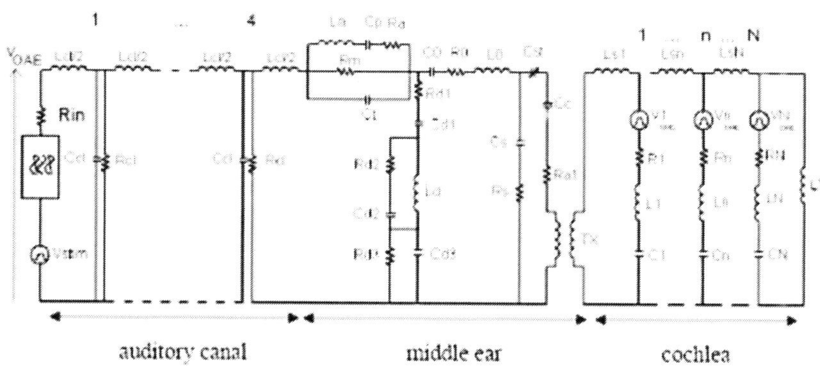

Figure 11. The electronic ear model.

The considered ear model is depicted in figure 11 and encompasses the human ear anatomy from the auditory canal to the OHC within the cochlea. The auditory canal is represented by a cascade of four T-sections, corresponding to the segmented form of a uniform transmission line, while the middle ear is modeled as a complex electrical network based on its functional anatomy [25]. An ideal transformer connects the middle ear to the cochlea, to represent the acoustic transformer ratio between the eardrum and the oval window [20-21]. Finally, the cochlea is modeled as a non-uniform and non-linear transmission line, divided into several sections from the base to the apex, each one consisting of a series inductor, a shunt resonant circuit (composed of a resistor, an inductor, and a capacitor), and a non-linear voltage source. In the electro-acoustic analogy, the series inductors represent the acoustic mass of the cochlear fluids; the resistors, inductors and capacitors forming the shunt resonant circuits represent the acoustic resistance, mass and stiffness of the basilar membrane, respectively, and the non linear voltage sources represent the OHC active processes. Finally, the helicotrema is modeled by the inductor LT. The initial values of the electric ear model components are those reported in Table 1 of [20] and also used in [22]. Correspondingly, the cochlea was represented by 128 and 64 partitions [19]

To verify the hypothesis that TEOAE are strongly modulated by the middle ear [17], some elements in the middle ear section were varied according to the experimental study of Avan and colleagues [4]. The first change is the addition of a stapes capacitor ($C\,st$) to the middle-ear section of the circuit, as already considered by [20] Giguère and Woodland (1994a) and by [25]. When $C\,st$ has a large value, its impedance is small, corresponding to small tension in the stapedius muscle ($C\,st$ equal to infinity corresponds to no stiffness in the resting condition). Conversely, when $C\,st$ is small, its impedance is large, corresponding to high muscle tension. Then, changes in the tympanic membrane stiffness ($C\,0, C\,d1$), to account for changes in the middle ear pressure, and in the tympanic membrane mass ($L\,0$, Ld), to simulate an additive mass, have been considered [4]. Furthermore, a dead cochlea condition has been simulated by de-activating the voltage sources in all cochlear sections.

The role of middle ear effects is a hot topic in the OAE field, and would be of high interest to audiology and hearing researchers.

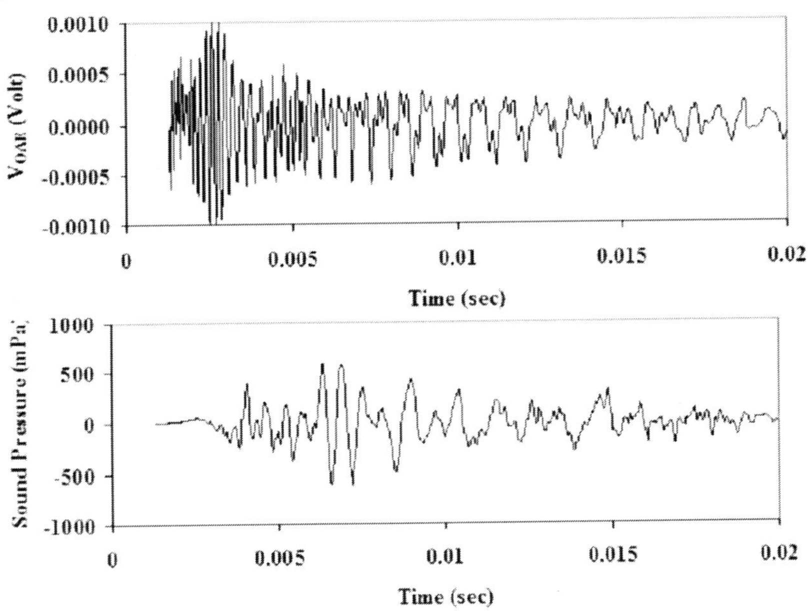

Figure 12. TEOAE Simulated (top) and real (bottom) in normal ear.

Figure 12 (top) reports a typical simulated signal, and a real TEOAE signal recorded from a normoacousic subject is reported in figure 12 (bottom). In both simulated and real signals, recording starts after 2.5 ms from the initial external excitation (t = 0), to get rid of the initial ringing. Both signals show oscillations lasting up to 20 ms, with higher frequencies having shorter latency than lower frequencies, in agreement with the latency-frequency relationship typical of TEOAEs. In fact, according to the place–frequency (tonotopic) effect characteristic of the basilar membrane, each element of the membrane acts as a resonator at a frequency inversely proportional to its distance from the oval window.

A very important goal in prevention and clinical applications is to improve the specificity of diagnostic tests and to reduce inter-subject variability in TEOAE signals. A new pass/fail test could be useful for screening but the quantification of cochlear damage is of great interest in research programs. To determine the amount of damage, an ear model can be used to simulate different levels of cochlear damage by silencing a growing number of cochlear partitions. The relation between a new parameter and the number of silenced partitions in the model was evaluated.

From the comparison between the real and simulated RAD2D values it is possible to extrapolate the corresponding number of "hypothetical silenced partitions". In this way, since each partition corresponded to a specific portion of uncoiled cochlea and to a specific number of outer hair cells, a descriptor of OHC integrity is obtained [26].

8. CONCLUSION

A very important goal in prevention and clinical applications is to improve the specificity of diagnostic tests and to reduce inter-subject variability in TEOAE signals. The availability of new micro-probes able to pick up both the pressure and the air particle velocity signals inside the ear canal, while allowing to update the standard multi-tonal tympanometry with the wideband implementation of p-v tympanometric non-invasive tests, points also to record and analyze TEOAEs as the non-linear by-product of DSP algorithms used in the ear-immitance function computing process. Furthermore, to prevent and to mitigate noise and aging effects on cochlea, a new post-processing procedure could be employed in *longitudinal studies* [27] as well as to test the efficacy of new pharmacological treatments and the opportunity to follow a subject over time.

REFERENCES

1. Kemp DT. Stimulated acoustic emissions from within the human auditory system," J Acous. Soc Am. 1978;64 1386-1391.
2. Keefe DH, Folson RC, Gorga MP, Vohr BR, Bulen JC, Norton et al. Identification of Neonatal Hearing Impairment: Ear-Canal Measurements of Acoustic Admittance and Reflectance in Neonates. Ear & Hearing 2000;21(5) 443-461.
3. Probst R, Lonsbury-Martin BL, Martin GK.. A review of otoacoustic emissions. Journal of the Acoustical Society of America 1991;89 2027–2067.
4. Avan P, Buki B, Maat B, Dordain M, and Wit HP. Middle ear influence on otoacoustic emissions. I: Non invasive investigation of the human transmission apparatus and comparison with model results Hearing Research 2000;140 189-201.
5. Avan P and Bonfils P.. Distortion-product otoacoustic emission spectra and high-resolution audiometry in noise-induced hearing loss. Hearing Research 2005;209 (1-2) 68-75.
6. Uchida Y Ando F , Shimokat, H et al. The Effects of Aging on Distortion-Product Otoacoustic Emissions in Adults with Normal Hearing. Ear & Hearing 2008;29(2) 176-184.
7. Hamdan AL, Abouchacra KS, Al Hazzouri AGZ.Transient-evoked otoacoustic emissions in a group of professional singers who have normal pure-tone hearing thresholds. Ear and Hearing 2008;29(3) 360-377.
8. Cianfrone G, Pentangelo D, Cianfrone F, Mazzei F, Turchetta R, Orlando MP, Altissimi G. Pharmacological Drugs Inducing ototoxicity, vestibular symtoms and tinnitus: a reasoned and updated guide. European Review for Medical and Pharmacological Sciences 2011;15 601-636.
9. Attias J, Furst M, Furman V et al. Noise-induced otoacoustic emission loss with or without hearing loss. Ear Hear 1995;16 612-8.
10. Shupak A, Tal D, Sharoni Z, et al. Otoacoustic Emissions in Early Noise-Induced Hearing Loss. Otology & Neurotology 2007;28 745-752.
11. Paglialonga A, Fiocchi S, Del Bo L, Ravazzani P and Tognola G.Quantitative analysis of cochlear active mechanisms in tinnitus subjects with normal hearing sensitivity: time–frequency analysis of transient evoked otoacoustic emissions and contralateral suppression. Auris Nasus Larynx 2011;38(1) 33-40.

12. Stanzial, 2011 Stanzial D., Sacchi G., Schiffrer G., Calibration of pressure-velocity probes using a progressive plane wave reference field and comparison with nominal calibration filters, J. Acoust. Soc. Am 2011;129 (6) 3745-3755.
13. Stanzial 2012 Stanzial D, Sacchi G, Schiffrer G. On the physical meaning of the power factor in acoustics, J. Acoust. Soc. Am 2012;131(1) 269-280.
14. Zimatore G, Giuliani A, Parlapiano C. et al.. Revealing deterministic structures in click-evoked otoacoustic emissions. Journal of Applied Physiology 2000;88(4) 1431-1437.
15. Zimatore G. The speaking ear. Analysis and modelling of Otoacoustic emissions. PhD thesis in Biophysics, Sapienza University, Rome, 2001.
16. Zimatore G , Hatzopoulos S, Giuliani et al. Comparison of transient otoacoustic emission responses from neonatal and adult ears. Journal of Applied Physiology 2002;92(6) 2521-2528.
17. Zimatore G , Hatzopoulos S, Giuliani et al. Otoacoustic emissions at different click intensities: invariant and subject dependent features. Journal of Applied Physiology 2003;95 2299-2305.
18. Zimatore G, Cavagnaro M, Giuliani A, Colosimo A. Human acoustic fingerprints. Biophysics & Bioengineering Letters 2008;1(2) 1-8.
19. Zimatore G, CavagnaroM, Giuliani A, Colosimo A. Reproducing Cochlear Signals by a Minimal Electroacoustic Model. Open Journal of Biophysics, 2012;2: 33-39 http://www.SciRP.org/journal/ojbiphy (accessed 7 April 2012)
20. Giguère C, and Woodland PC. A computational model of the auditory periphery for speech and hearing research. I. Ascending path J. Acoust. Soc. Am 1994;95(1) 331-342.
21. Giguère C and Woodland PC. A computational model of the auditory periphery for speech and hearing research. II. Descending paths J. Acoust. Soc. Am 1994;95(1) 343-349.
22. Zheng L, Zhang YT Yang FS et al. Synthesis and decomposition of transient-evoked otoacoustic emissions based on an active auditory model. IEEE Transaction on Biomedical Engineering 1999;46(9) 1098-1105.
23. PSPICE 1997 PSPICE A/D User's Guide. Irvine CA: Microsim Corp. Version 8.0
24. Suesserman MF and Spelman FA. Lumped-Parameter Model for In Vivo Cochlear Stimulation. IEEE Trans on Biomed Eng 1993;40(3) 237-245.
25. Lutman ME and Martin AM Development of an electroacoustic analogue model of the middle ear and acoustic reflex J. of Sound and Vibration 1979;64(1) 133-157.

26. Zimatore G, Fetoni AR, Paludetti G, Cavagnaro et al.. Post-processing analysis of transient-evoked otoacoustic emissions to detect 4 kHz-notch hearing impairment – a pilot study. Med Sci Monit 2011;17(6) MT41-49.
27. Helleman HW, Jansen EJ, Dreschler WA. Otoacoustic emissions in a hearing conservation program: general applicability in longitudinal monitoring and the relation to changes in pure-tone thresholds. Int J Audiol 2010;49(6) 410-9.

Chapter 13

Non-Destructive Evaluation for Corrosion Monitoring in Concrete: A Review and Capability of Acoustic Emission Technique

Ahmad Zaki [1], Hwa Kian Chai [1,*], Dimitrios G. Aggelis [2] and Ninel Alver [3]

[1] *Department of Civil Engineering, Faculty of Engineering, University of Malaya, Kuala Lumpur 50603, Malaysia*
[2] *Department of Mechanics of Materials and Constructions, Vrije Universiteit Brussel, Pleinlaan 2, Brussels 1050, Belgium*
[3] *Department of Civil Engineering, Faculty of Engineering, Ege University, Bornova, Izmir 35100, Turkey*

ABSTRACT

Corrosion of reinforced concrete (RC) structures has been one of the major causes of structural failure. Early detection of the corrosion process could help limit the location and the extent of necessary repairs or replacement, as well as reduce the cost associated with rehabilitation work. Non-destructive testing (NDT) methods have been found to be useful for *in-situ* evaluation of steel corrosion in RC, where the effect of steel corrosion and the integrity of the concrete structure can be assessed effectively. A complementary study of NDT methods for the investigation of corrosion is presented here. In this paper, acoustic emission (AE) effectively detects the corrosion of concrete structures at an early stage. The capability of the AE technique to detect corrosion occurring in real-time makes it a strong candidate for serving as an efficient NDT method, giving it an advantage over other NDT methods.

Keywords: reinforced concrete; corrosion; non-destructive testing; acoustic emission

1. INTRODUCTION

Corrosion of steel reinforcement is a global problem that leads to deterioration of RC structures [1]. Damage induced by steel corrosion usually requires proper repair, followed by maintenance [2,3]. It is reported that the costs of repair and maintenance of corroded structures exceed billions of dollars per year [4]. The costs vary depending on the condition of the concrete structures, including the cause of damage, degree of damage, and effect of damage on structural behavior [5]. A reliable inspection method is required at an early stage before the functionality of the RC structure is seriously damaged due to steel corrosion. The inspection method usually provides information about the current condition of the RC structures, so that their future performance can be predicted. Furthermore, the inspection method is almost a prerequisite for efficient and cost effective rehabilitation of existing RC structures. The inspection should be done without damaging the RC structures, both new and old ones.

The NDT methods are of many inspection methods for corrosion monitoring in RC structures. For new structures, the principal applications of NDT methods are likely to be quality control of the concrete conditions, while in old structures, the methods are expected to provide needed feedback on monitoring, detection, and identification of damage [6]. This paper aims to give a brief review of the NDT methods for monitoring and evaluating steel corrosion in RC structures, followed by a discussion and review of a technique based on AE developed to achieve the same objectives. Due to its principles of application, it is considered that AE has the capability of monitoring the steel corrosion in RC structures at an early stage.

2. NDT METHODS FOR CORROSION MONITORING

Recently, various methods have been implemented for corrosion monitoring in RC structures. They are classified into six main categories as follows: visual inspection, electrochemical methods (*i.e.*, open circuit potential (OCP) monitoring, resistivity method, polarization resistance, galvanostatic pulse method (GPM), electrochemical noise (EN)), elastic wave methods (*i.e.*, ultrasonic pulse velocity (UPV), acoustic emission (AE), and impact echo (IE)), electromagnetic (EM) methods (*i.e.*, ground penetrating radar (GPR)), optical sensing methods (*i.e.*, fiber Bragg grating (FBG)), and infrared thermography (IRT). Visual inspection, OCP, polarization resistance, and other electrochemical methods are more commonly used for corrosion monitoring in RC structures.

2.1. Visual Inspection

Visual inspection is a regular inspection method to assess corrosion damage on the surface of concrete structures. The appearance of the corroded area often provides valuable insight into the cause and the extent of corrosion. However, the method is very dependent on the inspector's experience. In addition, visual inspection is limited in its effectiveness to detect surface discontinuities due to steel corrosion and the unseen corrosion is difficult to spot [7,8,9,10].

2.2. Electrochemical Methods

Electrochemical methods are by far the most suitable for corrosion monitoring in RC structures. Electrochemical methods, in general, can provide fast and reliable information on the probability of corrosion, the corrosion rate of steel reinforcement, and the resistivity of the concrete structure [11,12,13,14,15,16,17,18,19,20,21,22]. Electrochemical methods are related to the interrelation of the electrical and the chemical effects. An electrochemical system measures the potential and current of oxidation and reduction reactions [23,24,25,26,27,28]. The principles of the five electrochemical methods are shown in Figure 1.

In the open circuit potential (OCP) monitoring, one of the electrochemical methods, the electrical potential (in mV or V) between a steel reinforcement and a reference electrode (*i.e.*, a copper/copper sulfate cell), in contact with the concrete surface, is measured [29,30,31,32]. The principle of OCP method is shown in Figure 1a. The OCP provides information pertaining to the probability and potential level of corrosion in RC structures [7,9]. Table 1 gives the potential ranges for different corrosion conditions of steel inside the reinforced concrete. The method is often unconvincing in terms of interpretation because the measurement depends on the condition of the RC structure. Moisture levels and amount of chloride concentration can affect the potential readings and give erroneous results [7,33]. The changes in moisture content (*i.e.,* wet condition of the concrete surface) lead to a shift in potential values so they become more negative value (*i.e.,* a shift of 100 mV was found on a bridge deck measured in the dry condition and the wet condition after rainfalls). The potential gradients and the local potential do not change; differences were found only in the magnitude of potential gradients. In a structure with a high chloride concentration, there is usually low resistivity in the concrete. The readings are often mismatched when there is low resistivity in the concrete, but it has readings with high passive potential. This is due the fact that chlorides are transported more easily in lower resistivity of concrete, high contents of chlorides depassivated the steel reinforcement, and active potentials were measured [7].

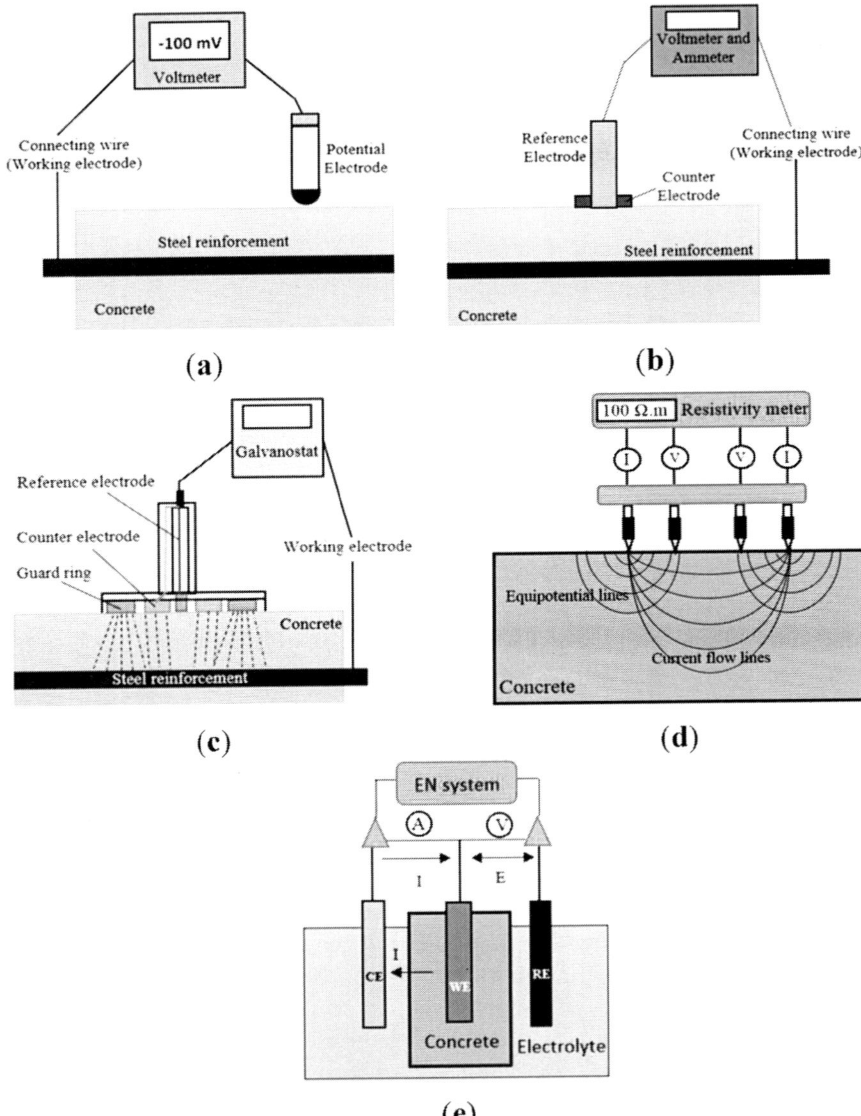

Figure 1. Principle of electrochemical methods: (**a**) open circuit potential monitoring (OCP) (Reproduced and modified from [7]); (**b**) polarization resistance (Reproduced and modified from [23]); (**c**) galvanostatic pulse method (GPM) (Reproduced and modified from [37]); (**d**) resistivity method (Reproduced and modified from [36]); and (**e**) electrochemical noise (EN) (Reproduced and modified from [38]).

On the other hand, the polarization resistance method is commonly used for measuring the corrosion rate in RC structures. The method records the current generated and consumed by anodic and cathodic reactions. The change in potential during reactions is known as polarization, which is used to evaluate the steel corrosion. However, this electrochemical method has some limitations, such as the method assumes uniform corrosion while pitting corrosion is a highly probable form of steel corrosion in concrete, which might lead to misleading results. In addition, the area of steel measured in concrete is not precisely known which creates some errors in the polarization resistance calculations. Another error in the polarization resistance method is the IR drop introduced by high resistive medium and high separation of the reference electrode from the steel reinforcement of the RC structures [34,35,36]. Figure 1b shows the principal of polarization resistance method.

The other practical method for measuring the corrosion rate is GPM. This generally involves impressing a small amplitude, short time anodic current pulse, to be applied galvanostatically on the steel reinforcement from the external counter electrode over the concrete surface [39,40]. The anodic current is usually in the range of 5 to 100 μA and the typical pulse duration is between 5 and 30 s [19]. The steel reinforcement is anodically polarized and there is a resulting change in the electrochemical potential. The potential is recorded by a reference electrode (usually in the center of the counter electrode) as a function of polarization response. Figure 1c shows a schematic setup for the GPM test. When the constant current I_{app} is applied to the system, the polarized potential of reinforcement (η_t), at given time t can be expressed as [39]:

$$\eta_t = I\text{app}[R_{ct}[1 - \exp(-t / R_{ct} \times C_{dl})] + R_\Omega] \quad (1)$$

where R_{ct} is polarization resistance, C_{dl} is double layer capacitance, and R_Ω is ohmic resistance.

The GPM results are much closer to the corrosion rate produced by the gravimetric method than those produced by polarization resistance, electrochemical impedance spectroscopy (EIS), the Tafel extrapolation method (TEM), and the harmonic analysis method (HAM) [13,14,19,41,42]. Uncertainty about the polarized area of steel reinforcement is one of the major sources of error when measuring the corrosion rate of steel reinforcement in RC structures affected by the electrical signal from the counter electrode and the non-uniform current distribution of the steel reinforcement [43].

The resistivity method is another electrochemical method, which relies upon the principle that corrosion is an electrochemical process. An ionic current must pass between the anode and cathode areas for corrosion monitoring of

RC structures [44]. The resistivity is an indirect indication of active corrosion of the steel reinforcement [45,46,47]. The corrosion process will be slower if the resistivity of the concrete is high. The resistivity of concrete exposed to chloride indicates the risk of early corrosion damage, because low resistivity is always associated with rapid chloride penetration [48]. Concrete resistivity is generally measured using the Wenner four probe method, as illustrated in Figure 1d. However, this method has limitations due to the heterogeneities (*i.e.*, steel reinforcement, resistivity layers, cracks, and large aggregates), which would influence the shape of the electric lines in the concrete. In addition, the low distance of the electrode probe (less than a few tens of centimeters) and weather conditions (*i.e.*, wet, temperature, and humidity) influence the resistivity, thus having a huge influence on the results [49,50].

On the other hand, the electrochemical noise (EN) is an emerging technique for monitoring the mechanisms and estimating the rate of corrosion in concrete structures. Electrochemical noise is used to describe fluctuations in the potential and current generated by corrosion reactions [51,52]. The phenomenon whereby the potential of the electrodes fluctuates due to current vibration is also referred to as noise [53]. The current vibration of electrodes is due to fluctuation of the oxide-reduction reaction on the surfaces of the electrodes [38]. A noise source is located within the probable corroding area. A noise signal is transformed from the time domain to the frequency domain displayed in the form of amplitude and frequency based on the fast Fourier transform (FFT). The noise resistance (Rn), which is believed to be similar to the polarization resistance, is given by [54,55,56]:

$$Rn = \sigma V / \sigma I \tag{2}$$

where σV is the standard deviation for the potential and σI, the standard deviation for the current obtained by statistical analysis of the noise data.

The main advantage of the EN technique is its lack of intrusiveness. It can avoid artificial disturbances to the system during measurement [57]. The EN measurement has been used to study the onset of localized corrosion [58,59]. However, only a few studies of EN for measuring corrosion in RC structures have been carried out [60,61]. Previous work using this technique on reinforcing bars has only studied the corrosion process of various metallic materials [62,63,64,65].

Table 1. Interpretation of corrosion activity of electrochemical methods.

Corrosion Activity	Potential Level (mV) [53]	Resistivity (Ω·m) [51]	Corrosion Rate	
			LPR (I_{corr} (A/cm^2)) [39]	GPM (R_{ct} (kΩ·cm^2)) [39]
Very High	-	-	10–100	0.25–2.5
High	<−350	<100	1.0–10	2.5–25
Moderate/Middle	−200 to −350	100–500	-	-
Low	>−200	500–1000	0.1–1	25–250
Negligible/passive	-	>1000	<0.1	>250

Table 1 provides a summary of the qualitative corrosion activity of the above electrochemical methods. However, there is no standardization for the evaluation of the corrosion activity by EN method. Therefore, most of the electrochemical methods require direct connection with the steel reinforcement in the RC structure as the electrode, which makes it an intrusive method [66]. It requires a localized breakdown in the concrete surface to provide the direct connection [67].

2.3. Electromagnetic (EM) Waves

Another NDT method of corrosion monitoring is ground penetrating radar (GPR), which is based on the propagation of an EM wave into an RC structure. Part of the EM wave is reflected back to the receiving antenna whenever it encounters an interface of two media with differing dielectric constants [68,69,70,71]. A GPR antenna receives direct and reflected waves, which are recorded as amplitude–time signals (*a*-scan) by the system. The direct wave signal (S_d) represents the EM energy transmitted directly to the receiving antenna, and the reflected wave (S_r) is the EM energy reflected from the steel reinforcement–concrete interface [70]. The propagation of EM waves depends on the corresponding dielectric permittivity of concrete, which is a quantity related to the ability of concrete to resist the flow of an electrical charge [69,72,73,74]. The permittivity in turn depends on the EM properties, which are influenced by temperature, moisture content, chloride content, pore structure, and deterioration (*i.e.*, corrosion) [75].

Despite its capability, GPR only presents a qualitative assessment of the corrosion damage. Moreover, the presence of chloride content and corrosion products would attenuate the GPR waves [74,76,77,78], which decrease the amplitude of the wave and the average velocity of reflected wave [79,80,81] increasing the travel time of the wave [82,83]. In addition, the GPR method is carried out periodically to monitor the corrosion process in RC structures and cannot be used for real-time monitoring [66]. Figure 2a shows a 3D image of GPR data in an RC slab with four different degrees of corrosion (*i.e.*, no corrosion, low corrosion, medium corrosion, and high corrosion from the left to the right side).

2.4. Infrared Thermography (IRT)

As a sub-surface inspection method, Infrared thermography (IRT) has been found to be capable of detecting corrosion in RC structures. The method is based on the resulting perturbations in the heat transfer characteristics of concrete materials [84,85,86,87,88,89,90]. Temperature is one of the most common indicators of the structural health of RC structures. Cracks, alkali aggregate reaction (AAR), corrosion, and other deterioration forms could cause abnormal temperature distribution [86]. With the advent of newer generations of IR cameras, IRT is becoming a more accurate, reliable, and cost effective technique for corrosion monitoring in RC structures [88]. However, like GPR method, the interpretation of IRT is also qualitative, *i.e.*, a higher degree of steel corrosion results in stronger IR thermal distribution on the concrete surface [89,91,92,93,94], which would exhibit higher peaks of IR intensity and faster rates of heating [95,96], and hence increase the temperature of the concrete materials [92]. Figure 2b shows an IRT image of the temperature distribution of a 10% degree of corrosion of the concrete cylinder.

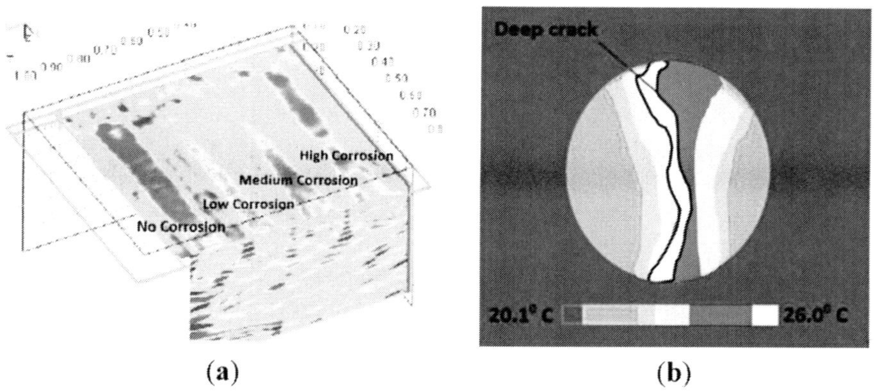

Figure 2. (**a**) 3D image of GPR of concrete slab with different level of steel corrosion, unit (m) (with permission from [68]) and (**b**) IRT image of concrete surface of cylinder, unit (°C) (Reproduced and modified from [89]).

2.5. Optical Sensing Methods

The optical sensing technology of the NDT methods, the fiber Bragg grating (FBG) method, involves creating periodic variations in the refractive index of the core of an optical fiber [97]. When light is made to pass through the grating at a particular wavelength, called the Bragg wavelength, the light reflected by the varying zones of refractive indices would be in phase and amplified [98,99]. During corrosion, the formation of corrosion products less dense than steel increases the volume and diameter of the bars, leads to an increase in fiber strain, which is measured by a shift in the wavelength of the FBG. The extent of corrosion is quantitatively evaluated through the change in the wavelength of the FBG [100,101,102]. Although it has advantages such as linear reaction, small volume, high anti-erosion capability, and automatic signal transmission [103,104], the FBG method also has limitations in corrosion monitoring in RC structures. The method can only conduct localized inspection of steel corrosion in RC structures. It is expected that a distributed and long-gauge FBG technique will be developed to solve this problem; until then this method is less effective for corrosion monitoring in RC structures [105].

2.6. Elastic Wave Methods

In order to enhance reliability, elastic wave methods should be adopted to complete the assessment of steel corrosion in RC structures. Elastic wave methods are essential when estimating mechanical properties and inhomogeneous characterization due to steel corrosion in RC structures. Wave-based damage detection excites transient waves to propagate into the concrete structure using sensors since the waves are reactive to damage such as cracks, voids, and also corrosion products [106]. There are three major wave-based methods for corrosion monitoring in RC structures, *i.e.*, ultrasonic pulse velocity (UPV), impact echo (IE), and acoustic emission (AE).

The UPV test is an NDT method that involves measuring the speed of sound through concrete in order to detect the condition of the concrete and the presence of steel corrosion [31,107,108,109,110,111,112] in chloride and oxide environments [112,113,114,115]. The principle of the UPV method is shown in Figure 3a. Ultrasonic waves can propagate a long distance along the steel reinforcement and have been found to be sensitive to the interface conditions between the steel reinforcement and the concrete [112,116]. Cracking due to steel corrosion results in wave attenuation and a decrease in UPV. It has been reported that the amplitude attenuation has good a correlation with the damage due to corrosion [31,117,118]. In addition, UPV as measured by the first wave peak could describe the process of steel corrosion. As the corrosion damage level increases, the relative variation for the first wave peak value of UPV first increases and then decreases [119]. This condition occurs because the corrosion products are increased during the process of steel corrosion, resulting in an increase in the delamination degree between the steel reinforcement and the concrete. As the reinforced concrete corrosion level increases, the pit on the steel becomes larger. This condition would lead to great reflection in the first wave energy, and the direct transmission wave energy would become low. The first wave peak value then decreased slowly. However, the method needs advanced study. The study could come from the use of surface and other, guided waves, with features other than longitudinal pulse velocity, and stemming from advanced signal processing techniques [120].

The stress wave method, IE, is employed to detect corrosion in structural elements using mechanical impact and then monitoring the displacement (d), detected by sensors placed on the concrete surface. The impulse is reflected by the arrival of reflections of the pulse from the crack and other internal defects under investigation [121,122], as shown in Figure 3b. In an early study by Liang and Su [123] more than a decade ago, the IE method was certainly able to detect the development of micro-cracks due to corrosion in RC blocks. In the latest studies, Samarkova *et al.* [124,125,126,127] have indicated that the dominant frequencies of the response signal are the main criteria used to detect the occurrence and position of steel corrosion in RC structures. However, only a number of limited studies have attempted to monitor steel corrosion in RC structures because the main application of the IE method is in the detection of voids and delamination. In addition, the proposed method is not as mature as electrochemical methods for corrosion monitoring.

Therefore, AE is considered a good complementary method to UPV and IE. The AE technique is a unique, non-invasive, and passive NDT method. AE is a class of phenomena whereby transient elastic waves (ultrasonic frequency range) from a localized source within a material and conversion of the waves into electrical signals through coupled piezoelectric sensors [128,129,130,131,132]. The sources of AE are deformation processes such as crack growth, void closure, plastic deformation, corrosion, and other material degradation. Localized energy release gives rise to elastic waves that are detected by sensors placed on the concrete surface [133,134]. The principle of AE is shown in Figure 3c. The AE technique is often capable of detecting corrosion in the early stages, so that an early warning can be given to allow for repair work before the structural RC element is seriously damaged and the functionality is lost due to steel corrosion. This study reviews comprehensively the capability of the AE technique in monitoring the corrosion activity and it discusses the distinct advantages of the AE technique in the following sections. Table 2 shows a summary of NDT methods for corrosion evaluation.

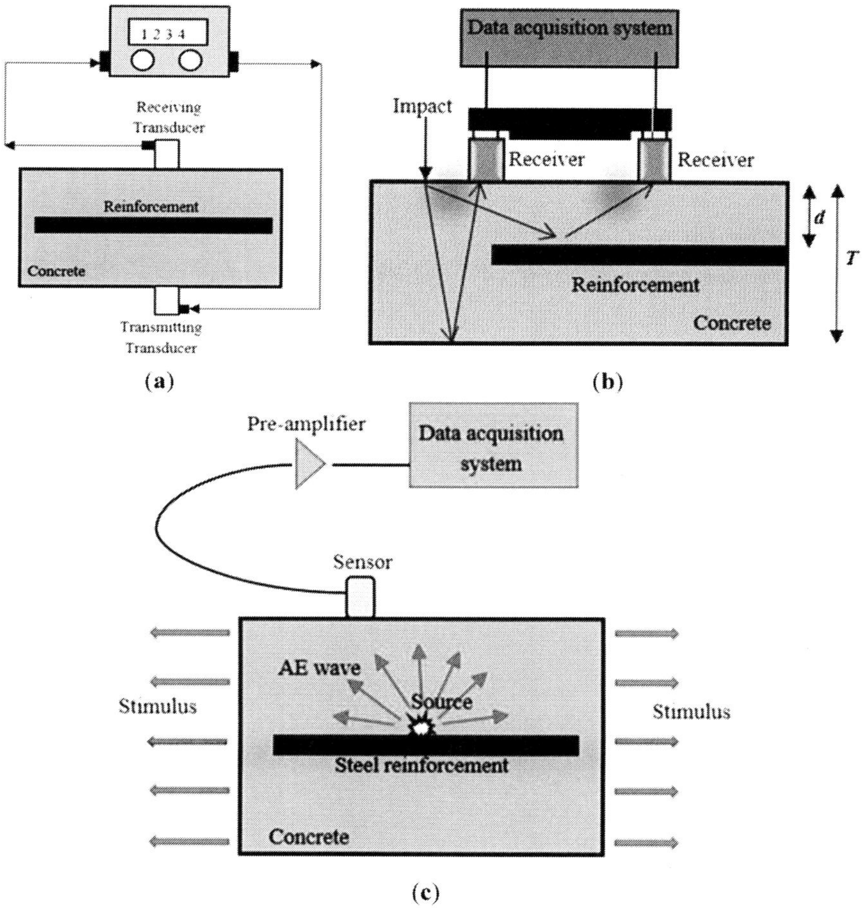

Figure 3. The principle of: (**a**) ultrasonic pulse velocity (Reproduced and modified from [120]); (**b**) impact echo (Reproduced and modified from [121]); and (**c**) acoustic emission (Reproduced and modified from [135]).

Table 2. Resume of NDT methods for corrosion evaluation.

No	NDT Methods	Principles	Advantages	Disadvantages	Corrosion Evaluation	Specific Equipment
1	**Electrochemical Method**					
	Open circuit potential (OCP) monitoring	Electrical potential value (in mV or V) is measured between steel reinforcement of RC and reference electrode (indicates corrosion potential of the steel inside RC)	The results are not in the form of equipotential contours, rather a single value that gives an indication of the steel condition.	Time consuming and need to be closed several hours during the inspection.	Potential level (mV or V)	Potential electrode, Voltmeter, and connecting wire (working electrode).
	Resistivity method	Resistivity (ρ) of RC, which the current can easily pass between anode and cathode areas of the concrete.	An easy, fast, portable, and inexpensive technique, which can be used for routine inspection.	Reinforcement in the test region can provide a "short-circuit" path and cause erroneous reduction in the measurement.	Resistivity (Ω·cm)	Current and potential electrodes, Voltmeter or resistivity unit, and insulated wire (working electrode).
	Polarization resistance	The change in potential during reactions (polarization) is recorded using an electrode plate on the concrete surface.	Short time for measurement and applies small perturbations that do not interfere with the existing electrochemical processes.	It takes time to obtain a full response because of the electrical capacitance across the steel and concrete interface. The voltage error introduced by IR drop in the concrete between working (steel rebars) and reference electrode	Corrosion current (I_{corr} (A/cm²))	Reference electrode, counter electrode, Voltmeter, Ammeter, and connecting wire (working electrode).
	Galvanostatic pulse method (GPM)	The anodic current pulse is applied galvanostatically on the steel reinforcement from counter electrode placed on the concrete surface.	A rapid device for determining the corrosion rate of steel reinforcement in RC. it enables display of corrosion rate, electrical resistance and potential value simultaneously.	Unstable reading due to parallel or crossing of the steel reinforcement, also cracks and delamination are often the reason for wrong readings	Potential resistance (R_{ct} (kΩ·cm²))	Reference electrode, counter electrode, guard ring, and connecting wire (working electrode).
	Electrochemical noise (EN)	EN describe the fluctuations of current and potential spontaneously generated by corrosion reactions.	Simple to use, no interference to the system, and measured signals can be analyzed by mathematical analysis.	The complicated kinds of noise (*i.e.*, physical origin) due to corrosion of steel reinforcement make mathematical analysis unsuccessful	Noise resistance (R_n (kΩ·cm²))	Electrodes (reference, counter, and working), Voltmeter, Ammeter, amplifier, and data acquisitions board.
2	**Elastic Wave Method**					
	Ultrasonic pulse velocity (UPV)	Mechanical energy propagates through the concrete as stress waves and is converted into electrical energy by a second transducer	A large penetration depth and it is easy to use for estimating the size, shape and nature of the concrete damage	The evaluation of UPV data is a highly specialized task, which requires careful data collection and expert analysis.	Pulse velocity (V)	Transducers (transmitter and receiver), amplifier, and oscillator.
	Acoustic emission (AE)	Elastic waves are generated due to rapid release of energy from a localized source within an RC structure.	A cost-effective and sensitive technique that can detect and locate the active defects.	Passive defects cannot be effectively detected.	AE parameters	Transducer, preamplifier, filter, amplifier, and storage equipment.
	Impact echo (IE)	Stress wave are propagated within the RC structure through vibrations and impact load.	A simple, fast, reliable method for inspecting the concrete is to impact the surface with a hammer and listen to the results.	The reliability of the IE method decreases with an increase in thickness.	Wave velocity (Vp)	Mechanical impactors, high-fidelity receiver, and data acquisition-signal analysis system.
3	**Electromagnetic Method**					
	Ground penetrating radar (GPR)	Transmission of electromagnetic (EM) waves into the RC structure under investigation.	Equipment portable and effective for investigating one large area from one surface.	Difficult interpretation of the results and needs post-processing analysis.	EM wave velocity (V)	Antennas (transmitter and receiver), a control unit, and computer.
4	**Optical Sensing Method**					
	Fiber Bragg grating (FBG)	The shift of FBG wavelength measures the increase in fiber strains with an increase in the cross section of steel reinforcement of corroded RC structures.	Small physical dimensions and suitable for embedding into structures.	The equipment has a high cost and there is no standardization of the procedure.	Bragg wavelength (λB)	Optical fiber sensor, Bragg meter, and computer.
5	**Infrared Thermography Method**					
	Infrared thermography (IRT)	IR radiation emitted by a concrete material is converted into an electrical signal and processed to create maps of the surface temperature.	Easy interpretation of the results and no radiation, rapid set-up, portable, and cost-effective technique.	There is no quantitative information on corrosion damage (*i.e.*, size or depth).	Radiation power (E)	Multi spectrum camera.

3. ACOUSTIC EMISSION (AE) FOR CORROSION MONITORING

The AE technique has been widely used in the field of civil engineering for structural health monitoring (SHM), especially the monitoring of corrosion [136,137,138]. The advantage of the AE technique is that can be used without intruding into any of the processes associated with the RC structures [139,140,141,142]. The first recorded application of the AE technique for corrosion monitoring was by Dunn *et al.* [143]. The AE technique was used to monitor and characterize the corrosion process in a series of controlled laboratory tests. The study illustrated the sensitivity of the AE technique to evaluate the ongoing corrosion process, suggesting its feasibility as a technique for monitoring corrosion in RC structures. There are two different approaches in analyzing AE data, one is the classical or parameter-based AE technique and the second is the quantitative or signal-based AE technique [142,144]. These approaches could be used for monitoring corrosion in RC structures, in the following sections.

3.1. AE Parameters for Corrosion Monitoring

The parameter-based technique is useful for better characterization of the AE source [145,146,147]. AE parameter analysis of hits or events, signal strength, and energy demonstrate that AE is readily applicable to the detection of corrosion in steel reinforcement, in order to identify the corrosion process, *i.e.*, initiation of cracks and propagation of cracks, and to locate the early stages of corrosion [148,149,150,151,152,153,154,155,156,157]. The AE sources are also classified in terms of RA value and average frequency to classify the type of failure and the *b*-value or *Ib*-value of AE amplitude distribution for assessing the damage severity [151,153,158,159]. The above AE parameters will be reviewed in the following sections. A simplified representation of an emitted signal as well as of commonly used parameters is shown in Figure 4. A summary of AE parameters and their contribution in providing information about the source event are listed in Table 3.

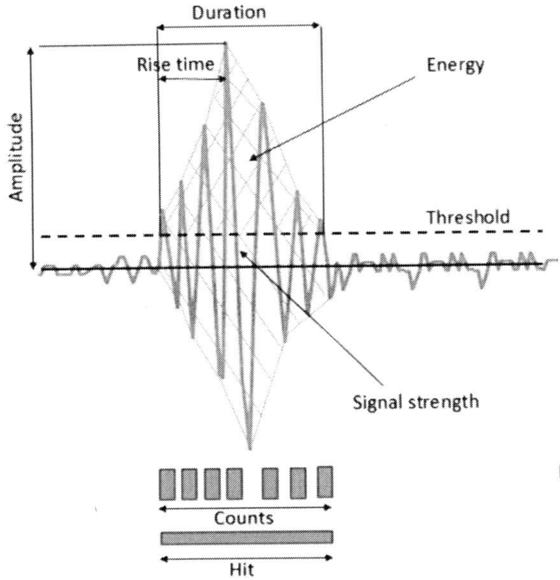

Figure 4. Parameters reflecting of an AE waveform.

Table 3. AE parameters and the applications in source events information (Reproduced and modified from [158]).

Variables	Parameter	Description	Variables	Parameter
Time domain	Hit	Detection of an AE signal	Hit	Detection of an AE signal
	Event	Local material change, an event is a number of hits	Event	Local material change, an event is a number of hits
	Amplitude	Largest voltage peak in the waveforms	Amplitude	Largest voltage peak in the waveforms
	Rise time	Time elapsed from signal start and peak amplitude	Rise time	Time elapsed from signal start and peak amplitude
	Duration	Time between signal start and signal end	Duration	Time between signal start and signal end
	Threshold	Electronic compactor such that signals with amplitude larger than this level will be recorded	Threshold	Electronic compactor such that signals with amplitude larger than this level will be recorded
	Counts	Number of times AE signal crosses threshold	Counts	Number of times AE signal crosses threshold
	Signal Strength	Area under the positive and negative envelope of linear voltage signal	Signal Strength	Area under the positive and negative envelope of linear voltage signal
Frequency domain	Frequency spectrum	Nature of source event	Frequency domain	Frequency spectrum
Time-frequency domain	Spectrogram	Energy distribution of source event through time	Time-frequency domain	Spectrogram

3.1.1. AE Hits

Many researchers have considered AE hits to be one of the AE parameters used to study the onset of corrosion and the nucleation of crack in RC structures [152,153,154,155]. Yoon *et al.* [152] carried out AE monitoring in RC beams subjected to four different degrees of corrosion. It was observed that AE hits increased with an increase in the degree of corrosion. This trend could provide important information for estimating the degree of corrosion. Ohtsu and Tomoda [155,156] investigated AE hits of RC specimens in sodium chloride (NaCl) concentration. Corresponding to two high AE activities, two periods of onset of corrosion and nucleation of cracks were observed. This could suggest that the AE activity observed corresponds to the corrosion loss of steel reinforcement in a marine environment observed by Melcher and Li [160], and shown in Figure 5a. At Phase 1, the onset of corrosion is initiated and the phase is dominated by the presence of oxygen and water. Then, corrosion loss decreases and stabilizes at Phase 2. The mass loss of corrosion increases again at Phases 3 and 4, where the corrosion penetrates inside and the expansion of corrosion products occurs due to anaerobic corrosion. Thus, based on the four phases of corrosion loss, two stages of corrosion activity are characterized, *i.e.*, the onset of corrosion and the growth of corrosion products (nucleation of cracks).

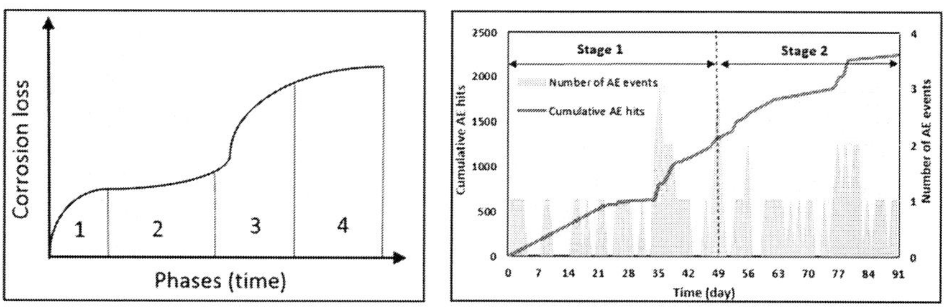

(a) (b)

Figure 5. (**a**) Typical corrosion loss of steel reinforcement due to chloride immersion (Reproduced and modified from ([160]); and (**b**) cumulative AE hits and number of AE events during corrosion test (Reproduced and modified from [153]).

On the other hand, Kawasaki *et al.* [153,154] showed cumulative AE hits and number of AE events in the corrosion process in a cyclic wet–dry test. The generating process of AE hits observed is classified into two stages, which refer to the process of corrosion loss of steel reinforcement due to chloride immersion as observed by Melcher and Li [160], and previously by Ohtsu and Tomoda [155,156]. Stage 1 (onset of corrosion) corresponds to Phases 1 and

2 and Stage 2 (nucleation of crack) corresponds to Phases 3 and 4. Thus, the AE technique could detect corrosion at an early stage and AE activities from the beginning, as in Figure 5b.

3.1.2. Signal Strength (SS) and Cumulative Signal Strength (CSS)

Signal strength (SS) is one of the AE parameters; it is defined as the area under the voltage signal of AE over the duration of the waveforms. Since it provides a measure of the waveform energy released by the specimen, it is a rational damage indicator [67,161]. Velez *et al.* [161] showed the results of plots of the SS and cumulative signal strength (CSS) of corroded specimens. The SS of the prestressed specimen (P1) is attributed to the nucleation of cracks caused by the accumulation of corrosion products at the steel-concrete interface [162]. A different analysis is conducted on the SS of the non-prestressed specimen (N3), which suggests that it could be an indicator of early crack formation due to corrosion. On the other hand, the results show that the cumulative signal strength (CSS) exhibits a clear rate change before the onset of corrosion according to electrochemical methods, which suggests that the AE technique could detect the onset of corrosion.

Patil *et al.* [67] showed the curve of CSS with time of a concrete specimen in Figure 6. The curve clearly distinguishes the AE activity recorded for the concrete specimen under active corrosion. The CSS rate increases slowly in Phase 1 indicating de-passivation of the layer surrounding the steel reinforcement and the onset of corrosion. The presence of a sudden rise at the end of Phase 1 might indicate crack initiation due to steel corrosion. The rise of CSS in Phase 2 indicates corrosion activity. The sudden rise at the end of Phase 2 seems to be crack propagation leading to a macro-crack. Further phases are a repetition of Phase 2 continuing in the same manner. According to the phenomenological model of corrosion loss in a marine environment, the CSS curve obtained could also be divided into four phases, but the curve shows two sudden rises at the end of Phases 1 and 2. If these sudden rises are excluded from the CSS curve, the curve shown by the dotted line will be obtained and if connected by a smooth line, the curve would be exactly in agreement with the conventional curve (as shown in Figure 5a).

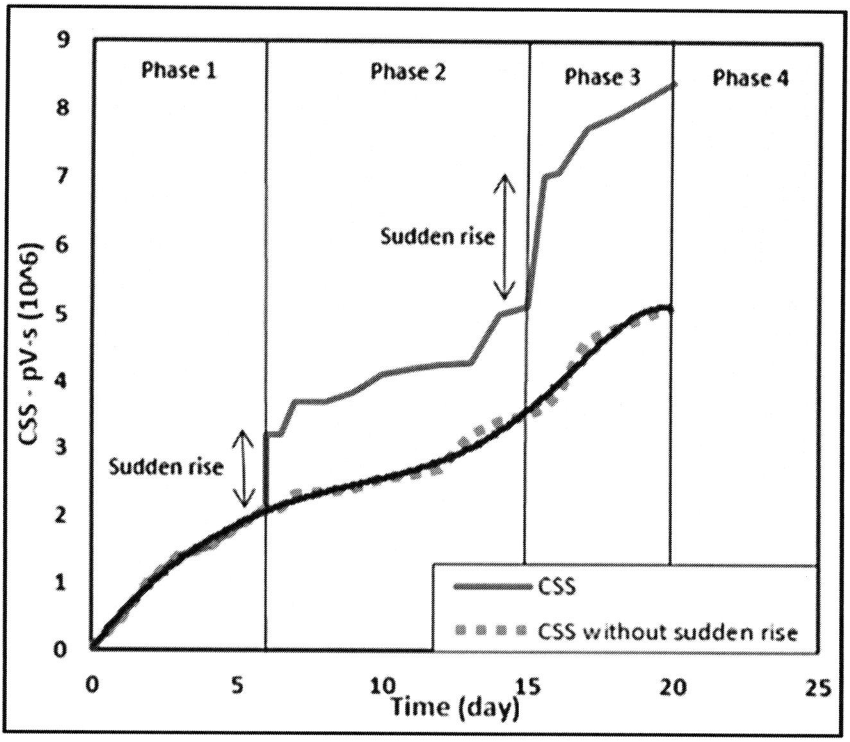

Figure 6. The variations in CSS parameter, which are similar to the curve of typical corrosion of steel reinforcement (Reproduced and modified from [68]).

3.1.3. Absolute (ABS) Energy

The feature of ABS energy is a quantifiable measurement of energy obtained for all the AE events or hits. Ing *et al.* [163] demonstrated the potential of ABS energy for identifying corrosion at an early stage before any external signs (*i.e.*, cracking) of corrosion occur. The thickness of the concrete cover is found to have a significant effect on ABS energy in the early stages of steel corrosion. An exponential relationship has been established between the compressive strength and ABS energy, which shows that AE detects the sudden release of micro-fractures in the RC structure. In addition, increasing the steel diameter due to corrosion is found to increase the ABS energy of AE data.

3.1.4. RA Values and Average Frequency

The characteristics of AE signals are estimated using two parameters, RA value and average frequency. The RA value and average frequency are defined from the AE parameters, *i.e.*, rise time, maximum amplitude, counts, and duration [151,156], as shown in Equations (3) and (4).

$$\text{RA value} = \text{Rise Time/Amplitude} \tag{3}$$

$$\text{Average frequency} = \text{Counts/Duration} \tag{4}$$

A crack type is classified by the relationship between RA value and average frequency as shown in Figure 7. A tensile-type crack is referred to as an AE signal with high average frequency and low RA value. A shear-type crack is identified by low average frequency and high RA value. This criterion is used to classify AE data detected in the corrosion process.

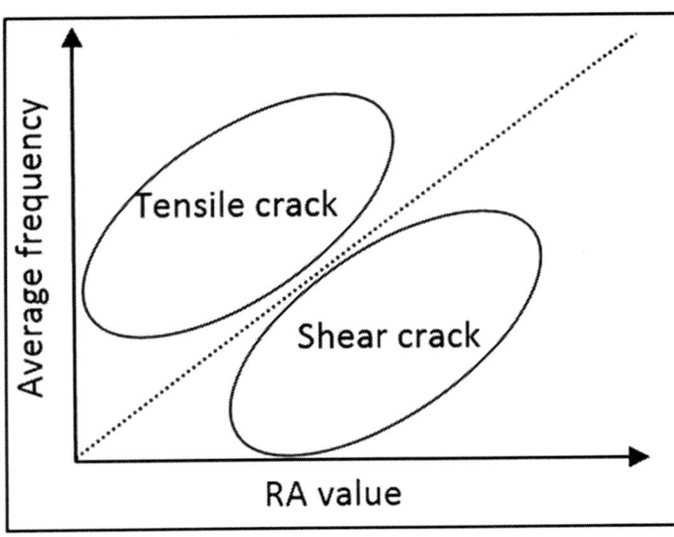

Figure 7. Classification of cracks by AE indexes (Reproduced and modified from [164]).

The RA values and the average frequency were used by Ohtsu and Tomoda [156] to classify AE sources, in two weeks of wet and dry tests in an RC slab. At 40 days, the RA value was high and the AF was low, indicating shear cracking and later the RA value decreased and AF increased showing tensile cracking, as shown in Figure 8a. However, on the other hand, Kawasaki *et al.* [153] showed different trends for RA values and average frequency, as shown in Figure 8b. The trend lines proposed classify the onset of corrosion and nucleation of cracks in an RC beam. At Stage 1, at 14 days to 21 days, RA values drastically increase and the average frequencies become smaller. The crack could be classified as a tensile crack due to the onset of corrosion in the RC beam. At Stage 2, an increase in RA values and a decrease in average frequencies were observed. This implies that generations of nucleation of crack due to corrosion were induced in the concrete specimen. From the above references, the trends for RA value and average frequency, which were proposed by Ohtsu and Tomoda [156] for an RC slab, seem similar to the trends for RA value and average frequency proposed by Kawasaki *et al.* [153] for an RC beam at Stage 1. The RA value is low and the average frequency is fairly high.

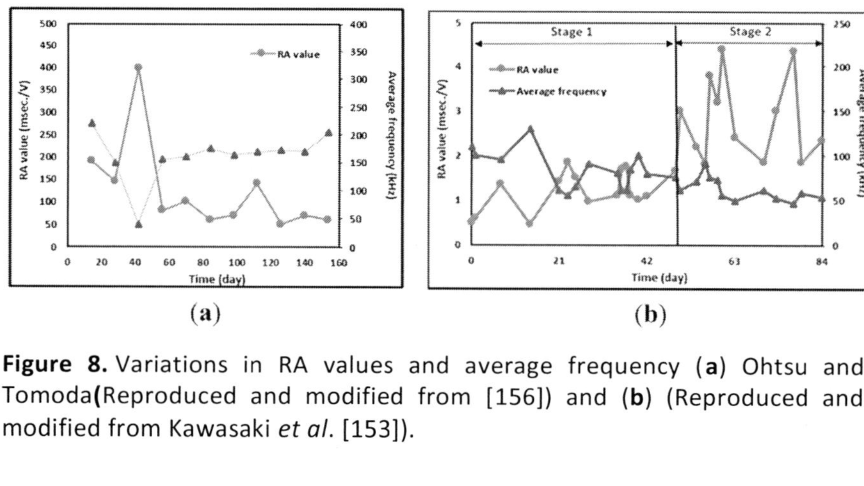

(a) (b)

Figure 8. Variations in RA values and average frequency (**a**) Ohtsu and Tomoda(Reproduced and modified from [156]) and (**b**) (Reproduced and modified from Kawasaki *et al.* [153]).

3.1.5. *b*-Value and *Ib*-Value

Gutenberg and Richter have developed the *b*-value in seismology to understand the relationship between the magnitude and frequency of earthquakes [140,165,166], as shown in Equation (5):

$$\mathrm{Log}_{10} \mathrm{N} = \mathrm{a} - \mathrm{bM} \quad (5)$$

where, M is the Richter magnitude of the event, N is the incremental frequency, "*a*" is an empirical constant and *b* is the *b*-value. The value of M is proportional to the logarithm of the maximum amplitude A_{max} recorded in a seismic trace.

In the AE method, the same principle can be applied to determine the scaling of the amplitude distribution of the AE waves during the fracture process. In term of AE technique, the formula was modified by Colombo *et al.* [140]. Where, $A_{max}$is peak amplitude of the AE events in decibels (dB).

$$\mathrm{Log}_{10}\, N = a - b(A\max) \tag{6}$$

However, to evaluate the slope failure and facture process, Shiotani *et al.* [167] improved the formula to improve b-value (*Ib*-value). This formulation is more based on statistical analysis, such as mean and standard deviation, for each of the AE amplitude sets. The formula is defined as:

$$Ib = \frac{Log\ N\ (\mu - \ \alpha 1\sigma) - Log\ N\ (\mu - \alpha 2\sigma)}{(\alpha 1 + \ \alpha 2)\sigma} \tag{7}$$

where, σ is the standard deviation, μ is the mean value of the amplitude distribution, $\alpha 1$ is the coefficient related to the smaller amplitude, and $\alpha 2$ is coefficient related to the fracture level.

In an RC structure application, the *b*-value is used as an indication and demarcation of degradation in the integrity of the RC specimen and is associated with cracks [140]. When cracks are forming, the large number of events increases causing a decrease in the *b*-value. When the distributed micro-crack occurs at an early stage of corrosion, the *b*-value is large and when the macro-crack begins to localize the *b*-value is small. Previous research has indicated that *b*-values below 1.0 correspond to the transition from micro-crack to macro-crack [168,169]. Figure 9a shows an example of *b*-value distribution in acyclic wet–dry test in an RC slab by Ohtsu and Tomoda [156]. The *b*-value becomes large in the first period and then the*b*-values keep fairly low. This result might imply the generation of small shear cracks in the first period. Then, nucleation of fairly large tensile cracks follows, leading to the second period.

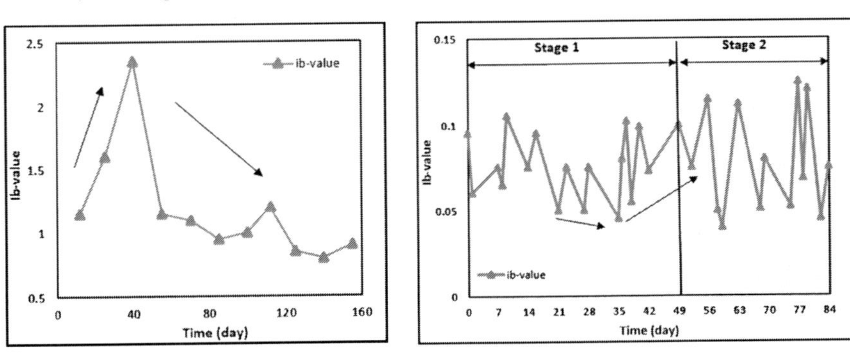

Figure 9. (**a**) *b*-value analysis (Reproduced and modified from [156]); and (**b**) *Ib*-value analysis (Reproduced and modified from [153]).

For the latter, the *Ib*-value is adopted for calculation, based on cumulative distribution as proposed by Shiotani [144]. In contrast, the case where the *Ib*-values become small implies nucleation of large AE hits. The variations in the *Ib*-value are given by Kawasaki *et al.* [153], as shown in Figure 9b. Large drops are observed between 21 days and 35 days, before the first dramatic increase in *Ib*-value. This might imply that the micro-cracks are generated at the onset of corrosion on the surface of the steel reinforcement. Due to high AE activity at Stage 2, the *Ib*-values decrease. Since the results of *Ib*-values at 56 and 84 days are comparatively lower than those of Stage 1, large-scale cracks are considered to be actively generated as corrosion-induced cracks in the RC beam. Furthermore, the fluctuations in *Ib*-values in Stage 2 are even bigger than in Stage 1. These results imply that the cracks are repeatedly generated due to corrosion products expansion.

3.1.6. Intensity Analysis

Intensity analysis (IA) is used to quantify corrosion rate and level [170,171,172,173]. Intensity analysis evaluates the structural significance of an AE event by calculating two values, called the historic index (*Hi*) and the severity index (*Sr*), from the signal strength [174,175,176], as shown in Equations (8) and (9). The historic index is used to estimate changes of the slope in the CSS plotted as a function of time. The severity index is the average of the large signal strength received at a sensor (*i.e.*, 50 events having the highest signal strength) [177]. An increase in severity index often corresponds to structural or material damage.

$$Hi = \frac{N}{N-K} + \left(\frac{\sum_{i=k+1}^{N} Soi}{\sum_{n=1}^{N} Soi}\right) \tag{8}$$

$$Sr = \frac{1}{J}\left(\sum_{m=1}^{J} Som\right) \tag{9}$$

where, Hi = Historic index, N = Number of hits up to time t, S_{oi} = Signal strength of the *i*th hit, K = empirically derived constant based on material, Sr = Severity index, J = empirically derived constant based on material, S_{om} = signal strength of the *m*th hit where the order of m is based on magnitude of the signal strength.

The K and J value are related to N by the relations: K = 0, N ≤ 50; K = N−30, 51 ≤ N ≤ 200; K = 0.85N, 201 ≤ N ≤ 500 and J = 0, N < 50; J = 50, J ≥ 50.

Velez *et al.* [161] developed IA-based criteria for assessing corrosion in prestressed concrete (PC) piles. An assessment chart divides the criteria into three areas, *i.e.*, no corrosion, early corrosion, and cracking, as shown in Figure 10. The figure shows that corroding and non-corroding specimens can be distinguished. In addition, the values of *Hi* and *Sr* are consistent with the levels of corrosion.

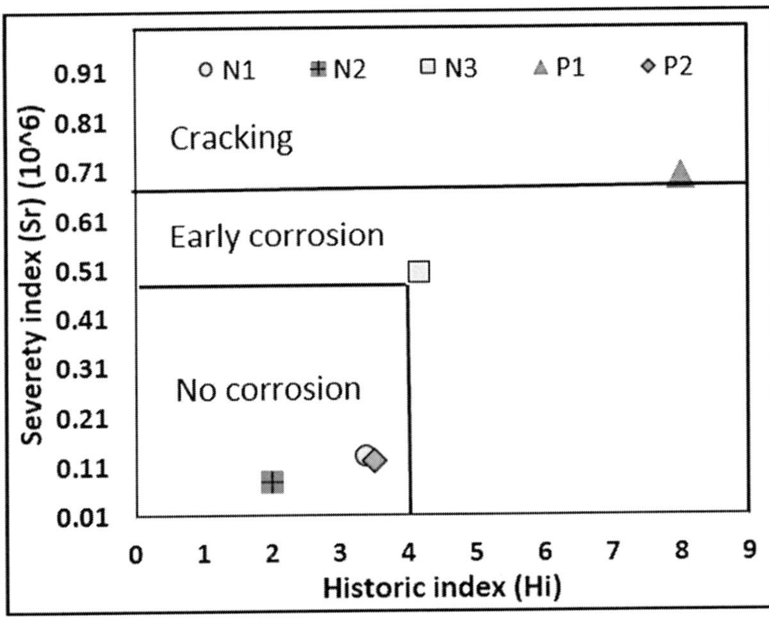

Figure 10. Intensity analysis results (Reproduced and modified from [161]).

3.1.7. Relaxation Ratio

This refers to the ratio of average energy recorded during the unloading and loading phases of a concrete test [177]. The method is based on the effect of the cracks present in RC structures. If the RC structure contains no cracks, the energy recorded during the unloading phase is low, however, if the cracks exist, the cracks emit significant energy during the unloading phase. Therefore, the greater amount of AE energy collected during the unloading phase compared to the loading phase could be used as an indicator of corrosion, if the relaxation ratio is more than 1 [159].

3.2. Signal-Based AE for Corrosion Monitoring

The signal-based technique involved a large number of waveforms that are recorded over a sufficiently short period time [142,144]. The most prominent feature of this approach compared to parametric analysis is that it performs better in filtering signal noise, thus offering a better interpretation of the data monitoring in RC structures [158].

3.2.1. AE Source Location

AE source location is performed to monitor the onset of corrosion, crack initiation due to corrosion, crack propagation, and location of corrosion in RC structures. The AE source location is determined using the velocity of the longitudinal wave computed by the time differences among the arrival times of the first longitudinal wave detected by AE sensor [178]. Figure 11 shows the AE source location of high corrosion activity in an RC slab by Ohtsu and Tomoda [155]. It is reasonable to assume that AE sources are located around the steel reinforcement. This implies that corrosion and activities *i.e.*, cracks and the expansion of corrosion products are readily detected and located by the AE technique.

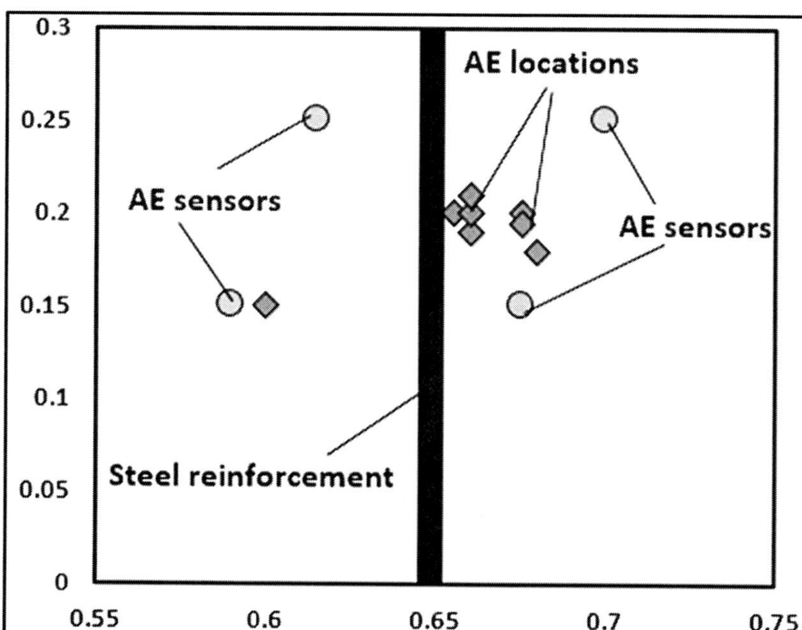

Figure 11. AE source location of corroded RC slab, unit (m) (Reproduced and modified from [155]).

3.2.2. SiGMA

In order to determine the moment tensor of an AE source, a simplified procedure has been developed and implemented as SiGMA (Simplified Green's functions for Moment tensor Analysis) by Ohtsu [179]. The analysis consists of a three-dimensional (3D) AE source location procedure and moment tensor analysis of AE sources. The location of the AE source is determined from the differences in arrival time [178]. Then, the components of the moment tensor are determined from the amplitudes of the first motions at the AE channels [153]. For this matter, SiGMA is a sophisticated method for estimating the size, orientation, crack type, location, and fracture mode of individual micro-cracking [180]. Farid Uddin and Ohtsu [181] and Kawasaki *et al*. [153] presented the results of SiGMA at corrosion test in an RC structures. By SiGMA, the cracking mechanisms due to corrosion of 594 events were identified close to steel reinforcement by [181], as shown in Figure 12. The legend of SiGMA results is shown in Figure 12b. In other hand, Thirty AE events of cracking mechanisms are detected in Stage 1 by [153]. These events are located around the steel reinforcement and at around the top side of the RC beam. This phenomenon could be related to the corrosion initiation as shear and mixed-mode cracks. In Stage 2, 19 AE events were determined close to the steel reinforcement with those of corrosion-induced crack in concrete and related to tensile cracks.

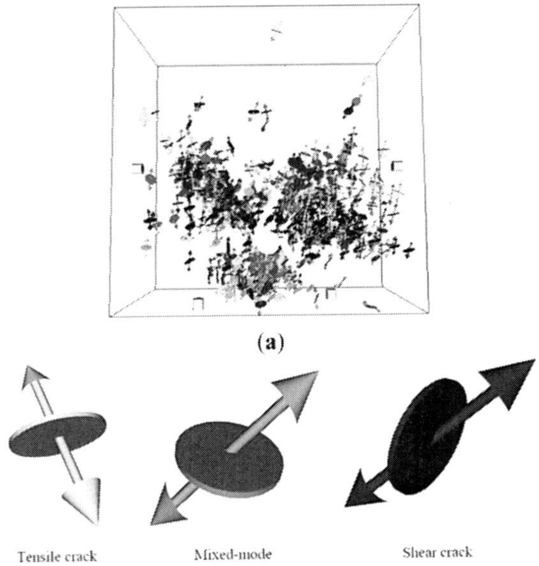

(a)

(b)

Figure 12. Results and models of SiGMA data (**a**) 594 events; and (**b**) models of tensile, mixed-mode, and shear crack (Reproduced and modified from [181]).

3.2.3. Noise Filtering

The noise of AE data can originate from external sources, *i.e.*, environment, traffic, man activities, *etc.* or be due to instrumental sources. Instrumental or electrical noise is generated by fluctuations occurring in the instrumentation including thermal noise, leakage current instability, and power supply voltage fluctuation [182]. Consequently, it is necessary to distinguish between real events related to real sources (such as crack propagation or fracture phenomena) and spurious events related to noise sources. Kouroussis and Anastassopoulos [183] proposed Unsupervised Pattern Recognition (UPR) to discriminate the signal and to distinguish real events from noise. The UPR applies mathematical or clustering algorithms in order to divide the set of AE hits into groups or clusters, which are close to one another in the same data set. This algorithm can discriminate the various sources of data that are related to the noise. Thus, Calabrese *et al.* [184] carried out an investigation to detect corrosion induced cracking in PC structures. It was focused on the use of multivariate analysis with the aim of establishing a procedure that would allow differentiation between steel corrosion based signals and background noise. Figure 13 shows the scheme for clustering methodology for noise removal. The three step models were implemented using Matlab software. The results show that more than 60% of detected signals were classified as noise.

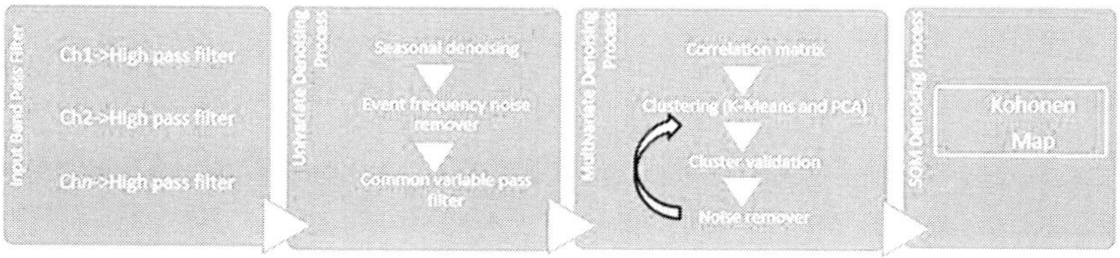

Figure 13. AE noise filtering procedure (Reproduced and modified from [183]).

4. CONCLUSIONS

A number of NDT methods for the corrosion monitoring of steel reinforcement have been reviewed in this paper. There are six major NDT methods for corrosion monitoring: *i.e.*, visual inspection, electrochemical methods (*i.e.*, HCP, resistivity method, LRP, and GPM), elastic wave methods (*i.e.*, UPV, AE, and IE), the electromagnetic (EM) method (*i.e.*, GPR), the optical sensing method (*i.e.*, FBG), and the IRT method. NDT methods for corrosion evaluation are summarized in Table 2. Each technique was reviewed in relation to principles, certain applications, and limitations. However, AE is more effective for monitoring and detecting steel corrosion in RC structures at an early stage.

1. The early number of cumulative AE hits can detect the corrosion at early stage. If there is significant increase of cumulative AE hits, it corresponds to onset of corrosion.
2. Other AE parameters, like the sudden rise in cumulative signal strength (CSS) and absolute energy (ABS) might indicate crack initiation due to steel corrosion.
3. The distribution of RA value and average frequency (AF) are also proposed as a means of classifying the onset of corrosion and nucleation of cracks. The onset of corrosion is identified by a drastic increase in RA value and a decrease in average frequency. In addition, the nucleation of cracks is implied by an increase in RA value and a decrease in average frequency at the next stage. Based on the above parameters, the steel corrosion of RC structures can be analyzed according to two stages, *i.e.*, the onset of corrosion and the nucleation of crack due to corrosion product.
4. In addition, *b*-value, *Ib*-value, and intensity analysis are developed to characterize damage in RC structures. Previous studies have indicated that a *b*-value below 1.0 indicates the transition from microcracking to macrocracking. The large fluctuations in *Ib*-value imply that these cracks are generally repeated due to expansion of corrosion product. On the other hand, intensity analysis uses several criteria to identify the condition of the RC structure (*i.e.*, no corrosion, early corrosion, and cracking).

Therefore, these capabilities make AE a strong candidate to become an efficient NDT method in the detection of corrosion occurring in real-time, giving it an advantage over other NDT methods. Although more research may be required to fully exploit current methods and devices, the future performance of AE in the corrosion evaluation of RC structures seems promising.

ACKNOWLEDGMENTS

The Authors would like to acknowledge the Minister of Higher Education (Malaysia) for providing the financial support for this project under Grant No. UM.C/625/1/HIR/MOHE/ENG/54.

REFERENCES

1. Badawi, M.; Soudki, K. Control of corrosion-induced damage in reinforced concrete beams using carbon fiber-reinforced polymer laminates. J. Compos. Constr. 2005, 9, 195–201.

2. Ahmad, S. Reinforcement corrosion in concrete structures, its monitoring and service life prediction—A review. Cem. Concr. Compos. 2003, 25, 459–471.
3. Yuan, Y.; Ji, Y.; Shah, S.P. Comparison of two accelerated corrosion techniques for concrete structures. ACI Struct. J.2007, 104, 344–347.
4. Song, G.; Shayan, A. Corrosion of Steel in Concrete: Causes, Detection and Prediction; Australian Road Research Board: Vermont South, Australia, 1999.
5. Almusallam, A.A. Effect of degree of corrosion on the properties of reinforcing steel bars. Constr. Build. Mater. 2001,15, 361–368.
6. Büyüköztürk, O. Imaging of concrete structures. NDT&E Int. 1998, 31, 233–243.
7. Elsener, B.; Andrade, C.; Gulikers, J.; Polder, R.; Raupach, M. Half-cell potential measurements—Potential mapping on reinforced concrete structures. Mater. Struct. 2003, 36, 461–471.
8. Wiggenhauser, H.; Reinhardt, H.W. NDT in civil engineering: Experience and results of the for 384 research group. In Proceedings of the AIP Conference, Melbourne, Australia, 15–18 Febraury 2010.
9. Arndt, R.; Jalinoos, F. NDE for corrosion detection in reinforced concrete structures-benchmark approach. In Proceedings of the Non-Destructive Testing In Civil Engineering (NDTCE'09), Nantes, France, 30 June–3 July 2009.
10. Rhazi, J.; Dous, O.; Laurens, S. A new application of the GPR technique to reinforced concrete bridge decks. In Proceeding of the 4th Middle East NDT conference and Exhibition, Manama, Kingdom of Bahrain, 2–5 December 2007.
11. Rodriguez, P.; Ramirez, E.; Gonzalez, J.A. Methods for studying corrosion in reinforced concrete. Mag. Concr. Res.1994, 46, 81–90.
12. Carino, N.J. Nondestructive techniques to investigate corrosion status in concrete structures. J. Perform. Constr. Facil1999, 13, 96–106.
13. Vedalakshmi, R.; Balamurugan, L.; Saraswathy, V.; Kim, S.H.; Ann, K.Y. Reliability of galvanostatic pulse technique in assessing the

corrosion rate of rebar in concrete structures: Laboratory vs field studies. KSCE J. Civ. Eng. 2010, 14, 867–877.

14. Holloway, L.; Karajayli, P.; Birbilis, N. Galvanostatic pulse corrosion rate monitoring for steel in concrete. In Proceeding of the 49th Annual Conference of the Australasian Corrosion Association: Corrosion and Prevention, Charleston, SC, USA, 13 January 2009.
15. Law, D.W.; Millard, S.G.; Bungey, J.H. Use of galvanostatic pulse measurements on active reinforcing steel in concrete to assess corrosion rates. Br. Corros. J. 2001, 36, 75–80.
16. Lu, C.; Peiyu, Y. An algorithm of galvanostatic pulse method to determine the corrosion status of reinforcement in concrete. Corros. Sci. 2000, 42, 675–686.
17. Xu, J.; Yao, W. Corrosion detection of steel in concrete by use of galvanostatic pulse technique. J. Chin. Soc. Corros. Prot. 2010, 30, 181–186.
18. Shi, J.J.; Sun, W.; Geng, G.Q. Steel corrosion in simulated concrete pore solutions using a galvanostatic pulse method.J. Univ. Sci. Technol. Beijing 2011, 33, 727–733.
19. Poursaee, A.; Hansson, C.M. Galvanostatic pulse technique with the current confinement guard ring: The laboratory and finite element analysis. Corros. Sci. 2008, 50, 2739–2746.
20. Holloway, L.; Birbilis, N.; Forsyth, M. Long term monitoring of a reinforced concrete remediation method in a marine environment. In Proceeding of the European Corrosion Conference: Long Term Prediction and Modelling of Corrosion (EUROCORR), Nice, France, 12–16 September 2004.
21. Gonzalez, J.A.; Miranda, J.M.; Birbilis, N.; Feliu, S. Electrochemical techniques for studying corrosion of reinforcing steel: Limitations and advantages. Corrosion 2005, 61, 37–50.
22. Koleva, D.A.; de Wit, J.H.W.; van Breugel, K.; Lodhi, Z.F.; van Westing, E. Investigation of corrosion and cathodic protection in reinforced concrete: I. Application of electrochemical techniques. J. Electrochem. Soc. 2007, 154, P52–P61.

23. Song, H.W.; Saraswathy, V. Corrosion monitoring of reinforced concrete structures—A review. Int. J. Electrochem. Sci.2007, 2, 1–28.
24. Andrade, C.; Gonzalez, J.A. Quantitative measurements of corrosion rate of reinforcing steels embedded in concrete using polarization resistance measurements. Werkst. Korros. 1978, 29, 515–519.
25. Baronio, G.; Berra, M.; Bertolini, L.; Pastore, T. Steel corrosion monitoring in normal and total-lightweight concretes exposed to chloride and sulphate solutions Part I: Potential measurements. Cem. Concr. Res. 1996, 26, 683–689.
26. Andrade, C.; Keddam, M.; Novoa, X.R.; Perez, M.C.; Rangel, C.M.; Takenouti, H. Electrochemical behaviour of steel rebars in concrete: Influence of environmental factors and cement chemistry. Electrochim. Acta 2001, 46, 3905–3912.
27. Fang, X.; Chen, L.; Pan, J. Comparative study of on-site corrosion monitoring for concrete by different methods. Adv. Mater. Res. 2014, 838–841, 2066–2070.
28. Swarup, J.; Sharma, P.C. Electrochemical techniques for the monitoring of corrosion of reinforcement in concrete structures. Bull. Electrochem. 1996, 12, 103–108.
29. Ohtsu, M.; Yamamoto, T. Compensation procedure for half-cell potential measurement. Constr. Build. Mater. 1997, 11, 395–402.
30. Assouli, B.; Ballivy, G.; Rivard, P. Influence of environmental parameters on application of standard ASTM C876–91: Half cell potential measurements. Corros. Eng. Sci. Technol. 2008, 43, 93–96.
31. Yeih, W.; Huang, R. Detection of the corrosion damage in reinforced concrete members by ultrasonic testing. Cem. Concr. Res. 1998, 28, 1071–1083.
32. Sadowski, L. Methodology for assessing the probability of corrosion in concrete structures on the basis of half-cell potential and concrete resistivity measurements. Sci. World J. 2013, 2013.
33. Ismail, M.A.; Soleymani, H.; Ohtsu, M. Early detection of corrosion activity in reinforced concrete slab by AE technique. In Proceeding of the 6th Asia-Pasific Structural Engineering and Construction

Conference (APSEC 2006), Kuala Lumpur, Malaysia, 5–6 September 2006.

34. Andrade, C.; Alonso, M.C.; Gonzalez, J.A. An initial effort to use corrosion rate measurements for estimating rebar durability corrosion rates of steel in concrete. Corros. Rat. Steel Concr. ASTM STP 1990, 1065, 29–37.
35. Clear, K.C. Measuring rate of corrosion of steel in field concrete structures. Trans. Res. Rec. 1989, 1211, 28–37.
36. Broomfield, J.P. Field measurement of the corrosion rate of steel in concrete using a microprocessor controlled unit with a monitored guard ring for signal confinement. ASTM Spec. Tech. Publ. 1996, 1276, 91–106.
37. Reichling, K.; Raupach, M.; Broomfield, J.; Gulikers, J.; Nygaard, P.V.; Schneck, U.; Sergi, G. Local detailed inspection methods regarding reinforcement corrosion of concrete structures. Mater. Corros. 2013, 64, 128–134.
38. Fukuyama, T.; Nagai, H.; Noguchi, T. Corrosion monitoring of reinforcing bars by electrochemical noise measurement. In Proceeding of the 4th International Conference on Structural Health Monitoring of Intelligent Infrastructure (SHMII-4), Zurich, Switzerland, 22–24 July 2009.
39. Newton, C.J.; Sykes, J.M. A galvanostatic pulse technique for investigation of steel corrosion in concrete. Corros. Sci.1988, 28, 1051–1074.
40. Elsener, B.; Wojtas, H.; Bohni, H. Inspection and monitoring of reinforced concrete structures—Electrochemical methods to detect corrosion. Insight 1994, 36, 502–506.
41. Sathiyanarayanan, S.; Natarajan, P.; Saravanan, K.; Srinivasan, S.; Venkatachari, G. Corrosion monitoring of steel in concrete by galvanostatic pulse technique. Cem. Concr. Compos. 2006, 28, 630–637.
42. Vedalakshmi, R.; Thangavel, K. Reliability of electrochemical techniques to predict the corrosion rate of steel in concrete structures. Arab. J. Sci. Eng. 2011, 36, 769–783.

43. Wojtas, H. Determination of corrosion rate of reinforcement with a modulated guard ring electrode: Analysis of errors due to lateral current distribution. Corros. Sci. 2004, 46, 1621–1632.
44. Bungey, J.H.; Millard, S.G. Testing of Concrete in Structures, 3rd ed.; Blackie Academic & Professional: New York, NY, USA, 1996.
45. Sadowski, L. New non-destructive method for linear polarisation resistance corrosion rate measurement. Arch. Civ. Mech. Eng. 2010, 10, 109–116.
46. Cavalier, P.G.; Vassie, P.R. Investigation and repair of reinforcement corrosion in a bridge deck. In Proceedings of the Institution of Civil Engineers (London) Part 1—Design & Construction, Birmingham, UK, 15–16 September 1981; pp. 461–480.
47. Clear, K.C. Time to Corrosion of Reinforcing Steel in Concrete Slabs; Federal Highway Administration: Washington, DC, USA, 1982.
48. Polder, R.B. Test methods for on site measurement of resistivity of concrete—A RILEM TC-154 technical recommendation. Constr. Build. Mater. 2001, 15, 125–131.
49. Polder, R.; Andrade, C.; Elsener, B.; Vennesland, O.; Gulikers, J.; Weidert, R.; Raupach, M. Test methods for on site measurement of resistivity of concrete. Mater. Struct. 2000, 33, 603–611.
50. Reichling, K.; Raupach, M.; Broomfield, J.; Gulikers, J.; L'Hostis, V.; Kessler, S.; Osterminski, K.; Pepenar, I.; Schneck, U.; Sergi, G.; et al. Full surface inspection methods regarding reinforcement corrosion of concrete structures. Mater. Corros. 2013, 64, 116–127.
51. Legat, A. Monitoring of steel corrosion in concrete by electrode arrays and electrical resistance probes. Electrochim. Acta 2007, 52, 7590–7598.
52. Legat, A.; Govekar, E. Detection of corrosion by analysis of electrochemical noise. Fractals 1994, 2, 241–244.
53. Inoue, H.; Moriya, R.; Nishimura, R. Extreme-value distribution of the metastable-pit depth estimated by potential noise method. J. Soc. Mater. Sci. Jpn 2008, 57, 1097–1100.
54. Bertocci, U.; Huet, F. Noise analysis applied to electrochemical systems. Corrosion 1995, 51, 131–144.

55. Xiao, H.; Mansfeld, F. Evaluation of coating degradation with electrochemical impedance spectroscopy and electrochemical noise analysis. J. Electrochem. Soc. 1994, 141, 2332–2337.
56. Mansfeld, F.; Xiao, H. Electrochemical noise analysis of iron exposed to NaCl solutions of different corrosivity. J. Electrochem. Soc. 1993, 140, 2205–2209.
57. Yu, J.G.; Luo, J.L.; Norton, P.R. Investigation of hydrogen induced pitting active sites. Electrochim. Acta 2002, 47, 4019–4025.
58. Mansfeld, F.; Sun, Z.; Hsu, C.H. Electrochemical noise analysis (ENA) for active and passive systems in chloride media. Electrochim. Acta 2001, 46, 3651–3664.
59. Hashimoto, M.; Miyajima, S.; Murata, T. An experimental study of potential fluctuation during passive film breakdown and repair on iron. Corros. Sci. 1992, 33, 905–915.
60. Mariaca, L.; Bautista, A.; Rodríguez, P.; González, J.A. Use of electrochemical noise for studying the rate of corrosion of reinforcements embedded in concrete. Mater. Struct. 1997, 30, 613–617.
61. Legat, A.; Leban, M.; Bajt, Ž. Corrosion processes of steel in concrete characterized by means of electrochemical noise.Electrochim. Acta 2004, 49, 2741–2751.
62. Zhao, B.; Li, J.H.; Hu, R.G.; Du, R.G.; Lin, C.J. Study on the corrosion behavior of reinforcing steel in cement mortar by electrochemical noise measurements. Electrochim. Acta 2007, 52, 3976–3984.
63. Aballe, A.; Bethencourt, M.; Botana, F.J.; Marcos, M.; Sánchez-Amaya, J.M. Influence of the degree of polishing of alloy AA 5083 on its behaviour against localised alkaline corrosion. Corros. Sci. 2004, 46, 1909–1920.
64. Mills, D.; Mabbutt, S.; Bierwagen, G. Investigation into mechanism of protection of pigmented alkyd coatings using electrochemical and other methods. Prog. Org. Coat. 2003, 46, 176–181.
65. Nagiub, A.; Mansfeld, F. Evaluation of corrosion inhibition of brass in chloride media using EIS and ENA. Corros. Sci.2001, 43, 2147–2171.
66. Hong, S.; Lai, W.L.; Helmerich, R. Experimental monitoring of chloride-induced reinforcement corrosion and chloride

contamination in concrete with ground-penetrating radar. Struct. Infrastruct. Eng. 2014, 15–26.

67. Patil, S.; Karkare, B.; Goyal, S. Acoustic emission vis-Ã-vis electrochemical techniques for corrosion monitoring of reinforced concrete element. Constr. Build. Mater. 2014, 68, 326–332.
68. Kabir, S.; Zaki, A. Detection and quantification of corrosion damage using ground penetrating radar (GPR). In Proceeding of the Progress in Electromagnetic Research Symposium (PIERS), Marrakesh, Marocco, 20–23 March 2011.
69. Halabe, U.B. Condition Assessment of Reinforced Concrete Structures Using Electromagnetic Waves. Ph.D. Thesis, Massachusetts Institute of Technology, Cambridge, MA, USA, 1990.
70. Sbartai, Z.M.; Laurens, S.; Rhazi, J.; Balayssac, J.P.; Arliguie, G. Using radar direct wave for concrete condition assessment: Correlation with electrical resistivity. J. Appl. Geophys. 2007, 62, 361–374.
71. Lai, W.L.; Kind, T.; Wiggenhauser, H. Frequency-dependent dispersion of high-frequency ground penetrating radar wave in concrete. NDT&E Int. 2011, 44, 267–273.
72. Robert, A. Dielectric permittivity of concrete between 50 MHz and 1 GHz and GPR measurements for building materials evaluation. J. Appl. Geophys. 1998, 40, 89–94.
73. Soutsos, M.N.; Bungey, J.H.; Millard, S.G.; Shaw, M.R.; Patterson, A. Dielectric properties of concrete and their influence on radar testing. NDT&E Int. 2001, 34, 419–425.
74. Sbartai, Z.M.; Laurens, S.; Balayssac, J.P.; Ballivy, G.; Arliguie, G. Effect of concrete moisture on radar signal amplitude. ACI Mater. J. 2006, 103, 419–426.
75. Maierhofer, C. Nondestructive evaluation of concrete infrastructure with ground penetrating radar. J. Mater. Civ. Eng.2003, 15, 287–297.
76. Kim, W. Ground Penetrating Radar Application for Non-Destructive Testing: Bridge Deck Inspection and Dowel Bar Detection. Ph.D, Thesis, University of Missouri–Rolla, Rolla, MO, USA, 2003.

77. Klysz, G.; Balayssac, J.P.; Laurens, S. Spectral analysis of radar surface waves for non-destructive evaluation of cover concrete. NDT&E Int. 2004, 37, 221–227.
78. Lai, W.L.; Kou, S.C.; Tsang, W.F.; Poon, C.S. Characterization of concrete properties from dielectric properties using ground penetrating radar. Cem. Concr. Res. 2009, 39, 687–695.
79. Sbartai, Z.M.; Laurens, S.; Balayssac, J.P.; Arliguie, G.; Ballivy, G. Ability of the direct wave of radar ground-coupled antenna for NDT of concrete structures. NDT&E Int. 2006, 39, 400–407.
80. Barnes, C.L.; Trottier, J.F.; Forgeron, D. Improved concrete bridge deck evaluation using GPR by accounting for signal depth-amplitude effects. NDT&E Int. 2008, 41, 427–433.
81. Kim, W.; Ismail, A.; Anderson, N.L.; Atekwana, E.A.; Buccellato, A. Non-destructive testing (NDT) for corrosion in bridge decks using GPR. In Proceedings of the 3rd International Conference on the Application of Geophysical Methodologies and NDT to Transportation Facilities and Infrastructure, Geophysics 2003, Orlando, FL, USA, 8–12 December 2003.
82. Lai, W.L.; Kind, T.; Wiggenhauser, H. Detection of accelerated reinforcement corrosion in concrete by ground penetrating radar. In Proceeding of the 13th International Conference on Ground Penetrating Radar (GPR), Lecce, Italy, 21–25 June 2010.
83. Hong, S.X.; Lai, W.L.; Helmerich, R. Monitoring accelerated corrosion in chloride contaminated concrete with ground penetrating radar. In Proceeding of the 14th International Conference on Ground Penetrating Radar, Shanghai, China, 4–8 June 2012.
84. Manning, D.G.; Holt, F.B. Detecting delamination in concrete bridge decks. Concr. Int. 1980, 2, 34–41.
85. Kordatos, E.Z.; Strantza, M.; Soulioti, D.V.; Matikas, T.E.; Aggelis, D.G. Combined NDT methods for characterization of subsurface cracks in concrete. In Proceeding of the SPIE—The International Society for Optical Engineering, San Diego, CA, USA, 6 March 2011.
86. Bagavathiappan, S.; Lahiri, B.B.; Saravanan, T.; Philip, J.; Jayakumar, T. Infrared thermography for condition monitoring—A review. Infrared Phys. Technol. 2013, 60, 35–55.

87. Washer, G.; Bolleni, N.; Fenwick, R. Thermographic imaging of subsurface deterioration in concrete bridges. Trans. Res. Rec. 2010, 2201, 27–33.
88. Pieper, D.; Donnell, K.M.; Ghasr, M.T.; Kinzel, E.C. Integration of microwave and thermographic NDT methods for corrosion detection. In Proceeding of AIP Conference, Canberra, Australia, 7–12 December 2014.
89. Kobayashi, K.; Banthia, N. Detection of reinforcement corrosion with a new non-destructive test method using induction heating. In Proceeding of the 6th International Conference on Concrete under Severe Conditions (CONSEC'10), Merida, Mexico, 7–9 June 2010.
90. Clark, M.R.; McCann, D.M.; Forde, M.C. Application of infrared thermography to the non-destructive testing of concrete and masonry bridges. NDT&E Int. 2003, 36, 265–275.
91. Chung, L.; Paik, I.K.; Roh, Y.S. Non-destructive evaluation techniques of reinforcing steel corrosion using infrared thermography. Key Eng. Mater. 2004, 270, 1592–1597.
92. Chung, L.; Paik, I.K.; Cho, S.H.; Roh, Y.S. Infrared thermographic technique to measure corrosion in reinforcing bar.Key Eng. Mater. 2006, 321, 821–824.
93. Washer, G. Advances in the use of thermographic imaging for the condition assessment of bridges. Bridg. Struct. Assess. Des. Constr. 2012, 8, 81–90.
94. Kobayashi, K.; Banthia, N. Corrosion detection in reinforced concrete using induction heating and infrared thermography. J. Civ. Struct. Health Monit. 2011, 1, 25–35.
95. Kwon, S.J.; Xue, H.; Feng, M.Q.; Baek, S. Nondestructive corrosion detection in concrete through integrated heat induction and IR thermography. In Proceeding of the SPIE—The International Society for Optical Engineering, San Diego, CA, USA, 6 March 2011.
96. Baek, S.; Xue, W.; Feng, M.Q.; Kwon, S. Nondestructive corrosion detection in RC through integrated heat induction and IR thermography. J. Nondestr. Eval. 2012, 31, 181–190.

97. Vallee, R.; Fredrick, S.; Asatryan, K.; Fischer, M.; Galstian, T. Real-time observation of Bragg grating formation in As2S3 chalcogenide ridge waveguides. Opt. Commun. 2004, 230, 301–307.
98. Hassan, M.R.A.; Bakar, M.H.A.; Dambul, K.; Adikan, F.R.M. Optical-based sensors for monitoring corrosion of reinforcement rebar via an etched Cladding Bragg grating. Sensors 2012, 12, 15820–15826.
99. Chen, W.; Dong, X. Modification of the wavelength-strain coefficient of FBG for the prediction of steel bar corrosion embedded in concrete. Opt. Fiber Technol. 2012, 18, 47–50.
100. Zheng, Z.; Sun, X.; Lei, Y. Monitoring corrosion of reinforcement in concrete structures via fiber Bragg grating sensors. Front. Mech. Eng. Chin. 2009, 4, 316–319.
101. Lee, J.R.; Yun, C.Y.; Yoon, D.J. A structural corrosion-monitoring sensor based on a pair of prestrained fiber Bragg gratings. Meas. Sci. Technol. 2010, 21.
102. Gao, J.; Wu, J.; Li, J.; Zhao, X. Monitoring of corrosion in reinforced concrete structure using Bragg grating sensing.NDT&E Int. 2011, 44, 202–205.
103. Lau, K.T. Fibre-optic sensors and smart composites for concrete applications. Mag. Concr. Res. 2003, 55, 19–34.
104. Geng, J.; Wu, J.; Zhao, X. Simulation of fiber Bragg grating sensor for rebar corrosion. In Proceeding of SPIE—The International Society for Optical Engineering, Weihai, China, 8 July 2009.
105. Lei, Y.; Zheng, Z.P. Review of physical based monitoring techniques for condition assessment of corrosion in reinforced concrete. Math. Probl. Eng. 2013, 2013.
106. Sohn, H.; Park, G.; Wait, J.R.; Limback, N.P.; Farrar, C.R. Wavelet-based active sensing for delamination detection in composite structures. Smart Mater. Struct. 2004, 13, 153–160.
107. ASTM. ASTM C-597–83: Standard Test Method for Pulse Velocity Through Concrete. Available online: http://www.astm.org/DATABASE.CART/HISTORICAL/C597-97.htm (accessed on 6 May 2015).
108. Bungey, J.H. Ultrasonic testing to identify alkali-silica reaction in concrete. Br. J. Non-Destr. Test. 1991, 33, 227–231.

109. John, G.; Hladky, K.; Gaydecki, P.; Dawson, J. Recent developments in inspection techniques for corrosion damaged concrete structures. Corros. Forms Control Infrastruct. 1992, 1137, 246–257.

110. Raad, J.; Aktan, H.; Usmen, M. Laboratory for non destructive evaluation of civil structures. In Proceeding of ASEE Annual Conference, Milwaukee, WI, USA, 15–18 June 1997.

111. Ervin, B.; Kuchma, D.; Bernhard, J.; Reis, H. Monitoring corrosion of rebar embedded in mortar using high-frequency guided ultrasonic waves. J. Eng. Mech. 2009, 135, 9–19.

112. Sharma, S.; Mukherjee, A. Longitudinal guided waves for monitoring chloride corrosion in reinforcing bars in concrete. Struct. Health Monit. 2010, 9, 555–567.

113. Jung, Y.-C.; Na, W.-B.; Kundu, T.; Ehsani, M. Damage detection in concrete using Lamb waves. In Proceeding of SPIE—The International Society for Optical Engineering, Newport Beach, CA, USA, 6 March 2000; pp. 448–458.

114. Sharma, S.; Mukherjee, A. Monitoring corrosion in oxide and chloride environments using ultrasonic guided waves. J. Mater. Civ. Eng. 2011, 23, 207–211.

115. Sharma, S.; Mukherjee, A. Nondestructive evaluation of corrosion in varying environments using guided waves. Res. Nondestr. Eval. 2013, 24, 63–88.

116. Miller, T.H.; Yanagita, T.; Kundua, T.; Grill, J.; Grill, W. Nondestructive inspection of reinforced concrete structures. In Proceeding of the SPIE—The International Society for Optical Engineering, San Diego, CA, USA, 8 March 2009.

117. Goueygou, M.; Piwakowski, B.; Fnine, A.; Kaczmarek, M.; Buyle-Bodin, F. NDE of two-layered mortar samples using high-frequency Rayleigh waves. Ultrasonics 2004, 42, 889–895.

118. Aggelis, D.G.; Kordatos, E.Z.; Soulioti, D.V.; Matikas, T.E. Combined use of thermography and ultrasound for the characterization of subsurface cracks in concrete. Constr. Build. Mater. 2010, 24, 1888–1897.

119. Li, D.; Zhang, S.; Yang, W.; Zhang, W. Corrosion monitoring and evaluation of reinforced concrete structures utilizing the ultrasonic guided wave technique. Int. J. Distrib. Sens. Netw. 2014, 2014.
120. Komlos, K.; Popovics, S.; NÃrnbergerovÃ, T.; BabÃl, B.; Popovics, J.S. Ultrasonic pulse velocity test of concrete properties as specified in various standards. Cem. Concr. Compos. 1996, 18, 357–364.
121. Sansalone, M.J.; Streett, W.B. Impact-echo. Nondestructive Evaluation of Concrete & Masonry; Bullbrier Press: Jersey Shore, PA, USA, 1997.
122. Carino, N.J.; Sansalone, M.; Hsu, N.N. Flaw detection in concrete by frequency spectrum analysis of impact-echo waveforms. Int. Adv. Nondestruct. Test. 1986, 12, 117–146.
123. Liang, M.T.; Su, P.J. Detection of the corrosion damage of rebar in concrete using impact-echo method. Cem. Concr. Res. 2001, 31, 1427–1436.
124. Samarkova, K.; Chobola, Z.; Stefkova, D. The corrosion status of reinforced concrete structure monitoring by impact echo method. Adv. Mater. Res. 2014, 875–877, 445–449.
125. Samarkova, K.; Chobola, Z.; Stefkova, D. Using of impact-echo methods to assessment of reinforced concrete structures corrosion. Appl. Mech. Mater. 2014, 446–447, 1400–1404.
126. Samarkova, K.; Chobola, Z.; Stefkova, D.; Kusak, I. Impact-echo methods to assessment corrosion of reinforced concrete structures. Appl. Mech. Mater. 2014, 627, 268–271.
127. Samarkova, K.; Stefkova, D.; Chobola, Z. Monitoring of reinforced concrete structure corrosion by using impact-echo method. Adv. Mater. Res. 2014, 1000, 239–242.
128. Miller, R.K.; Mc Intire, P. Acoustic emission testing. In Encyclopedic Dictionary of Polymers; Springer New York: New York, NY, USA, 1987; pp. 12–154.
129. Beattie, A.G. Acoustic emission, principles and instrumentation. J. Acoust. Emission 1983, 2, 95–128.
130. Shiotani, T. Evaluation of long-term stability for rock slope by means of acoustic emission technique. NDT&E Int.2006, 39, 217–228.

131. Alvarez, M.G.; Lapitz, P.; Ruzzante, J. Analysis of acoustic emission signals generated from SCC propagation. Corros. Sci. 2012, 55, 5–9.
132. Lapitz, P.; Ruzzante, J.; Alvarez, M.G. AE response of α-brass during stress corrosion crack propagation. Corros. Sci.2007, 49, 3812–3825.
133. Wells, D. An acoustic apparatus to record emissions from concrete under strain. Nucl. Eng. Des. 1970, 12, 80–88.
134. Luo, X.; Haya, H.; Inaba, T.; Shiotani, T.; Nakanishi, Y. Experimental study on evaluation of breakage in foundations using train-induced acoustic emission. In Proceeding of the Structural Engineering World Congress, Tokyo, Japan, 13–14 December 2002.
135. ASTM. ASTM C-610–98A: Definitions of Terms Relating to Acoustic Emission. Available online: http://www.astm.org/DATABASE.CART/WITHDRAWN/E610.htm (accessed on 6 May 2015).
136. Malhotra, V.M.; Carino, N.J. CRC Handbook on Nondestructive Testing of Concrete; CRC Press: Boca Raton, FL, USA, 1991.
137. Ohtsu, M.; Yuyama, S. Recommended practice for in-situ monitoring of concrete structures by acoustic emission. In Proceedings of the 15th International Acoustic Emission Symposium, Tokyo, Japan, 11–14 September 2000; pp. 263–268.
138. Carpinteri, A.; Lacidogna, G.; Niccolini, G. Damage analysis of reinforced concrete buildings by the acoustic emission technique. Struct. Control Health Monit. 2011, 18, 660–673.
139. Ohtsu, M. Mathematical theory of acustic emission and moment tensor solution. J. Soc. Mater. Sci. 1987, 36, 1025–1031.
140. Colombo, S.; Main, I.G.; Forde, M.C. Assessing damage of reinforced concrete beam using "b-value" analysis of acoustic emission signals. J. Mater. Civ. Eng. 2003, 15, 280–286.
141. Grosse, C.U.; Reinhardt, H.W.; Finck, F. Signal-based acoustic emission techniques in civil engineering. J. Mater. Civ. Eng. 2003, 15, 274–279.
142. Ohno, K.; Ohtsu, M. Crack classification in concrete based on acoustic emission. Constr. Build. Mater. 2010, 24, 2339–2346.

143. Dunn, S.E.; Young, J.D.; Hartt, W.H.; Brown, R.P. Acoustic emission characterization of corrosion induced damage in reinforced concrete. Corrosion 1984, 40, 339–343.

144. Shiotani, T.; Ohtsu, M.; Ikeda, K. Detection and evaluation of AE waves due to rock deformation. Constr. Build. Mater.2001, 15, 235–246.

145. Titus, T.N.K.; Reddy, D.V.; Dunn, S.E.; Hartt, W.H. Acoustic emission crack monitoring and prediction of remaining life of corroding reinforced concrete beams. In Proceeding of the 4th European Conference Non-destructive Testing, London, UK, 13–17 September 1988.

146. Ohtsu, M. The history and development of acoustic emission in concrete engineering. Mag. Concr. Res. 1996, 48, 321–330.

147. Ohtsu, M. Recommendation of RILEM TC 212-ACD: Acoustic emission and related NDE techniques for crack detection and damage evaluation in concrete. Mater. Struct. 2010, 43, 1187–1189.

148. Ismail, M.E.; Ismail, M.; Ohtsu, M. Non-destructive evaluation of corrosion activity in reinforced concrete slab. In Proceeding of the 4th International Structural Engineering and Construction Conference (ISEC-4)—Innovations in Structural Engineering and Construction, Melbourne, Australia, 26–28 September 2008.

149. Calabrese, L.; Campanella, G.; Proverbio, E. Identification of corrosion mechanisms by univariate and multivariate statistical analysis during long term acoustic emission monitoring on a pre-stressed concrete beam. Corros. Sci. 2013,73, 161–171.

150. Yu, A.; Zhao, Y.; Wang, L. Acoustic emission (AE) online monitoring test on corrosion in reinforced concrete. J. Build. Mater. 2014, 17, 291–297.

151. Ohtsu, M.; Tomoda, Y. Corrosion process in reinforced concrete identified by acoustic emission. Mater. Trans. 2007, 48, 1184–1189.

152. Yoon, D.J.; Weiss, W.J.; Shah, S.P. Assessing damage in corroded reinforced concrete using acoustic emission. J. Eng. Mech. 2000, 126, 273–283.

153. Kawasaki, Y.; Wakuda, T.; Kobarai, T.; Ohtsu, M. Corrosion mechanisms in reinforced concrete by acoustic emission.Constr. Build. Mater. 2013, 48, 1240–1247.

154. Kawasaki, Y.; Tomoda, Y.; Ohtsu, M. AE monitoring of corrosion process in cyclic wet–dry test. Constr. Build. Mater.2010, 24, 2353–2357.

155. Ohtsu, M.; Tomoda, Y. Acoustic emission techniques for rebar corrosion in reinforced concrete. In Advances in Construction Materials 2007; Springer-Verlag: Berlin, Germany, 2007.

156. Ohtsu, M.; Tomoda, Y. Phenomenological model of corrosion process in reinforced concrete identified by acoustic emission. ACI Mater. J. 2008, 105, 194–199.

157. Elfergani, H.A.; Pullin, R.; Holford, K.M. Damage assessment of corrosion in prestressed concrete by acoustic emission. Constr. Build. Mater. 2013, 40, 925–933.

158. Behnia, A.; Chai, H.K.; Shiotani, T. Advanced structural health monitoring of concrete structures with the aid of acoustic emission. Constr. Build. Mater. 2014, 65, 282–302.

159. ElBatanouny, M.K.; Ziehl, P.H.; Larosche, A.; Mangual, J.; Matta, F.; Nanni, A. Acoustic emission monitoring for assessment of prestressed concrete beams. Constr. Build. Mater. 2014, 58, 46–53.

160. Melchers, R.E.; Li, C.Q. Phenomenological modeling of reinforcement corrosion in marine environments. ACI Mater. J.2006, 103, 25–32.

161. Velez, W.; Matta, F.; Ziehl, P. Acoustic emission intensity analysis of corrosion in prestressed concrete piles. In Proceeding of the 10th International Conference on Barkhausen Noise and Micromagnetic Noise and Micromagnetic Testing, Melville, NY, USA, 26–28 August 2014.

162. Velez, W.; Matta, F.; Ziehl, P. Acoustic emission characterization of early corrosion in prestressed concrete exposed to salt water. In Proceeding of the 11th International Conference on Structural Safety and Reliability (ICOSSAR), New York, NY, USA, 16–20 June 2013.

163. Ing, M.; Austin, S.; Lyons, R. Cover zone properties influencing acoustic emission due to corrosion. Cem. Concr. Res.2005, 35, 284–295.

164. JCMS-IIIB5706. Monitoring Method for Active Cracks in Concrete by Acoustic Emission. Available online: http://framcos.org/FraMCoS-7/09-01.pdf (accessed on 6 May 2015).

165. Rao, M.V.M.S.; Prasanna Lakshmi, K.J. Analysis of b-value and improved b-value of acoustic emissions accompanying rock fracture. Curr. Sci. 2005, 89, 1577–1582.

166. Gutenberg, B.; Richter, C.F. Seismicity of the Earth and Associated Phenomena; Princeton University Press: Princeton, NJ, USA, 1954.

167. Shiotani, T.; Ohtsu, M. Prediction of Slope Failure Based on AE Activity; ASTM Special Technical Publication: West Conshohocken, PA, USA, 1999; pp. 156–172.

168. ElBatanouny, M.K.; Larosche, A.; Mazzoleni, P.; Ziehl, P.H.; Matta, F.; Zappa, E. Identification of cracking mechanisms in scaled FRP reinforced concrete beams using acoustic emission. Exp. Mech. 2012, 54, 69–82.

169. Ono, K. Application of acoustic emission for structure diagnosis. Konf. Naukowa 2010, 2, 317–341.

170. Yuyama, S.; Yokoyama, K.; Niitani, K.; Ohtsu, M.; Uomoto, T. Detection and evaluation of failures in high-strength tendon of prestressed concrete bridges by acoustic emission. Constr. Build. Mater. 2007, 21, 491–500.

171. ElBatanouny, M.K.; Mangual, J.; Velez, W.; Ziehl, P.H.; Matta, F.; Gonzalez, M. Monitoring corrosion in prestressed concrete beams using acoustic emission technique. In Proceeding of the SPIE—The International Society for Optical Engineering, San Diego, CA, USA, 11 March 2012.

172. Mangual, J.; ElBatanouny, M.K.; Velez, W.; Ziehl, P.; Matta, F.; Gonzalez, M. Characterization of corrosion damage in prestressed concrete using acoustic emission. In Proceeding of the Sensors and Smart Structures Technologies for Civil, Mechanical, and Aerospace Systems, Parts 1 and 2, San Diego, CA, USA, 12–15 March 2012.

173. ElBatanouny, M.K.; Mangual, J.; Ziehl, P.H.; Matta, F. Early corrosion detection in prestressed concrete girders using acoustic emission. J. Mater. Civ. Eng. 2014, 26, 504–511.

174. Fowler, T.J.; Blessing, J.A.; Strauser, F.E. Intensity analysis. In Proceedinds of the 4 International Symposia Acoustic Emission from Composite Materials, Columbus, OH, USA, 27–31 July 1992; pp. 16–27.

175. Gostautas, R.S.; Ramirez, G.; Peterman, R.J.; Meggers, D. Acoustic emission monitoring and analysis of glass fiber-reinforced composites bridge decks. J. Bridg. Eng. 2005, 10, 713–721.

176. Mangual, J.; ElBatanouny, M.K.; Ziehl, P.; Matta, F. Acoustic emission-based characterization of corrosion damage in cracked concrete with prestressing strand. ACI Mater. J. 2013, 110, 89–98.

177. Colombo, S.; Forde, M.C.; Main, I.G.; Shigeishi, M. Predicting the ultimate bending capacity of concrete beams from the “relaxation ratio” analysis of AE signals. Constr. Build. Mater. 2005, 19, 746–754.

178. Yun, H.D.; Choi, W.C.; Seo, S.Y. Acoustic emission activities and damage evaluation of reinforced concrete beams strengthened with CFRP sheets. NDT&E Int. 2010, 43, 615–628.

179. Ohtsu, M.; Okamoto, T.; Yuyama, S. Moment tensor analysis of acoustic emission for cracking mechanisms in concrete. ACI Struct. J. 1998, 95, 87–95.

180. Shigeishi, M.; Ohtsu, M. Acoustic emission moment tensor analysis: Development for crack identification in concrete materials. Constr. Build. Mater. 2001, 15, 311–319.

181. Farid Uddin, A.K.M.; Ohtsu, M. Micromechanics of corrosion cracking in concrete by AE-SIGMA. J. Acoustic Emission.2005, 23, 136–141.

182. Baranov, V.M.; Kapralov, E.Y.; Karasevich, V.A. Acoustic and electrochemical noise methods in extreme-condition fatigue tests. Meas. Tech. 2003, 46, 994–999.

183. Kouroussis, D.A.; Anastassopoulos, A.A.; Lenain, J.C.; Proust, A. Advances in Classification of Acoustic Emission Sources. Les Journées COFREND Reims 2001. Available online:

http://www.epandt.com/Biblio/EA2001-06A.pdf(accessed on 5 May 2015).

184. Calabrese, L.; Campanella, G.; Proverbio, E. Noise removal by cluster analysis after long time AE corrosion monitoring of steel reinforcement in concrete. Constr. Build. Mater. 2012, 34, 362–371.

Index

A

Acoustic Analogies, 127, 149

Acoustic Bathymetry, 101

acoustic emission, 1, 2, 3, 8, 10, 23, 24, 26, 27, 40, 41, 43, 44, 48, 56, 57, 64, 65, 67, 68, 69, 70, 72, 73, 74, 75, 76, 83, 84, 86, 89, 90, 153, 154, 155, 168, 169, 170, 171, 173, 185, 186, 187, 189, 190, 195, 213, 215, 218, 219, 222, 223, 224, 225, 226, 227, 229, 231, 235, 237, 240, 241, 242, 244, 245, 246, 247, 274, 275, 282, 284, 311, 312, 313, 314, 315, 316

Acoustic Energy, 48

Acoustic Profiling, 101

AE Descriptors, 190

aluminum alloy, 171, 172, 173, 174, 175, 177, 183, 184, 186, 187

Amplitude, 14, 15, 16, 17, 19

C

corrosion, 40, 102, 187, 274, 275, 276, 278, 279, 280, 281, 282, 283, 285, 286, 288, 289, 290, 291, 293, 294, 295, 296, 297, 298, 299, 300, 301, 302, 303, 304, 305, 306, 307, 308, 309, 310, 311, 312, 313, 314, 315, 316, 317

Crude Oil, 101, 102, 103, 125

Cumulative Energy, 11, 12, 13

D

diagnosis, 67, 68, 87, 255, 257, 262, 315

E

epoxy, 153, 154, 155, 157, 163, 164, 165, 166, 167, 168

F

feature vector, 23, 26, 32, 33, 36, 37, 39

L

Linear Source Location Technique, 89

M

macro-fiber composite, 67, 68

mechanical behavior, 43, 44, 51, 63, 169

Mechanical Jig, 1, 2, 4

O

OpenFOAM, 127, 128, 129, 130, 131, 132, 133, 134, 135, 137, 138, 139, 140, 141, 144, 145, 149, 150

Q

quantification model, 171, 176, 180, 184, 185

R

Radiographic Film, 8

Radiographic Source, 8

Recurrence plot, 263

Refraction Errors, 117

reinforced concrete, 168, 274, 275, 276, 282, 299, 300, 301, 302, 303, 306, 308, 309, 310, 311, 312, 313, 314, 315, 316

S

SEM micrograph, 46

SiGMA, 297

Sludge, 101, 110

Sound Propagation, 127, 151

Sound Sources, 127

static fatigue, 43, 45, 47, 52, 53, 54, 55, 57, 58, 60, 61, 62, 65

Storage Oil Tanks, 101

Structural Damage, 89

T

tectorial membrane, 253

Temperature Gradients, 113

Tungsten Inert Gas, 1, 7

U

ultrasound, 25, 40, 91, 154, 157, 166, 230, 231, 242, 310

USB-AE node, 3, 5

W

welding (TIG), 1

wind turbine, 67, 68, 69, 70, 77, 83, 84, 85

X

X-ray Radiography, 1